Two Bits

 EXPERIMENTAL FUTURES

Technological Lives, Scientific Arts, Anthropological Voices

A series edited by Michael M. J. Fischer and Joseph Dumit

2008 · DUKE UNIVERSITY PRESS · DURHAM AND LONDON

THE CULTURAL

SIGNIFICANCE OF

FREE SOFTWARE

Two Bits

CHRISTOPHER M. KELTY

© 2008 Duke University Press

Printed in the United States of America on acid-free paper ∞

Designed by C. H. Westmoreland

Typeset in Charis (an Open Source font) by Achorn International

Library of Congress Cataloging-in-Publication data and republication acknowledgments appear on the last printed pages of this book.

Duke University Press gratefully acknowledges the support of HASTAC (Humanities, Arts, Science, and Technology Advanced Collaboratory), which provided funds to help support the electronic interface of this book.

Two Bits is accessible on the Web at twobits.net.

To my parents, Anne and Ted

Contents

This is a book about Free Software, also known as Open Source Software, and is meant for anyone who wants to understand the cultural significance of Free Software. *Two Bits* explains how Free Software works and how it emerged in tandem with the Internet as both a technical and a social form. Understanding Free Software in detail is the best way to understand many contentious and confusing changes related to the Internet, to "commons," to software, and to networks. Whether you think first of e-mail, Napster, Wikipedia, MySpace, or Flickr; whether you think of the proliferation of databases, identity thieves, and privacy concerns; whether you think of traditional knowledge, patents on genes, the death of scholarly publishing, or compulsory licensing of AIDS medicine; whether you think of MoveOn.org or net neutrality or YouTube—the issues raised by these phenomena can be better understood by looking carefully at the emergence of Free Software.

Why? Because it is in Free Software and its history that the issues raised—from intellectual property and piracy to online political advocacy and "social" software—were first figured out and confronted. Free Software's roots stretch back to the 1970s and crisscross the histories of the personal computer and the Internet, the peaks and troughs of the information-technology and software industries, the transformation of intellectual property law, the innovation of organizations and "virtual" collaboration, and the rise of networked social movements. Free Software does not explain why these various changes have occurred, but rather how individuals and groups are responding: by creating new things, new practices, and new forms of life. It is these practices and forms of life—not the software itself—that are most significant, and they have in turn served as templates that others can use and transform: practices of sharing source code, conceptualizing openness, writing copyright (and copyleft) licenses, coordinating collaboration, and proselytizing for all of the above. There are explanations aplenty for *why* things are the way they are: it's globalization, it's the network society, it's an ideology of transparency, it's the virtualization of work, it's the new flat earth, it's Empire. We are drowning in the *why*, both popular and scholarly, but starving for the *how*.

Understanding how Free Software works is not just an academic pursuit but an experience that transforms the lives and work of participants involved. Over the last decade, in fieldwork with software programmers, lawyers, entrepreneurs, artists, activists, and other geeks I have repeatedly observed that understanding how Free Software works results in a revelation. People—even (or, perhaps, especially) those who do not consider themselves programmers, hackers, geeks, or technophiles—come out of the experience with something like religion, because Free Software is all about the practices, not about the ideologies and goals that swirl on its surface. Free Software and its creators and users are not, as a group, antimarket or anticommercial; they are not, as a group, anti–intellectual property or antigovernment; they are not, as a group, pro- or anti- anything. In fact, they are not really a group at all: not a corporation or an organization; not an NGO or a government agency; not a professional society or an informal horde of hackers; not a movement or a research project.

Free Software is, however, public; it is about making things public. This fact is key to comprehending its cultural significance, its appeal, and its proliferation. Free Software is public in a particular way: it is a self-determining, collective, politically independent mode of creating very complex technical objects that are made publicly and freely available to everyone—a "commons," in common parlance. It is a practice of working through the promises of equality, fairness, justice, reason, and argument in a domain of technically complex software and networks, and in a context of powerful, lopsided laws about intellectual property. The fact that something public in this grand sense emerges out of practices so seemingly arcane is why the first urge of many converts is to ask: how can Free Software be "ported" to other aspects of life, such as movies, music, science or medicine, civil society, and education? It is this proselytizing urge and the ease with which the practices are spread that make up the cultural significance of Free Software. For better or for worse, we may all be using Free Software before we know it.

Acknowledgments

Anthropology is dependent on strangers who become friends and colleagues—strangers who contribute the very essence of the work. In my case, these strangers are also hyperaware of issues of credit, reputation, acknowledgment, reuse, and modification of ideas and things. Therefore, the list is extensive and detailed.

Sean Doyle and Adrian Gropper opened the doors to this project, providing unparalleled insight, hospitality, challenge, and curiosity. Axel Roch introduced me to Volker Grassmuck, and to much else. Volker Grassmuck introduced me to Berlin's Free Software world and invited me to participate in the Wizards of OS conferences. Udhay Shankar introduced me to almost everyone I know, sometimes after the fact. Shiv Sastry helped me find lodging in Bangalore at his Aunt Anasuya Sastry's house, which is called "Silicon Valley" and which was truly a lovely place to stay. Bharath Chari and Ram Sundaram let me haunt their office and cat-5 cables during one of the more turbulent periods of their careers. Glenn Otis Brown visited, drank, talked, invited, challenged, entertained, chided, encouraged, drove, was driven, and gave and received advice. Ross Reedstrom welcomed me to the Rice Linux Users' Group and to Connexions. Brent Hendricks did yeoman's work, suffering my questions and intrusions. Geneva Henry, Jenn Drummond, Chuck Bearden, Kathy Fletcher, Manpreet Kaur, Mark Husband, Max Starkenberg, Elvena Mayo, Joey King, and Joel Thierstein have been welcoming and enthusiastic at every meeting. Sid Burris has challenged and respected my work, which has been an honor. Rich Baraniuk listens to everything I say, for better or for worse; he is a magnificent collaborator and friend.

James Boyle has been constantly supportive, for what feels like very little return on investment. Very few people get to read and critique and help reshape the argument and structure of a book, and to appear in it as well. Mario Biagioli helped me see the intricate strategy described in chapter 6. Stefan Helmreich read early drafts and transformed my thinking about networks. Manuel De-Landa explained the term *assemblage* to me. James Faubion corrected my thinking in chapter 2, helped me immeasurably with the Protestants, and has been an exquisitely supportive colleague and department chair. Mazyar Lotfalian and Melissa Cefkin provided their apartment and library, in which I wrote large parts of chapter 1. Matt Price and Michelle Murphy have listened patiently to me construct and reconstruct versions of this book for at least six years. Tom and Elizabeth Landecker provided hospitality and stunningly beautiful surroundings in which to rewrite parts of the book. Lisa Gitelman read carefully and helped explain issues about documentation and versioning that I discuss in chapter 4. Matt Ratto read and commented on chapters 4–7, convinced me to drop a useless distinction, and to clarify the conclusion to chapter 7. Shay David provided strategic insights about openness from his own work and pushed me to explain the point of recursive publics more clearly. Biella Coleman has been a constant interlocutor on the issues in this book—her contributions are too deep, too various, and too thorough to detail. Her own work on Free Software and hackers has been a constant sounding board and guide, and it has been a pleasure to work together on our respective texts. Kim Fortun helped me figure it all out.

George Marcus hired me into a fantastic anthropology department and has had immense faith in this project throughout its lifetime. Paul Rabinow, Stephen Collier, and Andrew Lakoff have provided an extremely valuable setting—the Anthropology of the Contemporary Research Collaboratory—within which the arguments of this book developed in ways they could not have as a solitary project. Joe Dumit has encouraged and prodded and questioned and brainstormed and guided and inspired. Michael Fischer is the best mentor and advisor *ever*. He has read everything, has written much that precedes and shapes this work, and has been an unwavering supporter and friend throughout.

Tish Stringer, Michael Powell, Valerie Olson, Ala Alazzeh, Lina Dib, Angela Rivas, Anthony Potoczniak, Ayla Samli, Ebru Kayaalp, Michael Kriz, Erkan Saka, Elise McCarthy, Elitza Ranova, Amanda Randall, Kris Peterson, Laura Jones, Nahal Naficy, Andrea Frolic, and Casey O'Donnell make my job rock. Scott McGill, Sarah Ellenzweig, Stephen Collier, Carl Pearson, Dan Wallach, Tracy Volz, Rich Doyle, Ussama Makdisi, Elora Shehabbudin, Michael Morrow, Taryn Kinney, Gregory Kaplan, Jane Greenberg, Hajime Nakatani, Kirsten Ostherr, Henning Schmidgen, Jason Danziger, Kayte Young, Nicholas King, Jennifer Fishman, Paul Drueke, Roberta Bivins, Sherri Roush, Stefan Timmermans, Laura Lark, and Susann Wilkinson either made Houston a wonderful place to be or provided an opportunity to escape it. I am especially happy that Thom Chivens has done both and more.

The Center for the Study of Cultures provided me with a Faculty Fellowship in the fall of 2003, which allowed me to accomplish much of the work in conceptualizing the book. The Harvard History of Science Department and the MIT Program in History, Anthropology, and Social Studies of Science and Technology hosted me in the spring of 2005, allowing me to write most of chapters 7, 8, and 9. Rice University has been extremely generous in all respects, and a wonderful place to work. I'm most grateful for a junior sabbatical that gave me the chance to complete much of this book. John Hoffman graciously and generously allowed the use of the domain name twobits.net, in support of Free Software. Ken Wissoker, Courtney Berger, and the anonymous reviewers for Duke University Press have made this a much, much better book than when I started.

My parents, Ted and Anne, and my brother, Kevin, have always been supportive and loving; though they claim to have no idea what I do, I nonetheless owe my small success to their constant support. Hannah Landecker has read and reread and rewritten every part of this work; she has made it and me better, and I love her dearly for it. Last, but not least, my new project, Ida Jane Kelty Landecker, is much cuter and smarter and funnier than *Two Bits*, and I love her for distracting me from it.

Introduction

Around 1998 Free Software emerged from a happily subterranean and obscure existence stretching back roughly twenty years. At the very pinnacle of the dotcom boom, Free Software suddenly populated the pages of mainstream business journals, entered the strategy and planning discussions of executives, confounded the radar of political leaders and regulators around the globe, and permeated the consciousness of a generation of technophile teenagers growing up in the 1990s wondering how people ever lived without e-mail. Free Software appeared to be something shocking, something that economic history suggested could never exist: a practice of creating software—good software—that was privately owned, but freely and publicly accessible. Free Software, as its ambiguous moniker suggests, is both free from constraints and free of charge. Such characteristics seem to violate economic logic and the principles of private ownership and individual autonomy, yet there are tens of millions of people creating this software and hundreds of millions more using it. Why? Why now? And most important: how?

Free Software is a set of practices for the distributed collaborative creation of software source code that is then made openly and freely available through a clever, unconventional use of copyright law.[1] But it is much more: Free Software exemplifies a considerable reorientation of knowledge and power in contemporary society—a reorientation of power with respect to the creation, dissemination, and authorization of knowledge in the era of the Internet. This book is about the cultural significance of Free Software, and by *cultural* I mean much more than the exotic behavioral or sartorial traits of software programmers, fascinating though they be. By *culture*, I mean an ongoing experimental system, a space of modification and modulation, of figuring out and testing; culture is an experiment that is hard to keep an eye on, one that changes quickly and sometimes starkly. Culture as an experimental system crosses economies and governments, networked social spheres, and the infrastructure of knowledge and power within which our world functions today— or fails to. Free Software, as a cultural practice, weaves together a surprising range of places, objects, and people; it contains patterns, thresholds, and repetitions that are not simple or immediately obvious, either to the geeks who make Free Software or to those who want to understand it. It is my goal in this book to reveal some of those complex patterns and thresholds, both historically and anthropologically, and to explain not just what Free Software is but also how it has emerged in the recent past and will continue to change in the near future.[2]

The significance of Free Software extends far beyond the arcane and detailed technical practices of software programmers and "geeks" (as I refer to them herein). Since about 1998, the practices and ideas of Free Software have extended into new realms of life and creativity: from software to music and film to science, engineering, and education; from national politics of intellectual property to global debates about civil society; from UNIX to Mac OS X and Windows; from medical records and databases to international disease monitoring and synthetic biology; from Open Source to open access. Free Software is no longer only about software—it exemplifies a more general reorientation of power and knowledge.

The terms *Free Software* and *Open Source* don't quite capture the extent of this reorientation or their own cultural significance. They refer, quite narrowly, to the practice of creating software—an activity many people consider to be quite far from their experience. However, creating Free Software is more than that: it includes a unique combination of more familiar practices that range from creating and policing intellectual property to arguing about the meaning of "openness" to organizing and coordinating people and machines across locales and time zones. Taken together, these practices make Free Software distinct, significant, and meaningful both to those who create it and to those who take the time to understand how it comes into being.

In order to analyze and illustrate the more general cultural significance of Free Software and its consequences, I introduce the concept of a "recursive public." A recursive public is *a public that is vitally concerned with the material and practical maintenance and modification of the technical, legal, practical, and conceptual means of its own existence as a public; it is a collective independent of other forms of constituted power and is capable of speaking to existing forms of power through the production of actually existing alternatives.* Free Software is one instance of this concept, both as it has emerged in the recent past and as it undergoes transformation and differentiation in the near future. There are other instances, including those that emerge from the practices of Free Software, such as Creative Commons, the Connexions project, and the Open Access movement in science. These latter instances may or may not be Free Software, or even "software" projects per se, but they are connected through the same practices, and what makes them significant is that they may also be "recursive publics" in the sense I explore in this book. Recursive publics, and publics generally, differ from interest groups, corporations, unions, professions, churches, and other forms of organization because of their focus on the radical technological modifiability of their own terms of existence. In any public there inevitably arises a moment when the question of how things are said, who controls the means of communication, or whether each and everyone is being properly heard becomes an issue. A legitimate public sphere is one that gives outsiders a way in: they may or may not be heard, but they do not have to appeal to any authority (inside or outside the organization) in order to have a voice.[3] Such publics are not inherently modifiable, but are made so—and maintained—through the practices of participants. It is possible for Free Software as we know it to cease to be public, or to become just one more settled form of power, but my focus is on the recent past and near future of something that is (for the time being) public in a radical and novel way.

The concept of a recursive public is not meant to apply to any and every instance of a public—it is not a replacement for the concept of a "public sphere"—but is intended rather to give readers a specific and detailed sense of the non-obvious, but persistent threads that form the warp and weft of Free Software and to analyze similar and related projects that continue to emerge from it as novel and unprecedented forms of publicity and political action.

At first glance, the thread tying these projects together seems to be the Internet. And indeed, the history and cultural significance of Free Software has been intricately mixed up with that of the Internet over the last thirty years. The Internet is a unique platform—an environment or an infrastructure—for Free Software. But the Internet looks the way it does because of Free Software. Free Software and the Internet are related like figure and ground or like system and environment; neither are stable or unchanging in and of themselves, and there are a number of practical, technical, and historical places where the two are essentially indistinguishable. The Internet is not itself a recursive public, but it is something vitally important to that public, something about which such publics care deeply and act to preserve. Throughout this book, I will return to these three phenomena: the Internet, a heterogeneous and diverse, though singular, infrastructure of technologies and uses; Free Software, a very specific set of technical, legal, and social practices that now require the Internet; and recursive publics, an analytic concept intended to clarify the relation of the first two.

Both the Internet and Free Software are historically specific, that is, not just any old new media or information technology. But the Internet is many, many specific things to many, many specific people. As one reviewer of an early manuscript version of this book noted, "For most people, the Internet is porn, stock quotes, Al Jazeera clips of executions, Skype, seeing pictures of the grandkids, porn, never having to buy another encyclopedia, MySpace, e-mail, online housing listings, Amazon, Googling potential romantic interests, etc. etc." It is impossible to explain all of these things; the meaning and significance of the proliferation of digital pornography is a very different concern than that of the fall of the print encyclopedia and the rise of Wikipedia. Yet certain underlying practices relate these diverse phenomena to one another and help explain why they have occurred at this time and in this technical, legal, and social context. By looking carefully at Free Software and its modulations, I suggest, one can come to a better understanding of the changes affecting pornography, Wikipedia, stock quotes, and many other wonderful and terrifying things.[4]

Two Bits has three parts. Part I of this book introduces the reader to the concept of recursive publics by exploring the lives, works, and discussions of an international community of geeks brought together by their shared interest in the Internet. Chapter 1 asks, in an ethnographic voice, "Why do geeks associate with one another?" The answer—told via the story of Napster in 2000 and the standards process at the heart of the Internet—is that they are making a recursive public. Chapter 2 explores the words and attitudes of geeks more closely, focusing on the strange stories they tell (about the Protestant Reformation, about their practical everyday polymathy, about progress and enlightenment), stories that make sense of contemporary political economy in sometimes surprising ways. Central to part I is an explication of the ways in which geeks argue about technology but also argue with and through it, by building, modifying, and maintaining the very software, networks, and legal tools within which and by which they associate with one another. It is meant to give the reader a kind of visceral sense of why certain arrangements of technology, organization, and law—specifically that of the Internet and Free Software—are so vitally important to these geeks.

Part II takes a step back from ethnographic engagement to ask, "What is Free Software and why has it emerged at this point in history?" Part II is a historically detailed portrait of the emergence of Free Software beginning in 1998–99 and stretching back in time as far as the late 1950s; it recapitulates part I by examining Free Software as an exemplar of a recursive public. The five chapters in part II tell a coherent historical story, but each is focused on a separate component of Free Software. The stories in these chapters help distinguish the figure of Free Software from the ground of the Internet. The diversity of technical practices, economic concerns, information technologies, and legal and organizational practices is huge, and these five chapters distinguish and describe the specific practices in their historical contexts and settings: practices of proselytizing and arguing, of sharing, porting, and forking source code, of conceptualizing openness and open systems, of creating Free Software copyright, and of coordinating people and source code.

Part III returns to ethnographic engagement, analyzing two related projects inspired by Free Software which modulate one or more of the five components discussed in part II, that is, which take the practices as developed in Free Software and experiment with making something new and different. The two projects are Creative Commons, a nonprofit organization that creates copyright licenses, and Connexions, a project to develop an online scholarly textbook commons. By tracing the modulations of practices in detail, I ask, "Are these projects still Free Software?" and "Are these projects still recursive publics?" The answer to the first questions reveals how Free Software's flexible practices are influencing specific forms of practice far from software programming, while the answer to the second question helps explain how Free Software, Creative Commons, Connexions, and projects like them are all related, strategic responses to the reorientation of power and knowledge. The conclusion raises a series of questions intended to help scholars looking at related phenomena.

Recursive Publics and the Reorientation of Power and Knowledge

Governance and control of the creation and dissemination of knowledge have changed considerably in the context of the Internet over the last thirty years. Nearly all kinds of media are easier to produce, publish, circulate, modify, mash-up, remix, or reuse. The number of such creations, circulations, and borrowings has exploded, and the tools of knowledge creation and circulation—software and networks—have also become more and more pervasively available. The results have also been explosive and include anxieties about validity, quality, ownership and control, moral panics galore, and new concerns about the shape and legitimacy of global "intellectual property" systems. All of these concerns amount to a *reorientation of knowledge and power* that is incomplete and emergent, and whose implications reach directly into the heart of the legitimacy, certainty, reliability and especially the finality and temporality of the knowledge and infrastructures we collectively create. It is a reorientation at once more specific and more general than the grand diagnostic claims of an "information" or "network" society, or the rise of knowledge work or knowledge-based economies; it is more specific because it concerns precise and detailed technical and legal practices, more general because it is a *cultural* reorientation, not only an economic or legal one.

Free Software exemplifies this reorientation; it is not simply a technical pursuit but also the creation of a "public," a collective that asserts itself as a check on other constituted forms of power—like states, the church, and corporations—but which remains independent of these domains of power.[5] Free Software is a response to this reorientation that has resulted in a novel form of democratic political action, a means by which publics can be created and maintained in forms not at all familiar to us from the past. Free Software is a public of a particular kind: a recursive public. Recursive publics are publics concerned with the ability to build, control, modify, and maintain the infrastructure that allows them to come into being in the first place and which, in turn, constitutes their everyday practical commitments and the identities of the participants as creative and autonomous individuals. In the cases explored herein, that specific infrastructure includes the creation of the Internet itself, as well as its associated tools and structures, such as Usenet, e-mail, the World Wide Web (www), UNIX and UNIX-derived operating systems, protocols, standards, and standards processes. For the last thirty years, the Internet has been the subject of a contest in which Free Software has been both a central combatant and an important architect.

By calling Free Software a recursive public, I am doing two things: first, I am drawing attention to the democratic and political significance of Free Software and the Internet; and second, I am suggesting that our current understanding (both academic and colloquial) of what counts as a self-governing public, or even as "the public," is radically inadequate to understanding the contemporary reorientation of knowledge and power. The first case is easy to make: it is obvious that there is something political about Free Software, but most casual observers assume, erroneously, that it is simply an ideological stance and that it is anti–intellectual property or technolibertarian. I hope to show how geeks do not start with ideologies, but instead come to them through their involvement in the practices of creating Free Software and its derivatives. To be sure, there are ideologues aplenty, but there are far more people who start out thinking of themselves as libertarians or liberators, but who become something quite different through their participation in Free Software.

The second case is more complex: why another contribution to the debate about the public and public spheres? There are two reasons I have found it necessary to invent, and to attempt to make precise, the concept of a recursive public: the first is to signal the need to include within the spectrum of political activity the creation, modification, and maintenance of software, networks, and legal documents. Coding, hacking, patching, sharing, compiling, and modifying of software are forms of political action that now routinely accompany familiar political forms of expression like free speech, assembly, petition, and a free press. Such activities are expressive in ways that conventional political theory and social science do not recognize: they can both express and "implement" ideas about the social and moral order of society. Software and networks can express ideas in the conventional written sense as well as create (express) infrastructures that allow ideas to circulate in novel and unexpected ways. At an analytic level, the concept of a recursive public is a way of insisting on the importance to public debate of the unruly technical materiality of a political order, not just the embodied discourse (however material) about that order. Throughout this book, I raise the question of how Free Software and the Internet are themselves a public, as well as what that public actually makes, builds, and maintains.

The second reason I use the concept of a recursive public is that conventional publics have been described as "self-grounding," as constituted only through discourse in the conventional sense of speech, writing, and assembly.[6] Recursive publics are "recursive" not only because of the "self-grounding" of commitments and identities but also because they are concerned with the depth or strata of this self-grounding: the layers of technical and legal infrastructure which are necessary for, say, the Internet to exist as the infrastructure of a public. Every act of self-grounding that constitutes a public relies in turn on the existence of a medium or ground through which communication is possible—whether face-to-face speech, epistolary communication, or net-based assembly—and recursive publics relentlessly question the status of these media, suggesting that they, too, must be independent for a public to be authentic. At each of these layers, technical and legal and organizational decisions can affect whether or not the infrastructure will allow, or even ensure, the continued existence of the recursive publics that are concerned with it. Recursive publics' independence from power is not absolute; it is provisional and structured in response to the historically constituted layering of power and control within the infrastructures of computing and communication.

For instance, a very important aspect of the contemporary Internet, and one that has been fiercely disputed (recently under the banner of "net neutrality"), is its *singularity*: there is only one Internet. This was not an inevitable or a technically determined outcome, but the result of a contest in which a series of decisions were made about layers ranging from the very basic physical configuration of the Internet (packet-switched networks and routing systems indifferent to data types), to the standards and protocols that make it work (e.g., TCP/IP or DNS), to the applications that run on it (e-mail, www, ssh). The outcome of these decisions has been to privilege the singularity of the Internet and to champion its standardization, rather than to promote its fragmentation into multiple incompatible networks. These same kinds of decisions are routinely discussed, weighed, and programmed in the activity of various Free Software projects, as well as its derivatives. They are, I claim, decisions embedded in imaginations of order that are simultaneously moral and technical.

By contrast, governments, corporations, nongovernmental organizations (NGOs), and other institutions have plenty of reasons—profit, security, control—to seek to fragment the Internet. But it is the check on this power provided by recursive publics and especially the practices that now make up Free Software that has kept the Internet whole to date. It is a check on power that is by no means absolute, but is nonetheless rigorously and technically concerned with its legitimacy and independence not only from state-based forms of power and control, but from corporate, commercial, and nongovernmental power as well. To the extent that the Internet is public and extensible (including the capability of creating private subnetworks), it is because of the practices discussed herein and their culmination in a recursive public.

Recursive publics respond to governance by directly engaging in, maintaining, and often modifying the infrastructure they seek, as a public, to inhabit and extend—and not only by offering opinions or protesting decisions, as conventional publics do (in most theories of the public sphere). Recursive publics seek to create what might be understood, enigmatically, as a constantly "self-leveling" level playing field. And it is in the attempt to make the playing field self-leveling that they confront and resist forms of power and control that seek to level it to the advantage of one or another large constituency: state, government, corporation, profession. It is important to understand that geeks do not simply want to level the playing field to their advantage—they have no affinity or identity as such. Instead, they wish to devise ways to give the playing field a certain kind of agency, effected through the agency of many different humans, but checked by its technical and legal structure and openness. Geeks do not wish to compete qua capitalists or entrepreneurs unless they can assure themselves that (qua public actors) that they can compete fairly. It is an ethic of justice shot through with an aesthetic of technical elegance and legal cleverness.

The fact that recursive publics respond in this way—through direct engagement and modification—is a key aspect of the reorientation of power and knowledge that Free Software exemplifies. They are reconstituting the relationship between liberty and knowledge in a technically and historically specific context. Geeks create and modify and argue about licenses and source code and protocols and standards and revision control and ideologies of freedom and pragmatism not simply because these things are inherently or universally important, but because they concern the relationship of governance to the freedom of expression and nature of consent. Source code and copyright licenses, revision control and mailing lists are the pamphlets, coffeehouses, and salons of the twenty-first century: Tischgesellschaften become Schreibtischgesellschaften.[7]

The "reorientation of power and knowledge" has two key aspects that are part of the concept of recursive publics: availability and modifiability (or adaptability). Availability is a broad, diffuse, and familiar issue. It includes things like transparency, open governance or transparent organization, secrecy and freedom of information, and open access in science. Availability includes the business-school theories of "disintermediation" and "transparency and accountability" and the spread of "audit culture" and so-called neoliberal regimes of governance; it is just as often the subject of suspicion as it is a kind of moral mandate, as in the case of open access to scientific results and publications.[8] All of these issues are certainly touched on in detailed and practical ways in the creation of Free Software. Debates about the mode of availability of information made possible in the era of the Internet range from digital-rights management and copy protection, to national security and corporate espionage, to scientific progress and open societies.

However, it is modifiability that is the most fascinating, and unnerving, aspect of the reorientation of power and knowledge. Modifiability includes the ability not only to access—that is, to reuse in the trivial sense of using something without restrictions—but to transform it for use in new contexts, to different ends, or in order to participate directly in its improvement and to redistribute or re-circulate those improvements within the same infrastructures while securing the same rights for everyone else. In fact, the core practice of Free Software is the practice of reuse and modification of software source code. Reuse and modification are also the key ideas that projects modeled on Free Software (such as Connexions and Creative Commons) see as their goal. Creative Commons has as its motto "Culture always builds on the past," and they intend that to mean "through legal appropriation and modification." Connexions, which allows authors to create online bits and pieces of textbooks explicitly encourages authors to reuse work by other people, to modify it, and to make it their own. Modifiability therefore raises a very specific and important question about *finality*. When is something (software, a film, music, culture) finished? How long does it remain finished? Who decides? Or more generally, what does its temporality look like, and how does that temporality restructure political relationships? Such issues are generally familiar only to historians and literary scholars who understand the transformation of canons, the interplay of imitation and originality, and the theoretical questions raised, for instance, in textual scholarship. But the contemporary meaning of modification includes both a vast increase in the speed and scope of modifiability and a certain automation of the practice that was unfamiliar before the advent of sophisticated, distributed forms of software.

Modifiability is an oft-claimed advantage of Free Software. It can be updated, modified, extended, or changed to deal with other changing environments: new hardware, new operating systems, unforeseen technologies, or new laws and practices. At an infrastructural level, such modifiability makes sense: it is a response to and an alternative to technocratic forms of planning. It is a way of *planning in* the ability to *plan out;* an effort to continuously secure the ability to deal with surprise and unexpected outcomes; a way of making flexible, modifiable infrastructures like the Internet as safe as permanent, inflexible ones like roads and bridges.

But what is the cultural significance of modifiability? What does it mean to plan in modifiability to culture, to music, to education and science? At a clerical level, such a question is obvious whenever a scholar cannot recover a document written in WordPerfect 2.0 or on a disk for which there are no longer disk drives, or when a library archive considers saving both the media and the machines that read that media. Modifiability is an imperative for building infrastructures that can last longer. However, it is not only a solution to a clerical problem: it creates new possibilities and new problems for long-settled practices like publication, or the goals and structure of intellectual-property systems, or the definition of the finality, lifetime, monumentality, and especially, the identity of a work. Long-settled, seemingly unassailable practices—like the authority of published books or the power of governments to control information—are suddenly confounded and denaturalized by the techniques of modifiability.

Over the last ten to fifteen years, as the Internet has spread exponentially and insinuated itself into the most intimate practices of all kinds of people, the issues of availability and modifiability and the reorientation of knowledge and power they signify have become commonplace. As this has happened, the significance and practices associated with Free Software have also spread—and been modulated in the process. These practices provide a material and meaningful starting point for an array of recursive publics who play with, modulate, and transform them as they debate and build new ways to share, create, license, and control their respective productions. They do not all share the same goals, immediate or long-term, but by engaging in the technical, legal, and social practices pioneered in Free Software, they do in fact share a "social imaginary" that defines a particular relationship between technology, organs of governance (whether state, corporate, or nongovernmental), and the Internet. Scientists in a lab or musicians in a band; scholars creating a textbook or social movements contemplating modes of organization and protest; government bureaucrats issuing data or journalists investigating corruption; corporations that manage

personal data or co-ops that monitor community development—
all these groups and others may find themselves adopting, modulating, rejecting, or refining the practices that have made up Free
Software in the recent past and will do so in the near future.

Experiment and Modulation

What exactly is Free Software? This question is, perhaps surprisingly, an incredibly common one in geek life. Debates about definition and discussions and denunciations are ubiquitous. As an
anthropologist, I have routinely participated in such discussions
and debates, and it is through my immediate participation that
Two Bits opens. In part I I tell stories about geeks, stories that are
meant to give the reader that classic anthropological sense of being thrown into another world. The stories reveal several general
aspects of what geeks talk about and how they do so, without getting into what Free Software is in detail. I start in this way because
my project started this way. I did not initially intend to study Free
Software, but it was impossible to ignore its emergence and mani-
fest centrality to geeks. The debates about the definition of Free
Software that I participated in online and in the field eventually
led me away from studying geeks per se and turned me toward the
central research concern of this book: what is the cultural significance of Free Software?

In part II what I offer is not a definition of Free Software, but
a history of how it came to be. The story begins in 1998, with an
important announcement by Netscape that it would give away the
source code to its main product, Netscape Navigator, and works
backward from this announcement into the stories of the UNIX
operating system, "open systems," copyright law, the Internet, and
tools for coordinating people and code. Together, these five stories
constitute a description of how Free Software works as a practice.
As a cultural analysis, these stories highlight just how experimental
the practices are, and how individuals keep track of and modulate
the practices along the way.

Netscape's decision came at an important point in the life of Free
Software. It was at just this moment that Free Software was becoming aware of itself as a coherent *movement* and not just a diverse amalgamation of projects, tools, or practices. Ironically, this

recognition also betokened a split: certain parties started to insist that the movement be called "Open Source" software instead, to highlight the practical over the ideological commitments of the movement. The proposal itself unleashed an enormous public discussion about what defined Free Software (or Open Source). This enigmatic event, in which a movement became aware of itself at the same time that it began to question its mission, is the subject of chapter 3. I use the term *movement* to designate one of the five core components of Free Software: the practices of argument and disagreement about the meaning of Free Software. Through these practices of discussion and critique, the other four practices start to come into relief, and participants in both Free Software and Open Source come to realize something surprising: for all the ideological distinctions at the level of discourse, they are *doing exactly the same thing* at the level of practice. The affect-laden histrionics with which geeks argue about the definition of what makes Free Software free or Open Source open can be matched only by the sober specificity of the detailed practices they share.

The second component of Free Software is just such a mundane activity: sharing source code (chapter 4). It is an essential and fundamentally routine practice, but one with a history that reveals the goals of software portability, the interactions of commercial and academic software development, and the centrality of source code (and not only of abstract concepts) in pedagogical settings. The details of "sharing" source code also form the story of the rise and proliferation of the UNIX operating system and its myriad derivatives.

The third component, conceptualizing openness (chapter 5), is about the specific technical and "moral" meanings of openness, especially as it emerged in the 1980s in the computer industry's debates over "open systems." These debates concerned the creation of a particular infrastructure, including both technical standards and protocols (a standard UNIX and protocols for networks), and an ideal market infrastructure that would allow open systems to flourish. Chapter 5 is the story of the failure to achieve a market infrastructure for open systems, in part due to a significant blind spot: the role of intellectual property.

The fourth component, applying copyright (and copyleft) licenses (chapter 6), involves the problem of intellectual property as it faced programmers and geeks in the late 1970s and early 1980s. In this chapter I detail the story of the first Free Software license—the GNU General Public License (GPL)—which emerged out of a controversy around a very famous piece of software called EMACS. The controversy is coincident with changing laws (in 1976 and 1980) and changing practices in the software industry—a general drift from trade secret to copyright protection—and it is also a story about the vaunted "hacker ethic" that reveals it in its native practical setting, rather than as a rarefied list of rules.

The fifth component, the practice of coordination and collaboration (chapter 7), is the most talked about: the idea of tens or hundreds of thousands of people volunteering their time to contribute to the creation of complex software. In this chapter I show how novel forms of coordination developed in the 1990s and how they worked in the canonical cases of Apache and Linux; I also highlight how coordination facilitates the commitment to adaptability (or modifiability) over against planning and hierarchy, and how this commitment resolves the tension between individual virtuosity and the need for collective control.

Taken together, these five components make up Free Software— but they are not a definition. Within each of these five practices, many similar and dissimilar activities might reasonably be included. The point of such a redescription of the practices of Free Software is to conceptualize them as a kind of *collective technical experimental system*. Within each component are a range of differences in practice, from conventional to experimental. At the center, so to speak, are the most common and accepted versions of a practice; at the edges are more unusual or controversial versions. Together, the components make up an experimental system whose infrastructure is the Internet and whose "hypotheses" concern the reorientation of knowledge and power.

For example, one can hardly have Free Software without source code, but it need not be written in C (though the vast majority of it is); it can be written in Java or perl or TeX. However, if one stretches the meaning of source code to include music (sheet music as source and performance as binary), what happens? Is this still Free Software? What happens when both the sheet and the performance are "born digital"? Or, to take a different example, Free Software requires Free Software licenses, but the terms of these licenses are often changed and often heatedly discussed and vigilantly policed by geeks. What degree of change removes a license from the realm of Free Software and why? How much flexibility is allowed?

Conceived this way, Free Software is a system of thresholds, not of classification; the excitement that participants and observers sense comes from the modulation (experimentation) of each of these practices and the subsequent discovery of where the thresholds are. Many, many people have written their own "Free Software" copyright licenses, but only some of them remain within the threshold of the practice as defined by the system. Modulations happen whenever someone learns how some component of Free Software works and asks, "Can I try these practices out in some other domain?"

The reality of constant modulation means that these five practices do not define Free Software once and for all; they define it with respect to its constitution in the contemporary. It is a set of practices defined "around the point" 1998–99, an intensive coordinate space that allows one to explore Free Software's components prospectively and retrospectively: into the near future and the recent past. Free Software is a machine for charting the (re)emergence of a problematic of power and knowledge as it is filtered through the technical realities of the Internet and the political and economic configuration of the contemporary. Each of these practices has its own temporality of development and emergence, but they have recently come together into this full house called either Free Software or Open Source.[9]

Viewing Free Software as an experimental system has a strategic purpose in *Two Bits*. It sets the stage for part III, wherein I ask what kinds of modulations might no longer qualify as Free Software per se, but still qualify as recursive publics. It was around 2000 that talk of "commons" began to percolate out of discussions about Free Software: commons in educational materials, commons in biodiversity materials, commons in music, text, and video, commons in medical data, commons in scientific results and data.[10] On the one hand, it was continuous with interest in creating "digital archives" or "online collections" or "digital libraries"; on the other hand, it was a conjugation of the digital collection with the problems and practices of intellectual property. The very term *commons*—at once a new name and a theoretical object of investigation—was meant to suggest something more than simply a collection, whether of digital objects or anything else; it was meant to signal the public interest, collective management, and legal status of the collection.[11]

In part III, I look in detail at two "commons" understood as modulations of the component practices of Free Software. Rather than treating commons projects simply as metaphorical or inspirational uses of Free Software, I treat them as modulations, which allows me to remain directly connected to the changing practices involved. The goal of part III is to understand how commons projects like Connexions and Creative Commons breach the thresholds of these practices and yet maintain something of the same orientation. What changes, for instance, have made it possible to imagine new forms of free content, free culture, open source music, or a science commons? What happens as new communities of people adopt and modulate the five component practices? Do they also become recursive publics, concerned with the maintenance and expansion of the infrastructures that allow them to come into being in the first place? Are they concerned with the implications of availability and modifiability that continue to unfold, continue to be figured out, in the realms of education, music, film, science, and writing?

The answers in part III make clear that, so far, these concerns are alive and well in the modulations of Free Software: Creative Commons and Connexions each struggle to come to terms with new ways of creating, sharing, and reusing content in the contemporary legal environment, with the Internet as infrastructure. Chapters 8 and 9 provide a detailed analysis of a technical and legal experiment: a modulation that begins with source code, but quickly requires modulations in licensing arrangements and forms of coordination. It is here that *Two Bits* provides the most detailed story of figuring out set against the background of the reorientation of knowledge and power. This story is, in particular, one of *reuse*, of modifiability and the problems that emerge in the attempt to build it into the everyday practices of pedagogical writing and cultural production of myriad forms. Doing so leads the actors involved directly to the question of the existence and ontology of norms: norms of scholarly production, borrowing, reuse, citation, reputation, and ownership. These last chapters open up questions about the stability of modern knowledge, not as an archival or a legal problem, but as a social and normative one; they raise questions about the invention and control of norms, and the forms of life that may emerge from these practices. Recursive publics come to exist where it is clear that such invention and control need to be widely shared, openly examined, and carefully monitored.

Three Ways of Looking at *Two Bits*

Two Bits makes three kinds of scholarly contributions: empirical, methodological, and theoretical. Because it is based largely on fieldwork (which includes historical and archival work), these three contributions are often mixed up with each other. Fieldwork, especially in cultural and social anthropology in the last thirty years, has come to be understood less and less as one particular tool in a methodological toolbox, and more and more as distinctive mode of epistemological encounter.[12] The questions I began with emerged out of science and technology studies, but they might end up making sense to a variety of fields, ranging from legal studies to computer science.

Empirically speaking, the actors in my stories are figuring something out, something unfamiliar, troubling, imprecise, and occasionally shocking to everyone involved at different times and to differing extents.[13] There are two kinds of figuring-out stories: the contemporary ones in which I have been an active participant (those of Connexions and Creative Commons), and the historical ones conducted through "archival" research and rereading of certain kinds of texts, discussions, and analyses-at-the-time (those of UNIX, EMACS, Linux, Apache, and Open Systems). Some are stories of technical figuring out, but most are stories of figuring out a problem that appears to have emerged. Some of these stories involve callow and earnest actors, some involve scheming and strategy, but in all of them the figuring out is presented "in the making" and not as something that can be conveniently narrated as obvious and uncontested with the benefit of hindsight. Throughout this book, I tell stories that illustrate what geeks are like in some respects, but, more important, that show them in the midst of figuring things out—a practice that can happen both in discussion and in the course of designing, planning, executing, writing, debugging, hacking, and fixing.

There are also myriad ways in which geeks narrate their own actions to themselves and others, as they figure things out. Indeed, there is no crisis of representing the other here: geeks are vocal, loud, persistent, and loquacious. The superalterns can speak for themselves. However, such representations should not necessarily be taken as evidence that geeks provide adequate analytic or critical explanations of their own actions. Some of the available writing provides excellent description, but distracting analysis. Eric Raymond's work is an example of such a combination.[14] Over the course of my fieldwork, Raymond's work has always been present as an excellent guide to the practices and questions that plague geeks—much like a classic "principal informant" in anthropology. And yet his analyses, which many geeks subscribe to, are distracting. They are fanciful, occasionally enjoyable and enlightening— but they are not about the cultural significance of Free Software. As such I am less interested in treating geeks as natives to be explained and more interested in arguing with them: the people in *Two Bits* are a *sine qua non* of the ethnography, but they are not the objects of its analysis.[15]

Because the stories I tell here are in fact recent by the standards of historical scholarship, there is not much by way of comparison in terms of the empirical material. I rely on a number of books and articles on the history of the early Internet, especially Janet Abbate's scholarship and the single historical work on UNIX, Peter Salus's *A Quarter Century of Unix*.[16] There are also a couple of excellent journalistic works, such as Glyn Moody's *Rebel Code: Inside Linux and the Open Source Revolution* (which, like *Two Bits*, relies heavily on the novel accessibility of detailed discussions carried out on public mailing lists). Similarly, the scholarship on Free Software and its history is just starting to establish itself around a coherent set of questions.[17]

Methodologically, *Two Bits* provides an example of how to study *distributed phenomena* ethnographically. Free Software and the Internet are objects that do not have a single geographic site at which they can be studied. Hence, this work is multisited in the simple sense of having multiple sites at which these objects were investigated: Boston, Bangalore, Berlin, Houston. It was conducted among particular people, projects, and companies and at conferences and online gatherings too numerous to list, but it has not been a study of a single Free Software project distributed around the globe. In all of these places and projects the geeks I worked with were randomly and loosely affiliated people with diverse lives and histories. Some identified as Free Software hackers, but most did not. Some had never met each other in real life, and some had. They represented multiple corporations and institutions, and came from diverse nations, but they nonetheless shared a certain set of ideas and idioms that made it possible for me to travel from Boston to Berlin to Bangalore and pick up an ongoing conversation with different people, in very different places, without missing a beat.

The study of distributed phenomena does not necessarily imply the detailed, local study of each instance of a phenomenon, nor does it necessitate visiting every relevant geographical site—indeed, such a project is not only extremely difficult, but confuses map and territory. As Max Weber put it, "It is not the 'actual' inter-connection of 'things' but the *conceptual* inter-connection of *problems* that define the scope of the various sciences."[18] The decisions about where to go, whom to study, and how to think about Free Software are arbitrary in the precise sense that because the phenomena are so widely distributed, it is possible to make any given node into a source of rich and detailed knowledge about the distributed phenomena itself, not only about the local site. Thus, for instance, the Connexions project would probably have remained largely unknown to me had I not taken a job in Houston, but it nevertheless possesses precise, identifiable connections to the other sites and sets of people that I have studied, and is therefore recognizable as part of *this* distributed phenomena, rather than some other. I was actively looking for something like Connexions in order to ask questions about what was becoming of Free Software and how it was transforming. Had there been no Connexions in my back yard, another similar field site would have served instead.

It is in this sense that the ethnographic object of this study is not geeks and not any particular project or place or set of people, but Free Software and the Internet. Even more precisely, the ethnographic object of this study is "recursive publics"—except that this concept is also the *work* of the ethnography, not its preliminary object. I could not have identified "recursive publics" as the object of the ethnography at the outset, and this is nice proof that ethnographic work is a particular kind of epistemological encounter, an encounter that requires considerable conceptual work during and after the material labor of fieldwork, and throughout the material labor of writing and rewriting, in order to make sense of and reorient it into a question that will have looked deliberate and answerable in hindsight. Ethnography of this sort requires a long-term commitment and an ability to see past the obvious surface of rapid transformation to a more obscure and slower temporality of cultural significance, yet still pose questions and refine debates about the near future.[19] Historically speaking, the chapters of part II can be understood as a contribution to a history of scientific infrastructure—or perhaps to an understanding of large-scale, collective experimentation.[20] The Internet and Free Software are each an important practical transformation that will have effects on the practice of science and a kind of complex technical practice for which there are few existing models of study.

A methodological note about the peculiarity of my subject is also in order. The Attentive Reader will note that there are very few fragments of conventional ethnographic material (i.e., interviews or notes) transcribed herein. Where they do appear, they tend to be "publicly available"—which is to say, accessible via the Internet—and are cited as such, with as much detail as necessary to allow the reader to recover them. Conventional wisdom in both anthropology and history has it that what makes a study interesting, in part, is the work a researcher has put into gathering that which is not already available, that is, primary sources as opposed to secondary sources. In some cases I provide that primary access (specifically in chapters 2, 8, and 9), but in many others it is now literally impossible: *nearly everything is archived.* Discussions, fights, collaborations, talks, papers, software, articles, news stories, history, old software, old software manuals, reminiscences, notes, and drawings—it is all saved by someone, somewhere, and, more important, often made instantly available by those who collect it. The range of conversations and interactions that count as private (either in the sense of disappearing from written memory or of being accessible only to the parties involved) has shrunk demonstrably since about 1981.

Such obsessive archiving means that ethnographic research is stratified in time. Questions that would otherwise have required "being there" are much easier to research after the fact, and this is most evident in my reconstruction from sources on USENET and mailing lists in chapters 1, 6, and 7. The overwhelming availability of quasi-archival materials is something I refer to, in a play on the EMACS text editor, as "self-documenting history." That is to say, one of the activities that geeks love to participate in, and encourage, is the creation, analysis, and archiving of their own roles in the development of the Internet. No matter how obscure or arcane, it seems most geeks have a well-developed sense of possibility—their contribution could turn out to have been transformative, important, originary. What geeks may lack in social adroitness, they make up for in archival hubris.

Finally, the theoretical contribution of *Two Bits* consists of a refinement of debates about publics, public spheres, and social imaginaries that appear troubled in the context of the Internet and Free Software. Terminology such as *virtual community*, *online community*, *cyberspace*, *network society*, or *information society* are generally not theoretical constructs, but ways of designating a subgenre of disciplinary research having to do with electronic networks. The need for a more precise analysis of the kinds of association that take place on and through information technology is clear; the first step is to make precise which information technologies and which specific practices make a difference.

There is a relatively large and growing literature on the Internet as a public sphere, but such literature is generally less concerned with refining the concept through research and more concerned with pronouncing whether or not the Internet fits Habermas's definition of the bourgeois public sphere, a definition primarily conceived to account for the eighteenth century in Britain, not the twenty-first-century Internet.[21] The facts of technical and human life, as they unfold through the Internet and around the practices of Free Software, are not easy to cram into Habermas's definition. The goal of *Two Bits* is not to do so, but to offer conceptual clarity based in ethnographic fieldwork.

The key texts for understanding the concept of recursive publics are the works of Habermas, Charles Taylor's *Modern Social Imaginaries*, and Michael Warner's *The Letters of the Republic* and *Publics and Counterpublics*. Secondary texts that refine these notions are John Dewey's *The Public and Its Problems* and Hannah Arendt's *The Human Condition*. Here it is not the public sphere per se that is the center of analysis, but the "ideas of modern moral and social order" and the terminology of "modern social imaginaries."[22] I find these concepts to be useful as starting points for a very specific reason: to distinguish the meaning of moral order from the meaning of *moral and technical order* that I explore with respect to geeks. I do not seek to test the concept of social imaginary here, but to build something on top of it.

If recursive public is a useful concept, it is because it helps elaborate the general question of the "reorientation of knowledge and power." In particular it is meant to bring into relief the ways in which the Internet and Free Software are related to the political economy of modern society through the creation not only of new knowledge, but of new infrastructures for circulating, maintaining, and modifying it. Just as Warner's book *The Letters of the Republic* was concerned with the emergence of the discourse of republicanism and the simultaneous development of an American republic of letters, or as Habermas's analysis was concerned with the relationship of the bourgeois public sphere to the democratic revolutions of the eighteenth century, this book asks a similar series of questions: how are the emergent practices of recursive publics related to emerging relations of political and technical life in a world that submits to the Internet and its forms of circulation? Is there still a role for a republic of letters, much less a species of public that can seriously claim independence and autonomy from other constituted forms of power? Are Habermas's pessimistic critiques of the bankruptcy of the public sphere in the twentieth century equally applicable to the structures of the twenty-first century? Or is it possible that recursive publics represent a reemergence of strong, authentic publics in a world shot through with cynicism and suspicion about mass media, verifiable knowledge, and enlightenment rationality?

PART I ⋮ THE INTERNET

The concept of the state, like most concepts which are introduced by "The," is both too rigid and too tied up with controversies to be of ready use. It is a concept which can be approached by a flank movement more easily than by a frontal attack. The moment we utter the words "The State" a score of intellectual ghosts rise to obscure our vision. Without our intention and without our notice, the notion of "The State" draws us imperceptibly into a consideration of the logical relationship of various ideas to one another, and away from the facts of human activity. It is better, if possible, to start from the latter and see if we are not led thereby into an idea of something which will turn out to implicate the marks and signs which characterize political behavior.

—JOHN DEWEY, *The Public and Its Problems*

Geeks and Recursive Publics 1.

Since about 1997, I have been living with geeks online and off. I have been drawn from Boston to Bangalore to Berlin to Houston to Palo Alto, from conferences and workshops to launch parties, pubs, and Internet Relay Chats (IRCs). All along the way in my research questions of commitment and practice, of ideology and imagination have arisen, even as the exact nature of the connections between these people and ideas remained obscure to me: what binds geeks together? As my fieldwork pulled me from a Boston start-up company that worked with radiological images to media labs in Berlin to young entrepreneurial elites in Bangalore, my logistical question eventually developed into an analytical concept: geeks are bound together as a recursive public.

How did I come to understand geeks as a public constituted around the technical and moral ideas of order that allow them to associate with one another? Through this question, one can start to understand the larger narrative of *Two Bits*: that of Free Software as an exemplary instance of a recursive public and as a set of practices that allow such publics to expand and spread. In this chapter I describe, ethnographically, the diverse, dispersed, and novel forms of entanglements that bind geeks together, and I construct the concept of a recursive public in order to explain these entanglements.

A recursive public is a public that is constituted by a shared concern for maintaining the means of association through which they come together as a public. Geeks find affinity with one another because they share an abiding moral imagination of the technical infrastructure, the Internet, that has allowed them to develop and maintain this affinity in the first place. I elaborate the concept of recursive public (which is not a term used by geeks) in relation to theories of ideology, publics, and public spheres and social imaginaries. I illustrate the concept through ethnographic stories and examples that highlight geeks' imaginations of the technical and moral order of the Internet. These stories include those of the fate of Amicas, a Boston-based healthcare start-up, between 1997 and 2003, of my participation with new media academics and activists in Berlin in 1999–2001, and of the activities of a group of largely Bangalore-based information technology (IT) professionals on and offline, especially concerning the events surrounding the peer-to-peer file sharing application Napster in 2000–2001.

The phrase "moral and technical order" signals both technology—principally software, hardware, networks, and protocols—and an imagination of the proper order of collective political and commercial action, that is, how economy and society should be ordered collectively. Recursive publics are just as concerned with the moral order of markets as they are with that of commons; they are not anticommercial or antigovernment. They exist independent of, and as a check on, constituted forms of power, which include markets and corporations. Unlike other concepts of a public or of a public sphere, "recursive public" captures the fact that geeks' principal mode of associating and acting is through the medium of the Internet, and it is through this medium that a recursive public can come into being in the first place. The Internet is not itself a public sphere, a public, or a recursive public, but a complex, heterogeneous infrastructure that constitutes and constrains geeks' everyday practical commitments, their ability to "become public" or to compose a common world. As such, their participation qua recursive publics structures their identity as creative and autonomous individuals. The fact that the geeks described here have been brought together by mailing lists and e-mail, bulletin-board services and Web sites, books and modems, air travel and academia, and cross-talking and cross-posting in ways that were not possible before the Internet is at the core of their own reasoning about why they associate with each other. They are the builders and imaginers of this space, and the space is what allows them to build and imagine it.

Why recursive? I call such publics *recursive* for two reasons: first, in order to signal that this kind of public includes the activities of making, maintaining, and modifying software and networks, as well as the more conventional discourse that is thereby enabled; and second, in order to suggest the recursive "depth" of the public, the series of technical and legal layers—from applications to protocols to the physical infrastructures of waves and wires—that are the subject of this making, maintaining, and modifying. The first of these characteristics is evident in the fact that geeks use technology as a kind of argument, for a specific kind of order: they argue *about* technology, but they also argue *through* it. They express ideas, but they also express *infrastructures* through which ideas can be expressed (and circulated) in new ways. The second of these characteristics—regarding layers—is reflected in the ability of geeks to immediately see connections between, for example, Napster (a user application) and TCP/IP (a network protocol) and to draw out implications for both of them. By connecting these layers, Napster comes to represent the Internet in miniature. The question of where these layers stop (hardware? laws and regulations? physical constants? etc.) circumscribes the limits of the imagination of technical and moral order shared by geeks.

Above all, "recursive public" is a concept—not a thing. It is intended to make distinctions, allow comparison, highlight salient features, and relate two diverse kinds of things (the Internet and Free Software) in a particular historical context of changing relations of power and knowledge. The stories in this chapter (and throughout the book) give some sense of *how* geeks interact and what they do technically and legally, but the concept of a recursive public provides a way of explaining *why* geeks (or people involved in Free Software or its derivatives) associate with one another, as well as a way of testing whether other similar cases of contemporary, technologically mediated affinity are similarly structured.

Recursion

Recursion (or "recursive") is a mathematical concept, one which is a standard feature of any education in computer programming. The definition from the Oxford English Dictionary reads: "2. a. Involving or being a repeated procedure such that the required result at each step except the last is given in terms of the result(s) of the next step, until after a finite number of steps a terminus is reached with an outright evaluation of the result." It should be distinguished from simple iteration or repetition. Recursion is always subject to a limit and is more like a process of repeated deferral, until the last step in the process, at which point all the deferred steps are calculated and the result given.

Recursion is powerful in programming because it allows for the definition of procedures in terms of themselves—something that seems at first counterintuitive. So, for example,

```
(defun (factorial n)          ; This is the name of the function and its input n.
(if (=n 1)                    ; This is the final limit, or recursive depth
1                             ; if n=1, then return 1
(* n (factorial (- n 1)))))   ; otherwise return n times factorial of n-1;
                              ; call the procedure from within itself, and
                              ; calculate the next step of the result before
                              ; giving an answer.[1]
```

In *Two Bits* a recursive public is one whose existence (which consists solely in address through discourse) is only possible through discursive and technical reference to the means of creating this public. Recursiveness is always contingent on a limit which determines the depth of a recursive procedure. So, for instance, a Free Software project may depend on some other kind of software or operating system, which may in turn depend on particular open protocols or a particular process, which in turn depend on certain kinds of hardware that implement them. The "depth" of recursion is determined by the openness necessary for the project itself.

James Boyle has also noted the recursive nature, in particular, of Free Software: "What's more, and this is a truly fascinating twist, when the production process does need more centralized coordination, some governance that guides how the sticky modular bits are put together, it is at least theoretically possible that we can come up with the control system *in exactly the same way*. In this sense, distributed production is potentially recursive."[2]

---

1. Abelson and Sussman, *The Structure and Interpretation of Computer Programs*, 30.
2. Boyle, "The Second Enclosure Movement and the Construction of the Public Domain," 46.

From the Facts of Human Activity

Boston, May 2003. Starbucks. Sean and Adrian are on their way to pick me up for dinner. I've already had too much coffee, so I sit at the window reading the paper. Eventually Adrian calls to find out where I am, I tell him, and he promises to show up in fifteen minutes. I get bored and go outside to wait, watch the traffic go by. More or less right on time (only post-dotcom is Adrian ever on time), Sean's new blue VW Beetle rolls into view. Adrian jumps out of the passenger seat and into the back, and I get in. Sean has been driving for a little over a year. He seems confident, cautious, but meanders through the streets of Cambridge. We are destined for Winchester, a township on the Charles River, in order to go to an Indian restaurant that one of Sean's friends has recommended. When I ask how they are doing, they say, "Good, good." Adrian offers, "Well, Sean's better than he has been in two years." "Really?" I say, impressed.

Sean says, "Well, happier than at least the last year. I, well, let me put it this way: forgive me father for I have sinned, I still have unclean thoughts about some of the upper management in the company, I occasionally think they are not doing things in the best interest of the company, and I see them as self-serving and sometimes wish them ill." In this rolling blue confessional Sean describes some of the people who I am familiar with whom he now tries very hard not to think about. I look at him and say, "Ten Hail Marys and ten Our Fathers, and you will be absolved, my child." Turning to Adrian, I ask, "And what about you?" Adrian continues the joke: "I, too, have sinned. I have reached the point where I can see absolutely nothing good coming of this company but that I can keep my investments in it long enough to pay for my children's college tuition." I say, "You, my son, I cannot help." Sean says, "Well, funny thing about tainted money . . . there just taint enough of it."

I am awestruck. When I met Sean and Adrian, in 1997, their start-up company, Amicas, was full of spit, with five employees working out of Adrian's living room and big plans to revolutionize the medical-imaging world. They had connived to get Massachusetts General Hospital to install their rudimentary system and let it compete with the big corporate sloths that normally stalked back offices: General Electric, Agfa, Siemens. It was these behemoths, according to Sean and Adrian, that were bilking hospitals and healthcare providers with promises of cure-all technologies and horribly designed "silos," "legacy systems," and other closed-system monsters of corporate IT harkening back to the days of IBM mainframes. These beasts obviously did not belong to the gleaming future of Internet-enabled scalability. By June of 2000, Amicas had hired new "professional" management, moved to Watertown, and grown to about a hundred employees. They had achieved their goal of creating an alternative Picture Archiving and Communication System (PACS) for use in hospital radiology departments and based on Internet standards.

At that point, in the spring of 2000, Sean could still cheerfully introduce me to his new boss—the same man he would come to hate, inasmuch as Sean hates anyone. But by 2002 he was frustrated by the extraordinary variety of corner-cutting and, more particularly, by the complacency with which management ignored his recommendations and released software that was almost certainly going to fail later, if not sooner. Sean, who is sort of permanently callow about things corporate, could find no other explanation than that the new management was evil.

But by 2003 the company had succeeded, having grown to more than 200 employees and established steady revenue and a stable presence throughout the healthcare world. Both Sean and Adrian were made rich—not wildly rich, but rich enough—by its success. In the process, however, it also morphed into exactly what Sean and Adrian had created it in order to fight: a slothlike corporate purveyor of promises and broken software. Promises Adrian had made and software Sean had built. The failure of Amicas to transform healthcare was a failure too complex and technical for most of America to understand, but it rested atop the success of Amicas in terms more readily comprehensible: a growing company making profit. Adrian and Sean had started the company not to make money, but in order to fix a broken healthcare system; yet the system stayed broken while they made money.

In the rolling confessional, Sean and Adrian did in fact see me, however jokingly, as a kind of redeemer, a priest (albeit of an order with no flock) whose judgment of the affairs past was essential to their narration of their venture as a success, a failure, or as an unsatisfying and complicated mixture of both. I thought about this strange moment of confession, of the combination of recognition and denial, of Adrian's new objectification of the company as an investment opportunity, and of Sean's continuing struggle to make his life and his work harmonize in order to produce good in the world. Only the promise of the next project, the next mission (and the ostensible reason for our dinner meeting) could possibly have mitigated the emotional disaster that their enterprise might otherwise be. Sean's and Adrian's endless, arcane fervor for the promise of new technologies did not cease, even given the quotidian calamities these technologies leave in their wake. Their faith was strong, and continuously tested.

Adrian's and Sean's passion was not for money—though money was a powerful drug—it was for the Internet: for the ways in which the Internet could replace the existing infrastructure of hospitals and healthcare providers, deliver on old promises of telemedicine and teleradiology, and, above all, level a playing field systematically distorted and angled by corporate and government institutions that sought secrecy and private control, and stymied progress. In healthcare, as Adrian repeatedly explained to me, this skewed playing field was not only unfair but malicious and irresponsible. It was costing lives. It slowed the creation and deployment of technologies and solutions that could lower costs and thus provide more healthcare for more people. The Internet was not part of the problem; it was part of the solution to the problems that ailed 1990s healthcare.

At the end of our car trip, at the Indian restaurant in Winchester, I learned about their next scheme, a project called MedCommons, which would build on the ideals of Free Software and give individuals a way to securely control and manage their own healthcare data. The rhetoric of commons and the promise of the Internet as an infrastructure dominated our conversation, but the realities of funding and the question of whether MedCommons could be pursued without starting another company remained unsettled. I tried to imagine what form a future confession might take.

Geeks and Their Internets

Sean and Adrian are geeks. They are entrepreneurs and idealists in different ways, a sometimes paradoxical combination. They are certainly obsessed with technology, but especially with the Internet, and they clearly distinguish themselves from others who are obsessed with technology of just any sort. They aren't quite representative—they do not stand in for all geeks—but the way they think about the Internet and its possibilities might be. Among the rich story of their successes and failures, one might glimpse the outlines of a question: where do their sympathies lie? Who are they *with*? Who do they recognize as being like them? What might draw them together with other geeks if not a corporation, a nation, a language, or a cause? What binds these two geeks to any others?

Sean worked for the Federal Reserve in the 1980s, where he was introduced to UNIX, C programming, EMACS, Usenet, Free Software, and the Free Software Foundation. But he was not a Free Software hacker; indeed, he resisted my attempts to call him a hacker at all. Nevertheless, he started a series of projects and companies with Adrian that drew on the repertoire of practices and ideas familiar from Free Software, including their MedCommons project, which was based more or less explicitly in the ideals of Free Software. Adrian has a degree in medicine and in engineering, and is a serial entrepreneur, with Amicas being his biggest success— and throughout the last ten years has attended all manner of conferences and meetings devoted to Free Software, Open Source, open standards, and so on, almost always as the lone representative from healthcare. Both graduated from the MIT (Sean in economics, Adrian in engineering), one of the more heated cauldrons of the Internet and the storied home of hackerdom, but neither were MIT hackers, nor even computer-science majors.

Their goals in creating a start-up rested on their understanding of the Internet as an infrastructure: as a *standardized infrastructure* with certain extremely powerful properties, not the least of which was its flexibility. Sean and Adrian talked endlessly about open systems, open standards, and the need for the Internet to remain open and standardized. Adrian spoke in general terms about how it would revolutionize healthcare; Sean spoke in specific terms about how it structured the way Amicas's software was being designed and written. Both participated in standards committees and in the online and offline discussions that are tantamount to policymaking in the Internet world. The company they created was a "virtual" company, that is, built on tools that depended on the Internet and allowed employees to manage and work from a variety of locations, though not without frustration, of course: Sean waited years for broadband access in his home, and the hospitals they served hemmed themselves in with virtual private networks, intranets, and security firewalls that betrayed the promises of openness that Sean and Adrian heralded.

The Internet was not the object of their work and lives, but it did represent in detail a kind of moral or social order embodied in a technical system and available to everyone to use as a platform whereby they might compete to improve and innovate in any realm. To be sure, although not all Internet entrepreneurs of the 1990s saw the Internet in the same way, Sean and Adrian were hardly alone in their vision. Something about the particular way in which they understood the Internet as representing a moral order— simultaneously a network, a market, a public, and a technology— was shared by a large group of people, those who I now refer to simply as geeks.

The term *geek* is meant to be inclusive and to index the problematic of a recursive public. Other terms may be equally useful, but perhaps semantically overdetermined, most notably *hacker*, which regardless of its definitional range, tends to connote someone subversive and/or criminal and to exclude geek-sympathetic entrepreneurs and lawyers and activists.[1] *Geek* is meant to signal, like the *public* in "recursive public," that geeks stand outside power, at least in some aspects, and that they are not capitalists or technocrats, even if they start businesses or work in government or industry.[2] *Geek* is meant to signal a mode of thinking and working, not an identity; it is a mode or quality that allows people to find each other, for reasons other than the fact that they share an office, a degree, a language, or a nation.

Until the mid-1990s, *hacker, geek,* and *computer nerd* designated a very specific type: programmers and lurkers on relatively underground networks, usually college students, computer scientists, and "amateurs" or "hobbyists." A classic mock self-diagnostic called the Geek Code, by Robert Hayden, accurately and humorously detailed the various ways in which one could be a geek in 1996—UNIX/Linux skills, love/hate of *Star Trek*, particular eating and clothing habits—but as Hayden himself points out, the geeks of the early 1990s exist no longer. The elite subcultural, relatively homogenous group it once was has been overrun: "The Internet of 1996 was still a wild untamed virgin paradise of geeks and eggheads unpopulated by script kiddies, and the denizens of AOL. When things changed, I seriously lost my way. I mean, all the 'geek' that was the Internet was gone and replaced by Xfiles buzzwords and politicians passing laws about a technology they refused to comprehend."[3]

For the purists like Hayden, geeks were there first, and they understood something, lived in a way, that simply cannot be comprehended by "script kiddies" (i.e., teenagers who perform the hacking equivalent of spray painting or cow tipping), crackers, or AOL users, all of whom are despised by Hayden-style geeks as unskilled users who parade around the Internet as if they own it. While certainly elitist, Hayden captures the distinction between those who are legitimately allowed to call themselves geeks (or hackers) and those who aren't, a distinction that is often formulated recursively, of course: "You are a hacker when another hacker calls you a hacker."

However, since the explosive growth of the Internet, *geek* has become more common a designation, and my use of the term thus suggests a role that is larger than programmer/hacker, but not as large as "all Internet users." Despite Hayden's frustration, geeks are still bound together as an elite and can be easily distinguished from "AOL users." Some of the people I discuss would not call themselves geeks, and some would. Not all are engineers or programmers: I have met businessmen, lawyers, activists, bloggers, gastroenterologists, anthropologists, lesbians, schizophrenics, scientists, poets, people suffering from malaria, sea captains, drug dealers, and people who keep lemurs, many of whom refer to themselves as geeks, some of the time.[4] There are also lawyers, politicians, sociologists, and economists who may not refer to themselves as geeks, but who care about the Internet just as other geeks do. By contrast "users" of the Internet, even those who use it eighteen out of twenty-four hours in a day to ship goods and play games, are not necessarily geeks by this characterization.

Operating Systems and Social Systems

Berlin, November 1999. I am in a very hip club in Mitte called WMF. It's about eight o'clock—five hours too early for me to be a hipster, but the context is extremely cool. WMF is in a hard-to-find, abandoned building in the former East; it is partially converted, filled with a mixture of new and old furnishings, video projectors, speakers, makeshift bars, and dance-floor lighting. A crowd of around fifty people lingers amid smoke and Beck's beer bottles, sitting on stools and chairs and sofas and the floor. We are listening to an academic read a paper about Claude Shannon, the MIT engineer credited with the creation of information theory. The author is smoking and reading in German while the audience politely listens. He speaks for about seventy minutes. There are questions and some perfunctory discussion. As the crowd breaks up, I find myself, in halting German that quickly converts to English, having a series of animated conversations about the GNU General Public License, the Debian Linux Distribution, open standards in net radio, and a variety of things for which Claude Shannon is the perfect ghostly technopaterfamilias, even if his seventy-minute invocation has clashed heavily with the surroundings.

Despite my lame German, I still manage to jump deeply into issues that seem extremely familiar: Internet standards and open systems and licensing issues and namespaces and patent law and so on. These are not businesspeople, this is not a start-up company. As I would eventually learn, there was even a certain disdain for *die Krawattenfaktor*, the suit-and-tie factor, at these occasional, hybrid events hosted by Mikro e.V., a nonprofit collective of journalists, academics, activists, artists, and others interested in new media, the Internet, and related issues. Mikro's constituency included people from Germany, Holland, Austria, and points eastward. They took some pride in describing Berlin as "the farthest East the West gets" and arranged for a group photo in which, facing West, they stood behind the statue of Marx and Lenin, who face East and look eternally at the iconic East German radio tower (*Funkturm*) in Alexanderplatz. Mikro's members are resolutely activist and see the issues around the Internet-as-infrastructure not in terms of its potential for business opportunities, but in urgently political and unrepentantly aesthetic terms—terms that are nonetheless similar to those of Sean and Adrian, from whom I learned the language that allows me to mingle with the Mikro crowd at WMF. I am now a geek.

Before long, I am talking with Volker Grassmuck, founding member of Mikro and organizer of the successful "Wizards of OS" conference, held earlier in the year, which had the very intriguing subtitle "Operating Systems and Social Systems." Grassmuck is inviting me to participate in a planning session for the next WOS, held at the Chaos Computer Congress, a hacker gathering that occurs each year in December in Berlin. In the following months I will meet a huge number of people who seem, uncharacteristically for artists and activists, strangely obsessed with configuring their Linux distributions or hacking the http protocol or attending German Parliament hearings on copyright reform. The political lives of these folks have indeed mixed up operating systems and social systems in ways that are more than metaphorical.

The Idea of Order at the Keyboard

If intuition can lead one from geek to geek, from start-up to nightclub, and across countries, languages, and professional orientations, it can only be due to a shared set of ideas of how things fit together in the world. These ideas might be "cultural" in the traditional sense of finding expression among a community of people who share backgrounds, homes, nations, languages, idioms, ethnos, norms, or other designators of belonging and co-presence. But because the Internet—like colonialism, satellite broadcasting, and air travel, among other things—crosses all these lines with abandon that the shared idea of order is better understood as part of a public, or public sphere, a vast republic of letters and media and ideas circulating in and through our thoughts and papers and letters and conversations, at a planetary scope and scale.

"Public sphere" is an odd kind of thing, however. It is at once a concept—intended to make sense of a space that is not the here and now, but one made up of writings, ideas, and discussions—and a set of ideas that people have about themselves and their own participation in such a space. I must be able to imagine myself speaking and being spoken to in such a space and to imagine a great number of other people also doing so according to unwritten rules we share. I don't need a complete theory, and I don't need to call it a public sphere, but I must somehow share an idea of order with all those other people who also imagine themselves participating in and subjecting themselves to that order. In fact, if the public sphere exists as more than just a theory, then it has no other basis than just such a shared imagination of order, an imagination which provides a guide against which to make judgments and a map for changing or achieving that order. Without such a shared imagination, a public sphere is otherwise nothing more than a cacophony of voices and information, nothing more than a stream of data, structured and formatted by and for machines, whether paper or electronic.

Charles Taylor, building on the work of Jürgen Habermas and Michael Warner, suggests that the public sphere (both idea and thing) that emerged in the eighteenth century was created through practices of communication and association that reflected a moral order in which the public stands outside power and guides or checks its operation through shared discourse and enlightened discussion. Contrary to the experience of bodies coming together into a common space (Taylor calls them "topical spaces," such as conversation, ritual, assembly), the crucial component is that the public sphere "transcends such topical spaces. We might say that it knits a plurality of spaces into one larger space of non-assembly. The same public discussion is deemed to pass through our debate today, and someone else's earnest conversation tomorrow, and the newspaper interview Thursday and so on. . . . The public sphere that emerges in the eighteenth century is a meta-topical common space."[5]

Because of this, Taylor refers to his version of a public as a "social imaginary," a way of capturing a phenomena that wavers between having concrete existence "out there" and imagined rational existence "in here." There are a handful of other such imagined spaces—the economy, the self-governing people, civil society—and in Taylor's philosophical history they are related to each through the "ideas of moral and social order" that have developed in the West and around the world.[6]

Taylor's social imaginary is intended to do something specific: to resist the "spectre of idealism," the distinction between ideas and practices, between "ideologies" and the so-called material world as "rival causal agents." Taylor suggests, "Because human practices are the kind of thing that makes sense, certain ideas are internal to them; one cannot distinguish the two in order to ask the question Which causes which?"[7] Even if materialist explanations of cause are satisfying, as they often are, Taylor suggests that they are so "at the cost of being implausible as a universal principle," and he offers instead an analysis of the rise of the modern imaginaries of moral order.[8]

The concept of recursive public, like that of Taylor's public sphere, is understood here as a kind of social imaginary. The primary reason is to bypass the dichotomy between ideas and material practice. Because the creation of software, networks, and legal documents are precisely the kinds of activities that trouble this distinction—they are at once ideas and things that have material effects in the world, both expressive and performative—it is extremely difficult to identify the properly material materiality (source code? computer chips? semiconductor manufacturing plants?). This is the first of the reasons why a recursive public is to be distinguished from the classic formulae of the public sphere, that is, that it requires a kind of imagination that includes the writing and publishing and speaking and arguing we are familiar with, as well as the making of new kinds of software infrastructures for the circulation, archiving, movement, and modifiability of our enunciations.

The concept of a social imaginary also avoids the conundrums created by the concept of "ideology" and its distinction from material practice. Ideology in its technical usage has been slowly and surely overwhelmed by its pejorative meaning: "The ideological is never one's own position; it is always the stance of someone else, always *their* ideology."[9] If one were to attempt an explanation of any particular ideology in nonpejorative terms, there is seemingly nothing that might rescue the explanation from itself becoming ideological.

The problem is an old one. Clifford Geertz noted it in "Ideology as a Cultural System," as did Karl Mannheim before him in *Ideology and Utopia*: it is the difficulty of employing a non-evaluative concept of ideology.[10] Of all the versions of struggle over the concept of a scientific or objective sociology, it is the claim of exploring ideology objectively that most rankles. As Geertz put it, "Men do not care to have beliefs to which they attach great moral significance examined dispassionately, no matter for how pure a purpose; and if they are themselves highly ideologized, they may find it simply impossible to believe that a disinterested approach to critical matters of social and political conviction can be other than a scholastic sham."[11]

Mannheim offered one response: a version of epistemological relativism in which the analysis of ideology included the ideological position of the analyst. Geertz offered another: a science of "symbolic action" based in Kenneth Burke's work and drawing on a host of philosophers and literary critics.[12] Neither the concept of ideology, nor the methods of cultural anthropology have been the same since. "Ideology" has become one of the most widely deployed (some might say, most diffuse) tools of critique, where critique is understood as the analysis of cultural patterns given in language and symbolic structures, for the purposes of bringing

GEEKS AND RECURSIVE PUBLICS

to light systems of hegemony, domination, authority, resistance, and/or misrecognition.[13] However, the practices of critique are just as (if not more) likely to be turned on critical scholars themselves, to show how the processes of analysis, hidden assumptions, latent functions of the university, or other unrecognized features the material, non-ideological real world cause the analyst to fall into an ideological trap.

The concept of ideology takes a turn toward "social imaginary" in Paul Ricoeur's *Lectures on Ideology and Utopia*, where he proposes ideological and utopian thought as two components of "social and cultural imagination." Ricoeur's overview divides approaches to the concept of ideology into three basic types—the distorting, the integrating, and the legitimating—according to how actors deal with reality through (symbolic) imagination. Does the imagination distort reality, integrate it, or legitimate it vis-à-vis the state? Ricoeur defends the second, Geertzian flavor: ideologies integrate the symbolic structure of the world into a meaningful whole, and "only because the structure of social life is already symbolic can it be distorted."[14]

For Ricoeur, the very substance of life begins in the interpretation of reality, and therefore ideologies (as well as utopias—and perhaps conspiracies) could well be treated as systems that integrate those interpretations into the meaningful wholes of political life. Ricoeur's analysis of the integration of reality though social imagination, however, does not explicitly address how imagination functions: what exactly is the nature of this symbolic action or interpretation, or imagination? Can one know it from the outside, and does it resist the distinction between ideology and material practice? Both Ricoeur and Geertz harbor hope that ideology can be made scientific, that the integration of reality through symbolic action requires only the development of concepts adequate to the job.

Re-enter Charles Taylor. In *Modern Social Imaginaries* the concept of social imaginary is distinctive in that it attempts to capture the specific integrative imaginations of modern moral and social order. Taylor stresses that they are *imaginations*—not necessarily theories—of modern moral and social order: "By social imaginary, I mean something much broader and deeper than the intellectual schemes people may entertain when they think about social reality in a disengaged mode. I am thinking, rather, of the ways in which people imagine their social existence, how they fit together with others, how things go on between them and their fellows, the expectations that are normally met, and the deeper normative notions and images that underlie these expectations."[15] Social imaginaries develop historically and result in both new institutions and new subjectivities; the concepts of public, market, and civil society (among others) are located in the imaginative faculties of actors who recognize the shared, common existence of these ideas, even if they differ on the details, and the practices of those actors reflect a commitment to working out these shared concepts.

Social imaginaries are an extension of "background" in the philosophical sense: "a wider grasp of our whole predicament."[16] The example Taylor uses is that of marching in a demonstration: the action is in our imaginative repertoire and has a meaning that cannot be reduced to the local context: "We know how to assemble, pick up banners and march. . . . [W]e understand the ritual. . . . [T]he immediate sense of what we are doing, getting the message to our government and our fellow citizens that the cuts must stop, say, makes sense in a wider context, in which we see ourselves standing in a continuing relation with others, in which it is appropriate to address them in this manner."[17] But we also stand "internationally" and "in history" against a background of stories, images, legends, symbols, and theories. "The background that makes sense of any given act is wide and deep. It doesn't include everything in our world, but the relevant sense-giving features can't be circumscribed. . . . [It] draws on our whole world, that is, our sense of our whole predicament in time and space, among others and in history."[18]

The social imaginary is not simply the norms that structure our actions; it is also a sense of what makes norms achievable or "realizable," as Taylor says. This is the idea of a "moral order," one that we expect to exist, and if it doesn't, one that provides a plan for achieving it. For Taylor, there is such a thing as a "modern idea of order," which includes, among other things, ideas of what it means to be an individual, ideas of how individual passions and desires are related to collective association, and, most important, ideas about living in time together (he stresses a radically secular conception of time—secular in a sense that means more than simply "outside religion"). He by no means insists that this is the only such definition of modernity (the door is wide open to understanding alternative modernities), but that the modern idea of moral order is one that dominates and structures a very wide array of institutions and individuals around the world.

The "modern idea of moral order" is a good place to return to the question of geeks and their recursive publics. Are the ideas of order shared by geeks different from those Taylor outlines? Do geeks like Sean and Adrian, or activists in Berlin, possess a distinctive social imaginary? Or do they (despite their planetary dispersal) participate in this common modern idea of moral order? Do the stories and narratives, the tools and technologies, the theories and imaginations they follow and build on have something distinctive about them? Sean's and Adrian's commitment to transforming healthcare seems to be, for instance, motivated by a notion of moral order in which the means of allocation of healthcare might become more just, but it is also shot through with technical ideas about the role of standards, the Internet, and the problems with current technical solutions; so while they may seem to be simply advocating for better healthcare, they do so through a technical language and practice that are probably quite alien to policymakers, upper management, and healthcare advocacy groups that might otherwise be in complete sympathy.

The affinity of geeks for each other is processed through and by ideas of order that are both moral *and* technical—ideas of order that do indeed mix up "operating systems and social systems." These systems include the technical means (the infrastructure) through which geeks meet, assemble, collaborate, and plan, as well as how they talk and think about those activities. The infrastructure—the Internet—allows for a remarkably wide and diverse array of people to encounter and engage with each other. That is to say, the idea of order shared by geeks is shared because they are geeks, because they "get it," because the Internet's structure and software have taken a particular form through which geeks come to understand the moral order that gives the fabric of their political lives warp and weft.

Internet Silk Road

Bangalore, March 2000. I am at another bar, this time on one of Bangalore's trendiest streets. The bar is called Purple Haze, and I have been taken there, the day after my arrival, by Udhay Shankar

N. Inside it is dark and smoky, purple, filled with men between eighteen and thirty, and decorated with posters of Jimi Hendrix, Black Sabbath, Jim Morrison (Udhay: "I hate that band"), Led Zeppelin, and a somewhat out of place Frank Zappa (Udhay: "One of my political and musical heroes"). All of the men, it appears, are singing along with the music, which is almost without exception heavy metal.

I engage in some stilted conversation with Udhay and his cousin Kirti about the difference between Karnatic music and rock-and-roll, which seems to boil down to the following: Karnatic music decreases metabolism and heart rate, leading to a relaxed state of mind; rock music does the opposite. Given my aim of focusing on the Internet and questions of openness, I have already decided not to pay attention to this talk of music. In retrospect, I understand this to have been a grave methodological error: I underestimated the extent to which the subject of music has been one of the primary routes into precisely the questions about the "reorientation of knowledge and power" I was interested in. Over the course of the evening and the following days, Udhay introduced me, as promised, to a range of people he either knew or worked with in some capacity. Almost all of the people I met appeared to sincerely love heavy-metal music.

I met Udhay Shankar N. in 1999 through a newsletter, distributed via e-mail, called *Tasty Bits from the Technology Front*. It was one of a handful of sources I watched closely while in Berlin, looking for such connections to geek culture. The newsletter described a start-up company in Bangalore, one that was devoted to creating a gateway between the Internet and mobile phones, and which was, according to the newsletter, an entirely Indian operation, though presumably with U.S. venture funds. I wanted to find a company to compare to Amicas: a start-up, run by geeks, with a similar approach to the Internet, but halfway around the world and in a "culture" that might be presumed to occupy a very different kind of moral order. Udhay invited me to visit and promised to introduce me to everyone he knew. He described himself as a "random networker"; he was not really a programmer or a designer or a Free Software geek, despite his extensive knowledge of software, devices, operating systems, and so on, including Free and Open Source Software. Neither was he a businessman, but rather described himself as the guy who "translates between the suits and the techs."

Udhay "collects interesting people," and it was primarily through his zest for collecting that I met all the people I did. I met cosmopolitan activists and elite lawyers and venture capitalists and engineers and cousins and brothers and sisters of engineers. I met advertising executives and airline flight attendants and consultants in Bombay. I met journalists and gastroenterologists, computer-science professors and musicians, and one mother of a robot scientist in Bangalore. Among them were Muslims, Hindus, Jains, Jews, Parsis, and Christians, but most of them considered themselves more secular and scientific than religious. Many were self-educated, or like their U.S. counterparts, had dropped out of university at some point, but continued to teach themselves about computers and networks. Some were graduates or employees of the Indian Institute of Science in Bangalore, an institution that was among the most important for Indian geeks (as Stanford University is to Silicon Valley, many would say). Among the geeks to whom Udhay introduced me, there were only two commonalities: the geeks were, for the most part, male, and they all loved heavy-metal music.[19]

While I was in Bangalore, I was invited to join a mailing list run by Udhay called Silk-list, an irregular, unmoderated list devoted to "intelligent conversation." The list has no particular focus: long, meandering conversations about Indian politics, religion, economics, and history erupt regularly; topics range from food to science fiction to movie reviews to discussions on Kashmir, Harry Potter, the singularity, or nanotechnology. Udhay started Silk-list in 1997 with Bharath Chari and Ram Sundaram, and the recipients have included hundreds of people around the world, some very well-known ones, programmers, lawyers, a Bombay advertising executive, science-fiction authors, entrepreneurs, one member of the start-up Amicas, at least two transhumanists, one (diagnosed) schizophrenic, and myself. Active participants usually numbered about ten to fifteen, while many more lurked in the background.

Silk-list is an excellent index of the relationship between the network of people in Bangalore and their connection to a worldwide community on the Internet—a fascinating story of the power of heterogeneously connected networks and media. Udhay explained that in the early 1990s he first participated in and then taught himself to configure and run a modem-based networking system known as a Bulletin Board Service (BBS) in Bangalore. In 1994 he heard about a book by Howard Rheingold called *The Virtual*

Community, which was his first introduction to the Internet. A couple of years later when he finally had access to the Internet, he immediately e-mailed John Perry Barlow, whose work he knew from *Wired* magazine, to ask for Rheingold's e-mail address in order to connect with him. Rheingold and Barlow exist, in some ways, at the center of a certain kind of geek world: Rheingold's books are widely read popular accounts of the social and community aspects of new technologies that have often had considerable impact internationally; Barlow helped found the Electronic Frontier Foundation and is responsible for popularizing the phrase "information wants to be free."[20] Both men had a profound influence on Udhay and ultimately provided him with the ideas central to running an online community. A series of other connections of similar sorts—some personal, some precipitated out of other media and other channels, some entirely random—are what make up the membership of Silk-list.[21]

Like many similar communities of "digerati" during and after the dot.com boom, Silk-list constituted itself more or less organically around people who "got it," that is, people who claimed to understand the Internet, its transformative potential, and who had the technical skills to participate in its expansion. Silk-list was not the only list of its kind. Others such as the *Tasty Bits* newsletter, the FoRK (Friends of Rohit Khare) mailing list (both based in Boston), and the Nettime and Syndicate mailing lists (both based in the Netherlands) ostensibly had different reasons for existence, but many had the same subscribers and overlapping communities of geeks. Subscription was open to anyone, and occasionally someone would stumble on the list and join in, but most were either invited by members or friends of friends, or they were connected by virtue of cross-posting from any number of other mailing lists to which members were subscribed.

/pub

Silk-list is public in many senses of the word. Practically speaking, one need not be invited to join, and the material that passes through the list is publicly archived and can be found easily on the Internet. Udhay does his best to encourage everyone to speak and to participate, and to discourage forms of discourse that he thinks might silence participants into lurking. Silk-list is not a government, corporate, or nongovernmental list, but is constituted only through the activity of geeks finding each other and speaking to each other on this list (which can happen in all manner of ways: through work, through school, through conferences, through fame, through random association, etc.). Recall Charles Taylor's distinction between a topical and a metatopical space. Silk-list is not a conventionally topical space: at no point do all of its members meet face-to-face (though there are regular meet-ups in cities around the world), and they are not all online at the same time (though the volume and tempo of messages often reflect who is online "speaking" to each other at any given moment). It is a topical space, however, if one considers it from the perspective of the machine: the list of names on the mailing list are all assembled together in a database, or in a file, on the server that manages the mailing list. It is a stretch to call this an "assembly," however, because it assembles only the avatars of the mailing-list readers, many of whom probably ignore or delete most of the messages.

Silk-list is certainly, on the other hand, a "metatopical" public. It "knits together" a variety of topical spaces: my discussion with friends in Houston, and other members' discussions with people around the world, as well as the sources of multiple discussions like newspaper and magazine articles, films, events, and so on that are reported and discussed online. But Silk-list is not "The" public—it is far from being the only forum in which the public sphere is knitted together. Many, many such lists exist.

In *Publics and Counterpublics* Michael Warner offers a further distinction. "The" public is a social imaginary, one operative in the terms laid out by Taylor: as a kind of vision of order evidenced through stories, images, narratives, and so on that constitute the imagination of what it means to be part of the public, as well as plans necessary for creating the public, if necessary. Warner distinguishes, however, between a concrete, embodied audience, like that at a play, a demonstration, or a riot (a topical public in Taylor's terms), and an audience brought into being by discourse and its circulation, an audience that is not metatopical so much as it is a public that is concrete in a different way; it is concrete not in the face-to-face temporality of the speech act, but in the sense of calling a public into being through an address that has a different temporality. It is a public that is concrete in a media-specific manner: it depends on the structures of creation, circulation, use, performance, and reuse of particular kinds of discourse, particular objects or instances of discourse.

Warner's distinction has a number of implications. The first, as Warner is careful to note, is that the existence of particular media is not sufficient for a public to come into existence. Just because a book is printed does not mean that a public exists; it requires also that the public take corresponding action, that is, that they read it. To be part of a particular public is to choose to pay attention to those who choose to address those who choose to pay attention . . . and so on. Or as Warner puts it, "The circularity is essential to the phenomenon. A public might be real and efficacious, but its reality lies in just this reflexivity by which an addressable object is conjured into being in order to enable the very discourse that gives it existence."[22]

This "autotelic" feature of a public is crucial if one is to understand the *function* of a public as standing outside of power. It simply cannot be organized by the state, by a corporation, or by any other social totality if it is to have the legitimacy of an independently functioning public. As Warner puts it, "A public organizes itself independently of state institutions, law, formal frameworks of citizenship, or preexisting institutions such as the church. If it were not possible to think of the public as organized independently of the state or other frameworks, the public could not be sovereign with respect to the state. . . . Speaking, writing, and thinking involve us—actively and immediately—in a public, and thus in the being of the sovereign."[23]

Warner's description makes no claim that any public or even The Public actually takes this form in the present: it is a description of a social imaginary or a "faith" that allows individuals to make sense of their actions according to a modern idea of social order. As Warner (and Habermas before him) suggests, the existence of such autonomous publics—and certainly the idea of "public opinion"— does not always conform to this idea of order. Often such publics turn out to have been controlled all along by states, corporations, capitalism, and other forms of social totality that determine the nature of discourse in insidious ways. A public whose participants have no faith that it is autotelic and autonomous is little more than a charade meant to assuage opposition to authority, to transform political power and equality into the negotiation between unequal parties.

Is Silk-list a public? More important, is it a sovereign one? Warner's distinction between different media-specific forms of assembly is crucial to answering this question. If one wants to know whether a mailing list on the Internet is more or less likely to be a sovereign public than a book-reading public or the nightly-news-hearing one, then one needs to approach it from the specificity of the form of discourse. This specificity not only includes whether the form is text or video and audio, or whether the text is ASCII or Unicode, or the video PAL or NTSC, but it also includes the means of creation, circulation, and reuse of that discourse as well.

For example, consider the differences between a book published in a conventional fashion, by a conventional, corporate press, distributed to bookstores or via Amazon.com, and a book published by an Internet start-up which makes an electronic copy freely available with a copyleft license, yet charges (a lower price) for a print-on-demand hardcopy. Both books might easily enter the metatopical space of The Public: discussed in homes, schools, on mailing lists, glowingly reviewed or pilloried, perhaps having effects on corporate behavior, state, or public policy. The former, however, is highly constrained in terms of who will author such a book, how it will be distributed, marketed, edited, and revised, and so on. Copyright law will restrict what readers can do with it, including how they might read it or subsequently circulate it or make derivative use of it. However, a traditionally published book is also enriched by its association with a reputable corporation: it is treated more or less immediately as authoritative, perhaps as meeting some standard of accuracy, precision, or even truth, and its quality is measured primarily by sales.

The on-demand, Internet-mediated book, by contrast, will have a much different temporality of circulation: it might languish in obscurity due to lack of marketing or reputable authority, or it might get mentioned somewhere like the *New York Times* and suddenly become a sensation. For such a book, copyright law (in the form of a copyleft license) might allow a much wider range of uses and reuses, but it will restrict certain forms of commercialization of the text. The two publics might therefore end up looking quite different, overlapping, to be sure, but varying in terms of their control and the terms of admittance. What is at stake is the power of one or the other such public to appear as an independent and sovereign entity—free from suspect constraints and control—whose function is to argue with other constituted forms of power.

The conventionally published book may well satisfy all the criteria of being a public, at least in the colloquial sense of making a set of ideas and a discourse widely available and expecting to influence, or receive a response from, constituted forms of sovereign power. However, it is only the latter "on-demand" scheme for publishing that satisfies the criteria of being a *recursive* public. The differences in this example offer a crude indication of why the Internet is so crucially important to geeks, so important that it draws them together, in its defense, as an infrastructure that enables the creation of publics that are thought to be autonomous, independent, and autotelic. Geeks share an idea of moral and technical order when it comes to the Internet; not only this, but they share a commitment to maintaining that order because it is what allows them to associate as a recursive public in the first place. They discover, or rediscover, through their association, the power and possibility of occupying the position of independent public—one not controlled by states, corporations, or other organizations, but open (they claim) through and through—and develop a desire to defend it from encroachment, destruction, or refeudalization (to use Habermas's term for the fragmentation of the public sphere).

The recursive public is thus not only the book and the discourse around the book. It is not even "content" expanded to include all kinds of media. It is also the technical structure of the Internet as well: its software, its protocols and standards, its applications and software, its legal status and the licenses and regulations that govern it. This captures both of the reasons why recursive publics are distinctive: (1) they include not only the discourses of a public, but the ability to make, maintain, and manipulate the infrastructures of those discourses as well; and (2) they are "layered" and include both discourses and infrastructures, to a specific technical extent (i.e., not all the way down). The meaning of which layers are important develops more or less immediately from direct engagement with the medium. In the following example, for instance, Napster represents the potential of the Internet in miniature—as an application—but it also connects immediately to concerns about the core protocols that govern the Internet and the process of standardization that governs the development of these protocols: hence recursion through the layers of an infrastructure.

These two aspects of the recursive public also relate to a concern about the fragmentation or refeudalization of the public sphere: *there is only one Internet.* Its singularity is not technically determined or by any means necessary, but it is what makes the Internet so valuable to geeks. It is a contest, the goal of which is to maintain the Internet as an infrastructure for autonomous and autotelic publics to emerge as part of The Public, understood as part of an imaginary of moral and technical order: operating systems and social systems.

From Napster to the Internet

On 27 July 2000 Eugen Leitl cross-posted to Silk-list a message with the subject line "Prelude to the Singularity." The message's original author, Jeff Bone (not at the time a member of Silk-list), had posted the "op-ed piece" initially to the FoRK mailing list as a response to the Recording Industry Association of America's (RIAA) actions against Napster. The RIAA had just succeeded in getting U.S. district judge Marilyn Hall Patel, Ninth Circuit Court of Appeals, to issue an injunction to Napster to stop downloads of copyrighted music. Bone's op-ed said,

> Popular folklore has it that the Internet was designed with decentralized routing protocols in order to withstand a nuclear attack. That is, the Internet "senses damage" and "routes around it." It has been said that, on the 'Net, censorship is perceived as damage and is subsequently routed around. The RIAA, in a sense, has cast itself in a censor's role. Consequently, the music industry will be perceived as damage—and it will be routed around. There is *no doubt* that this will happen, and that technology will evolve more quickly than businesses and social institutions can; there are numerous highly-visible projects already underway that attempt to create technology that is invulnerable to legal challenges of various kinds. Julian Morrison, the originator of a project (called Fling) to build a fully anonymous/untraceable suite of network protocols, expresses this particularly eloquently.[24]

Bone's message is replete with details that illustrate the meaning and value of the Internet to geeks, and that help clarify the concept of a recursive public. While it is only one message, it nonetheless condenses and expresses a variety of stories, images, folklore, and technical details that I elaborate herein.

The Napster shutdown in 2000 soured music fans and geeks alike, and it didn't really help the record labels who perpetrated it either. For many geeks, Napster represented the Internet in miniature, an innovation that both demonstrated something on a scope and scale never seen before, and that also connected people around something they cared deeply about—their shared interest in music. Napster raised interesting questions about its own success: Was it successful because it allowed people *to develop new musical interests* on a scope and scale they had never experienced before? Or was it successful because it gave people with already existing musical interests a way *to share music* on a scope and scale they had never experienced before? That is to say, was it an innovation in marketing or in distribution? The music industry experienced it as the latter and hence as direct competition with their own means of distribution. Many music fans experienced it as the former, what Cory Doctorow nicely labeled "risk-free grazing," meaning the ability to try out an almost unimaginable diversity of music before choosing what to invest one's interests (and money) in. To a large extent, Napster was therefore a recapitulation of what the Internet already meant to geeks.

Bone's message, the event of the Napster shutdown, and the various responses to it nicely illustrate the two key aspects of the recursive public: first, the way in which geeks argue not only about rights and ideas (e.g., is it legal to share music?) but also about the infrastructures that allow such arguing and sharing; second, the "layers" of a recursive public are evidenced in the immediate connection of Napster (an application familiar to millions) to the "decentralized routing protocols" (TCP/IP, DNS, and others) that made it possible for Napster to work the way it did.

Bone's message contains four interrelated points. The first concerns the concept of autonomous technical progress. The title "Prelude to the Singularity" refers to a 1993 article by Vernor Vinge about the notion of a "singularity," a point in time when the speed of autonomous technological development outstrips the human capacity to control it.[25] The notion of singularity has the status of a kind of colloquial "law" similar to Moore's Law or Metcalfe's Law, as well as signaling links to a more general literature with roots in libertarian or classically liberal ideas of social order ranging from John Locke and John Stuart Mill to Ayn Rand and David Brin.[26]

Bone's affinity for transhumanist stories of evolutionary theory, economic theory, and rapid innovation sets the stage for the rest of his message. The crucial rhetorical gambit here is the appeal to inevitability (as in the emphatic "there is *no doubt* that this will happen"): Bone establishes that he is speaking to an audience that is accustomed to hearing about the inevitability of technical progress and the impossibility of legal maneuvering to change it, but his audience may not necessarily agree with these assumptions. Geeks occupy a spectrum from "polymath" to "transhumanist," a spectrum that includes their understandings of technological progress and its relation to human intervention. Bone's message clearly lands on the far transhumanist side.

A second point concerns censorship and the locus of power: according to Bone, power does not primarily reside with the government or the church, but comes instead from the private sector, in this case the coalition of corporations represented by the RIAA. The significance of this has to do with the fact that a "public" is expected to be its own sovereign entity, distinct from church, state, or corporation, and while censorship by the church or the state is a familiar form of aggression against publics, censorship by corporations (or consortia representing them), as it strikes Bone and others, is a novel development. Whether the blocking of file-sharing can legitimately be called censorship is also controversial, and many Silk-list respondents found the accusation of censorship untenable.

Proving Bone's contention, over the course of the subsequent years and court cases, the RIAA and the Motion Picture Association of America (MPAA) have been given considerably more police authority than even many federal agencies—especially with regard to policing networks themselves (an issue which, given its technical abstruseness, has rarely been mentioned in the mainstream mass media). Both organizations have not only sought to prosecute file-sharers but have been granted rights to obtain information from Internet Service Providers about customer activities and have consistently sought the right to secretly disable (hack into, disable, or destroy) private computers suspected of illegal activity. Even if these practices may not be defined as censorship per se, they are nonetheless fine examples of the issues that most exercise geeks: the use of legal means by a few (in this case, private corporations) to suppress or transform technologies in wide use by the many. They also index the problems of monopoly, antitrust, and technical control that are not obvious and often find expression, for example, in allegories of reformation and the control of the music-sharing laity by papal authorities.

Third, Bone's message can itself be understood in terms of the reorientation of knowledge and power. Although what it means to call his message an "op-ed" piece may seem obvious, Bone's message was not published anywhere in any conventional sense. It doesn't appear to have been widely cited or linked to. However, for one day at least, it was a heated discussion topic on three mailing lists, including Silk-list. "Publication" in this instance is a different kind of event than getting an op-ed in the *New York Times*.

The material on Silk-list rests somewhere between private conversation (in a public place, perhaps) and published opinion. No editor made a decision to "publish" the message—Bone just clicked "send." However, as with any print publication, his piece was theoretically accessible by anyone, and what's more, a potentially huge number of copies may be archived in many different places (the computers of all the participants, the server that hosts the list, the Yahoo! Groups servers that archive it, Google's search databases, etc.). Bone's message exemplifies the recursive nature of the recursive public: it is a public statement about the openness of the Internet, and it is an *example* of the new forms of publicness it makes possible through its openness.

The constraints on who speaks in a public sphere (such as the power of printers and publishers, the requirements of licensing, or issues of cost and accessibility) are much looser in the Internet era than in any previous one. The Internet gives a previously unknown Jeff Bone the power to dash off a manifesto without so much as a second thought. On the other hand, the ease of distribution belies the difficulty of actually being heard: the multitudes of other Jeff Bones make it much harder to get an audience. In terms of publics, Bone's message can constitute a public in the same sense that a *New York Times* op-ed can, but its impact and meaning will be different. His message is openly and freely available for as long as there are geeks and laws and machines that maintain it, but the *New York Times* piece will have more authority, will be less accessible, and, most important, will not be available to just anyone. Geeks imagine a space where anyone can speak with similar reach and staying power—even if that does not automatically imply authority—and they imagine that it should remain open at all costs. Bone is therefore interested precisely in a technical infrastructure that ensures his right to speak about that infrastructure and offer critique and guidance concerning it.

The ability to create and to maintain such a recursive public, however, raises the fourth and most substantial point that Bone's message makes clear. The leap to speaking about the "decentralized routing protocols" represents clearly the shared moral and technical order of geeks, derived in this case from the specific details of the Internet. Bone's post begins with a series of statements that are part of the common repertoire of technical stories and images among geeks. Bone begins by making reference to the "folklore" of the Internet, in which routing protocols are commonly believed to have been created to withstand a nuclear attack. In calling it folklore he suggests that this is not a precise description of the Internet, but an image that captures its design goals. Bone collapses it into a more recent bit of folklore: "The Internet treats censorship as damage and routes around it."[27] Both bits of folklore are widely circulated and cited; they encapsulate one of the core intellectual ideas about the architecture of the Internet, that is, its open and distributed interconnectivity. There is certainly a specific technical backdrop for this suggestion: the TCP/IP "internetting" protocols were designed to link up multiple networks without making them sacrifice their autonomy and control. However, Bone uses this technical argument more in the manner of a social imaginary than of a theory, that is, as a way of thinking about the technical (and moral) order of the Internet, of what the Internet is supposed to be like.

In the early 1990s this version of the technical order of the Internet was part of a vibrant libertarian dogma asserting that the Internet simply could not be governed by any land-based sovereign and that it was fundamentally a place of liberty and freedom. This was the central message of people such as John Perry Barlow, John Gilmore, Howard Rheingold, Esther Dyson, and a host of others who populated both the pre-1993 Internet (that is, before the World Wide Web became widely available) and the pages of magazines such as *Wired* and *Mondo 2000*—the same group of people, incidentally, whose ideas were visible and meaningful to Udhay Shankar and his friends in India even prior to Internet access there, not to mention to Sean and Adrian in Boston, and artists and activists in

Europe, all of whom often reacted more strongly against this libertarian aesthetic.

For Jeff Bone (and a great many geeks), the folkloric notion that "the net treats censorship as damage" is a very powerful one: it suggests that censorship is impossible because there is no central point of control. A related and oft-cited sentiment is that "trying to take something off of the Internet is like trying to take pee out of a pool." This is perceived by geeks as a virtue, not a drawback, of the Internet.

The argument is quite complex, however: on one side of a spectrum, there is the belief that the structure of the Internet ensures that censorship cannot happen, technically speaking, so long as the Internet's protocols and software remain open. Furthermore, that structure ensures that all attempts to regulate the Internet will also fail (e.g., the related sentiment that "the Internet treats Congress as damage and routes around it").

On the other side of the spectrum, however, this view of the unregulatable nature of the Internet has been roundly criticized, most prominently by Lawrence Lessig, who is otherwise often in sympathy with geek culture. Lessig suggests that just because the Internet has a particular structure does not mean that it must always be that way.[28] His argument has two prongs: first, that the Internet is structured the way it is because it is made of code that people write, and thus it could have been and will be otherwise, given that there are changes and innovations occurring all the time; second, that the particular structure of the Internet therefore governs or regulates behavior in particular ways: Code is Law. So while it may be true that no one can make the Internet "closed" by passing a law, it is also true that the Internet could become closed if the technology were to be altered for that purpose, a process that may well be nudged and guided by laws, regulations, and norms.

Lessig's critique is actually at the heart of Bone's concern, and the concern of recursive publics generally: the Internet is a *contest* and one that needs to be repeatedly and constantly replayed in order to maintain it as the legitimate infrastructure through which geeks associate with one another. Geeks argue in detail about what distinguishes technical factors from legal or social ones. Openness on the Internet is complexly intertwined with issues of availability, price, legal restriction, usability, elegance of design, censorship, trade secrecy, and so on.

However, even where openness is presented as a natural tendency for technology (in oft-made analogies with reproductive fitness and biodiversity, for example), it is only a partial claim in that it represents only one of the "layers" of a recursive public. For instance, when Bone suggests that the net is "invulnerable to legal attack" because "technology will evolve more quickly than businesses and social institutions can," he is not only referring to the fact that the Internet's novel technical configuration has few central points of control, which makes it difficult for a single institution to control it, but also talking about the distributed, loosely connected networks of people who have the right to write and rewrite software and deal regularly with the underlying protocols of the Internet—in other words, of geeks themselves.

Operating systems and social systems: the imagination of order shared by geeks is both moral and technical. It is not only about the technical structure of the Internet, however innovative that is, but also about the legal and social structure that has emerged with it, the kind of order that has made it possible for geeks to associate in a planetary public and to become aware of the value of the space they have made.

Many geeks, perhaps including Bone, discover the nature of this order by coming to understand how the Internet works—how it works technically, but also who created it and how. Some have come to this understanding through participation in Free Software (an exemplary "recursive public"), others through stories and technologies and projects and histories that illuminate the process of creating, growing, and evolving the Internet. The story of the process by which the Internet is standardized is perhaps the most well known: it is the story of the Internet Engineering Task Force and its Requests for Comments system.

Requests for Comments

For many geeks, the Internet Engineering Task Force (IETF) and its Requests for Comments (RFC) system exemplify key features of the moral and technical order they share, the "stories and practices" that make up a social imaginary, according to Charles Taylor. The IETF is a longstanding association of Internet engineers who try to help disseminate some of the core standards of the Internet through the RFC process. Membership is open to individuals, and the association has very little real control over the structure or growth of the Internet—only over the key process of Internet standardization. Its standards rarely have the kind of political legitimacy that one associates with international treaties and the standards bodies of Geneva, but they are nonetheless de facto legitimate. The RFC process is an unusual standards process that allows modifications to existing technologies to be made before the standard is finalized. Together Internet standards and the RFC process form the background of the Napster debate and of Jeff Bone's claims about "internet routing protocols."

A famous bit of Internet-governance folklore expresses succinctly the combination of moral and technical order that geeks share (attributed to IETF member David Clark): "We reject kings, presidents, and voting. We believe in rough consensus and running code."[29] This quote emphasizes the necessity of arguing with and through technology, the first aspect of a recursive public; the only argument that convinces is working code. If it works, then it can be implemented; if it is implemented, it will "route around" the legal damage done by the RIAA. The notion of "running code" is central to an understanding of the relationship between argument-by-technology and argument-by-talk for geeks. Very commonly, the response by geeks to people who argued about Napster that summer—and the courts' decisions regarding it—was to dismiss their complaints as mere talk. Many suggested that if Napster were shut down, thousands more programs like it would spring up in its wake. As one mailing-list participant, Ashish "Hash" Gulhati, put it, "It is precisely these totally unenforceable and mindless judicial decisions that will start to look like self-satisfied wanking when there's code out there which will make the laws worth less than the paper they're written on. When it comes to fighting this shit in a way that counts, everything that isn't code is just talk."[30]

Such powerful rhetoric often collapses the process itself, for someone has to write the code. It can even be somewhat paradoxical: there is a need to talk forcefully about the need for less talk and more code, as demonstrated by Eugen Leitl when I objected that Silk-listers were "just talking": "Of course we should talk. Did my last post consist of some kickass Python code adding sore-missed functionality to Mojonation? Nope. Just more meta-level waffle about the importance of waffling less, coding more. I lack the proper mental equipment upstairs for being a good coder, hence I attempt to corrupt young impressionable innocents into contributing to the cause. Unashamedly so. So sue me."[31]

Eugen's flippancy reveals a recognition that there is a political component to coding, even if, in the end, talk disappears and only code remains. Though Eugen and others might like to adopt a rhetoric that suggests "it will just happen," in practice none of them really act that way. Rather, the activities of coding, writing software, or improving and diversifying the software that exists are not inevitable or automatic but have specific characteristics. They require time and "the proper mental equipment." The inevitability they refer to consists not in some fantasy of machine intelligence, but in a social imaginary shared by many people in loosely connected networks who spend all their free time building, downloading, hacking, testing, installing, patching, coding, arguing, blogging, and proselytizing—in short, creating a recursive public enabled by the Internet.

Jeff Bone's op-ed piece, which is typically enthusiastic about the inevitability of new technologies, still takes time to reference one of thousands (perhaps tens of thousands) of projects as worthy of attention and support, a project called Fling, which is an attempt to rewrite the core protocols of the Internet.[32] The goal of the project is to write a software implementation of these protocols with the explicit goal of making them "anonymous, untraceable, and untappable." Fling is not a corporation, a start-up, or a university research project (though some such projects are); it is only a Web site. The core protocols of the Internet, contained in the RFCs, are little more than documents describing how computers should interact with each other. They are standards, but of an unusual kind.[33] Bone's leap from a discussion about Napster to one about the core protocols of the Internet is not unusual. It represents the second aspect of a recursive public: the importance of understanding the Internet as a set of "layers," each enabling the next and each requiring an openness that both prevents central control and leads to maximum creativity.

RFCs have developed from an informal system of memos into a formal standardization process over the life of the Internet, as the IETF and the Internet Society (ISOC) have become more bureaucratic entities. The process of writing and maintaining these documents is particular to the Internet, precisely because the Internet is the kind of network experiment that facilitates the sharing of resources across administratively bounded networks. It is a process that has allowed all the experimenters to both share the network and to propose changes to it, in a common space. RFCs are primarily suggestions, not demands. They are "public domain" documents and thus available to everyone with access to the Internet. As David Clark's reference to "consensus and running code" demonstrates, the essential component of setting Internet standards is a good, working *implementation* of the protocols. Someone must write software that behaves in the ways specified by the RFC, which is, after all, only a document, not a piece of software. Different implementations of, for example, the TCP/IP protocol or the File Transfer Protocol (ftp) depend initially on individuals, groups, and/or corporations building them into an operating-system kernel or a piece of user software and subsequently on the existence of a large number of people using the same operating system or application.

In many cases, subsequent to an implementation that has been disseminated and adopted, the RFCs have been amended to reflect these working implementations and to ordain them as standards. So the current standards are actually bootstrapped, through a process of writing RFCs, followed by a process of creating implementations that adhere loosely to the rules in the RFC, then observing the progress of implementations, and then rewriting RFCs so that the process begins all over again. The fact that geeks can have a discussion via e-mail depends on the very existence of both an RFC to define the e-mail protocol and implementations of software to send the e-mails.

This standardization process essentially inverts the process of planning. Instead of planning a system, which is then standardized, refined, and finally built according to specification, the RFC process allows plans to be proposed, implemented, refined, reproposed, rebuilt, and so on until they are adopted by users and become the standard approved of by the IETF. The implication for most geeks is that this process is permanently and fundamentally open: changes to it can be proposed, implemented, and adopted without end, and the better a technology becomes, the more difficult it becomes to improve on it, and therefore the less reason there is to subvert it or reinvent it. Counterexamples, in which a standard emerges but no one adopts it, are also plentiful, and they suggest that the standardization process extends beyond the proposal-implementation-proposal-standard circle to include the problem of actually convincing users to switch from one working technology to a better one. However, such failures of adoption are also seen as a kind of confirmation of the quality or ease of use of the *current* solution, and they are all the more likely to be resisted when some organization or political entity tries to force users to switch to the new standard—something the IETF has refrained from doing for the most part.

Conclusion: Recursive Public

Napster was a familiar and widely discussed instance of the "re-orientation of power and knowledge" (or in this case, power and music) wrought by the Internet and the practices of geeks. Napster was not, however, a recursive public or a Free Software project, but a dot-com-inspired business plan in which proprietary software was given away for free in the hopes that revenue would flow from the stock market, from advertising, or from enhanced versions of the software. Therefore, geeks did not defend Napster as much as they experienced its legal restriction as a wake-up call: the Internet enables Napster and will enable many other things, but laws, corporations, lobbyists, money, and governments can destroy all of it.

I started this chapter by asking what draws geeks together: what constitutes the chain that binds geeks like Sean and Adrian to hipsters in Berlin and to entrepreneurs and programmers in Bangalore? What constitutes their affinity if it is not any of the conventional candidates like culture, nation, corporation, or language? A colloquial answer might be that it is simply the Internet that brings them together: cyberspace, virtual communities, online culture. But this doesn't answer the question of why? Because they can? Because Community Is Good? If mere association is the goal, why not AOL or a vast private network provided by Microsoft?

My answer, by contrast, is that geeks' affinity with one another is structured by shared moral and technical understandings of order. They are a public, an independent public that has the ability to build, maintain, and modify itself, that is not restricted to the activities of speaking, writing, arguing, or protesting. Recursive publics form through their experience with the Internet precisely because the Internet is the kind of thing they can inhabit and transform. Two things make recursive publics distinctive: the ability to include the practice of creating this infrastructure as part of the activity of being public or contesting control; and the ability to "recurse" through the layers of that infrastructure, maintaining its publicness at each level without making it into an unchanging, static, unmodifiable thing.

The affinity constituted by a recursive public, through the medium of the Internet, creates geeks who understand clearly what association through the Internet means. This affinity structures their imagination of what the Internet is and enables: creation, distribution, modification of knowledge, music, science, software. The infrastructure—*this-infrastructure-here*, the Internet—must be understood as part of this imaginary (in addition to being a pulsating tangle of computers, wires, waves, and electrons).

The Internet is not the only medium for such association. A corporation, for example, is also based on a shared imaginary of the economy, of how markets, exchanges, and business cycles are supposed to work; it is the creation of a concrete set of relations and practices, one that is generally inflexible—even in this age of so-called flexible capitalism—because it requires a commitment of time, humans, and capital. Even in fast capitalism one needs to rent office space, buy toilet paper, install payroll software, and so on.

Software and networks can be equally concrete—connecting people, capital, and other resources over time and thus creating an infrastructure—but they are arguably more flexible, more changeable, and more reprogrammable—than a corporation, a sewage system, or a stock exchange. The Internet, in particular, especially in the stories of the IETF and the RFC process, represents a radicalization of this flexibility: not only can one create an application like Napster that takes clever advantage of the layers (protocols, routers, and routes) of the Internet, but one can actually rewrite the layers themselves, rendering possible a new class of Napsters. The difficulty of doing so increases with ever deeper layers, but the possibility is not (yet) arbitrarily restricted by any organization, person, law, or government. Affinity—membership in a recursive public—depends on adopting the moral and technical imaginations of this kind of order.

The urgency evidenced in the case of Napster (and repeated in numerous other instances, such as the debate over net neutrality) is linked to a moral idea of order in which there is a shared imaginary of The Public, and not only a vast multiplicity of competing publics. It is an urgency linked directly to the fact that the Internet provides geeks with a platform, an environment, an infrastructure through which they not only associate, but create, and do so in a manner that is widely felt to be autonomous, autotelic, and independent of at least the most conventional forms of power: states and corporations—independent enough, in fact, that both states and corporations can make widespread use of this infrastructure (can become geeks themselves) without necessarily endangering its independence.

2. Protestant Reformers, Polymaths, Transhumanists

Geeks talk a lot. They don't talk about recursive publics. They don't often talk about imaginations, infrastructures, moral or technical orders. But they do talk a lot. A great deal of time and typing is necessary to create software and networks: learning and talking, teaching and arguing, telling stories and reading polemics, reflecting on the world in and about the infrastructure one inhabits. In this chapter I linger on the stories geeks tell, and especially on stories and reflections that mark out contemporary problems of knowledge and power—stories about grand issues like progress, enlightenment, liberty, and freedom.

Issues of enlightenment, progress, and freedom are quite obviously still part of a "social imaginary," especially imaginations of the relationship of knowledge and enlightenment to freedom and autonomy so clearly at stake in the notion of a public or public sphere. And while the example of Free Software illuminates how issues of enlightenment, progress, and freedom are proposed, contested, and implemented in and through software and networks, this chapter contains stories that are better understood as "usable pasts"—less technical and more accessible narratives that make sense of the contemporary world by reflecting on the past and its difference from today.

Usable pasts is a more charitable term for what might be called modern myths among geeks: stories that the tellers know to be a combination of fact and fiction. They are told not in order to remember the past, but in order to make sense of the present and of the future. They make sense of practices that are not questioned in the doing, but which are not easily understood in available intellectual or colloquial terms. The first set of stories I relate are those about the Protestant Reformation: allegories that make use of Catholic and Protestant churches, laity, clergy, high priests, and reformation-era images of control and liberation. It might be surprising that geeks turn to the past (and especially to religious allegory) in order to make sense of the present, but the reason is quite simple: there are no "ready-to-narrate" stories that make sense of the practices of geeks today. Precisely because geeks are "figuring out" things that are not clear or obvious, they are of necessity bereft of effective ways of talking about it. The Protestant Reformation makes for good allegory because it separates power from control; it draws on stories of catechism and ritual, alphabets, pamphlets and liturgies, indulgences and self-help in order to give geeks a way to make sense of the distinction between power and control, and how it relates to the technical and political economy they occupy. The contemporary relationship among states, corporations, small businesses, and geeks is not captured by familiar oppositions like commercial/noncommercial, for/against private property, or capitalist/socialist—it is a relationship of reform and conversion, not revolution or overthrow.

Usable pasts are stories, but they are stories that reflect specific attitudes and specific ways of thinking about the relationship between past, present, and future. Geeks think and talk a lot about time, progress, and change, but their conclusions and attitudes are by no means uniform. Some geeks are much more aware of the specific historical circumstances and contexts in which they operate, others less so. In this chapter I pose a question via Michel

Foucault's famous short piece "What Is Enlightenment?" Namely, are geeks modern? For Foucault, rereading Kant's eponymous piece from 1784, the problem of being modern (or of an age being "enlightened") is not one of a period or epoch that people live through; rather, it involves a subjective relationship, an *attitude.* Kant's explanation of enlightenment does not suggest that it is itself a universal, but that it occurs through a form of reflection on what difference the changes of one's immediate historical past make to one's understanding of the supposed universals of a much longer history—that is, one must ask why it is necessary to think the way one does today about problems that have been confronted in ages past. For Foucault, such reflections must be rooted in the "historically unique forms in which the generalities of our relations . . . have been problematized."[1] Thus, I want to ask of geeks, how do they connect the historically unique problems they confront—from the Internet to Napster to intellectual property to sharing and re-using source code—to the generalities of relations in which they narrate them as problems of liberty, knowledge, power, and enlightenment? Or, as Foucault puts it, are they *modern* in this sense? Do they "despise the present" or not?

The attitudes that geeks take in responding to these questions fall along a spectrum that I have identified as ranging from "polymaths" to "transhumanists." These monikers are drawn from real discussions with geeks, but they don't designate a kind of person. They are "subroutines," perhaps, called from within a larger program of moral and technical imaginations of order. It is possible for the same person to be a polymath at work and a transhumanist at home, but generally speaking they are conflicting and opposite mantles. In polymath routines, technology is an intervention into a complicated, historically unique field of people, customs, organizations, other technologies, and laws; in transhumanist routines, technology is seen as an inevitable force—a product of human action, but not of human design—that is impossible to control or resist through legal or customary means.

Protestant Reformation

Geeks love allegories about the Protestant Reformation; they relish stories of Luther and Calvin, of popery and iconoclasm, of reformation over revolution. Allegories of Protestant revolt allow geeks to make sense of the relationship between the state (the monarchy), large corporations (the Catholic Church), the small start-ups, individual programmers, and adepts among whom they spend most of their time (Protestant reformers), and the laity (known as "lusers" and "sheeple"). It gives them a way to assert that they prefer reformation (to save capitalism from the capitalists) over revolution. Obviously, not all geeks tell stories of "religious wars" and the Protestant Reformation, but these images reappear often enough in conversations that most geeks will more or less instantly recognize them as a way of making sense of modern corporate, state, and political power in the arena of information technology: the figures of Pope, the Catholic Church, the Vatican, the monarchs of various nations, the laity, the rebel adepts like Luther and Calvin, as well as models of sectarianism, iconoclasm ("In the beginning was the Command Line"), politicoreligious power, and arcane theological argumentation.[2] The allegories that unfold provide geeks a way to make sense of a similarly complex modern situation in which it is not the Church and the State that struggle, but the Corporation and the State; and what geeks struggle over are not matters of church doctrine and organization, but matters of information technology and its organization as intellectual property and economic motor. I stress here that this is not an analogy that I myself am making (though I happily make use of it), but is one that is in wide circulation among the geeks I study. To the historian or religious critic, it may seem incomplete, or absurd, or bizarre, but it still serves a specific function, and this is why I highlight it as one component of the practical and technical ideas of order that geeks share.

At the first level are allegories of "religious war" or "holy war" (and increasingly, of "jihads"). Such stories reveal a certain cynicism: they describe a technical war of details between two pieces of software that accomplish the same thing through different means, so devotion to one or the other is seen as a kind of arbitrary theological commitment, at once reliant on a pure rationality and requiring aesthetic or political judgment. Such stories imply that two technologies are equally good and equally bad and that one's choice of sect is thus an entirely nonrational one based in the vicissitudes of background and belief. Some people are zealous proselytizers of a technology, some are not. As one Usenet message explains: "Religious 'wars' have tended to occur over theological and doctrinal technicalities of one sort or another. The parallels between that and the *computing* technicalities that result in 'computing wars' are pretty strong."[3]

Perhaps the most familiar and famous of these wars is that between Apple and Microsoft (formerly between Apple and IBM), a conflict that is often played out in dramatic and broad strokes that imply fundamental differences, when in fact the differences are extremely slight.[4] Geeks are also familiar with a wealth of less well-known "holy wars": EMACS versus vi; KDE versus Gnome; Linux versus BSD; Oracle versus all other databases.[5]

Often the language of the Reformation creeps playfully into otherwise serious attempts to make aesthetic judgments about technology, as in this analysis of the programming language tcl/tk:

> It's also not clear that the primary design criterion in tcl, perl, or Visual BASIC was visual beauty—nor, probably, should it have been. Ouster-hout said people will vote with their feet. This is important. While the High Priests in their Ivory Towers design pristine languages of stark beauty and balanced perfection for their own appreciation, the rest of the mundane world will in blind and contented ignorance go plodding along using nasty little languages like those enumerated above. These poor sots will be getting a great deal of work done, putting bread on the table for their kids, and getting home at night to share it with them. The difference is that the priests will shake their fingers at the laity, and the laity won't care, because they'll be in bed asleep.[6]

In this instance, the "religious war" concerns the difference between academic programming languages and regular programmers made equivalent to a distinction between the insularity of the Catholic Church and the self-help of a protestant laity: the heroes (such as tcl/tk, perl, and python—all Free Software) are the "nasty little languages" of the laity; the High Priests design (presumably) Algol, LISP, and other "academic" languages.

At a second level, however, the allegory makes precise use of Protestant Reformation details. For example, in a discussion about the various fights over the Gnu C Compiler (gcc), a central component of the various UNIX operating systems, Christopher Browne posted this counter-reformation allegory to a Usenet group.

> The EGCS project was started around two years ago when G++ (and GCC) development got pretty "stuck." EGCS sought to integrate to-

gether a number of the groups of patches that people were making to the GCC "family." In effect, there had been a "Protestant Reformation," with split-offs of:

a) The GNU FORTRAN Denomination;

b) The Pentium Tuning Sect;

c) The IBM Haifa Instruction Scheduler Denomination;

d) The C++ Standard Acolytes.

These groups had been unable to integrate their efforts (for various reasons) with the Catholic Version, GCC 2.8. The Ecumenical GNU Compiler Society sought to draw these groups back into the Catholic flock. The project was fairly successful; GCC 2.8 was succeeded by GCC 2.9, which was not a direct upgrade from 2.8, but rather the results of the EGCS project. EGCS is now GCC.[7]

In addition to the obvious pleasure with which they deploy the sectarian aspects of the Protestant Reformation, geeks also allow themselves to see their struggles as those of Luther-like adepts, confronted by powerful worldly institutions that are distinct but intertwined: the Catholic Church and absolutist monarchs. Sometimes these comparisons are meant to mock theological argument; sometimes they are more straightforwardly hagiographic. For instance, a 1998 article in *Salon* compares Martin Luther and Linus Torvalds (originator of the Linux kernel).

In Luther's Day, the Roman Catholic Church had a near-monopoly on the cultural, intellectual and spiritual life of Europe. But the principal source text informing that life—the Bible—was off limits to ordinary people. . . . Linus Torvalds is an information-age reformer cut from the same cloth. Like Luther, his journey began while studying for ordination into the modern priesthood of computer scientists at the University of Helsinki—far from the seats of power in Redmond and Silicon Valley. Also like Luther, he had a divine, slightly nutty idea to remove the intervening bureaucracies and put ordinary folks in a direct relationship to a higher power—in this case, their computers. Dissolving the programmer-user distinction, he encouraged ordinary people to participate in the development of their computing environment. And just as Luther sought to make the entire sacramental shebang—the wine, the bread and the translated Word—available to the hoi polloi, Linus seeks to revoke the developer's proprietary access to the OS, insisting that the full operating system source code be delivered—without cost—to every ordinary Joe at the desktop.[8]

Adepts with strong convictions—monks and priests whose initiation and mastery are evident—make the allegory work. Other uses of Christian iconography are less, so to speak, faithful to the sources. Another prominent personality, Richard Stallman, of the Free Software Foundation, is prone to dressing as his alter-ego, St. IGNUcius, patron saint of the church of EMACS—a church with no god, but intense devotion to a baroque text-processing program of undeniable, nigh-miraculous power.[9]

Often the appeal of Reformation-era rhetoric comes from a kind of indictment of the present: despite all this high tech, superfabulous computronic wonderfulness, we are no less feudal, no less violent, no less arbitrary and undemocratic; which is to say, geeks have progressed, have seen the light and the way, but the rest of society—and especially management and marketing—have not. In this sense, Reformation allegories are stories of how "things never change."

But the most compelling use of the Protestant Reformation as usable past comes in the more detailed understandings geeks have of the political economy of information technology. The allegorization of the Catholic Church with Microsoft, for instance, is a frequent component, as in this brief message regarding start-up key combinations in the Be operating system: "These secret handshakes are intended to reinforce a cabalistic high priesthood and should not have been disclosed to the laity. Forget you ever saw this post and go by [sic] something from Microsoft."[10]

More generally, large corporations like IBM, Oracle, or Microsoft are made to stand in for Catholicism, while bureaucratic congresses and parliaments with their lobbyists take on the role of absolutist monarchs and their cronies. Geeks can then see themselves as fighting to uphold Christianity (true capitalism) against the church (corporations) and to be reforming a way of life that is corrupted by church and monarchs, instead of overthrowing through revolution a system they believe to be flawed. There is a historically and technically specific component of this political economy in which it is in the interest of corporations like IBM and Microsoft to keep users "locked as securely to Big Blue as an manacled wretch in a medieval dungeon."[11]

Such stories appeal because they bypass the language of modern American politics (liberal, conservative, Democrat, Republican) in which there are only two sides to any issue. They also bypass an argument between capitalism and socialism, in which if you are not pro-capitalism you must be a communist. They are stories that allow the more pragmatist of the geeks to engage in intervention and reformation, rather than revolution. Though I've rarely heard it articulated so bluntly, the allegory often implies that one must "save capitalism from the capitalists," a sentiment that implies at least some kind of human control over capitalism.

In fact, the allegorical use of the Reformation and the church generates all kinds of clever comparisons. A typical description of such comparisons might go like this: the Catholic Church stands in for large, publicly traded corporations, especially those controlling large amounts of intellectual property (the granting of which might roughly be equated with the ceremonies of communion and confession) for which they depend on the assistance and support of national governments. Naturally, it is the storied excesses of the church—indulgences, liturgical complexity, ritualistic ceremony, and corruption—which make for easy allegory. Modern corporations can be figured as a small, elite papal body with theologians (executives and their lawyers, boards of directors and their lawyers), who command a much larger clergy (employees), who serve a laity (consumers) largely imagined to be sinful (underspending on music and movies—indeed, even "stealing" them) and thus in need of elaborate and ritualistic cleansing (advertising and lawsuits) by the church. Access to grace (the American Dream) is mediated only by the church and is given form through the holy acts of shopping and home improvement. The executives preach messages of damnation to the government, messages most government officials are all too willing to hear: do not tamper with our market share, do not affect our pricing, do not limit our ability to expand these markets. The executives also offer unaccountable promises of salvation in the guise of deregulation and the American version of "reform"—the demolition of state and national social services. Government officials in turn have developed their own "divine right of kings," which justifies certain forms of manipulation (once called "elections") of succession. Indulgences are sold left and right by lobbyists or industry associations, and the decrees of the papacy evidence little but full disconnection from the miserable everyday existence of the flock.

In fact, it is remarkable how easy such comparisons become the more details of the political economy of information one learns. But allegories of the Reformation and clerical power can lead easily to cynicism, which should perhaps be read in this instance as evidence of political disenfranchisement, rather than a lapse in faith. And yet the usable pasts of these reformation-minded modern monks and priests crop up regularly not only because they provide relief from technical chatter but because they explain a political, technical, legal situation that does not have ready-to-narrate stories. Geeks live in a world finely controlled by corporate organizations, mass media, marketing departments, and lobbyists, yet they share a profound distrust of government regulation—they need another set of just-so stories to make sense of it. The standard *unusable* pasts of the freeing of markets, the inevitability of capitalism and democracy, or more lately, the necessity of security don't do justice to their experience.

Allegories of Reformation are stories that make sense of the political economy of information. But they also have a more precise use: to make sense of the distinction between power and control. Because geeks are "closer to the machine" than the rest of the laity, one might reasonably expect them to be the ones in power. This is clearly not the case, however, and it is the frustrations and mysteries by which states, corporations, and individuals manipulate technical details in order to shift power that often earns the deepest ire of geeks. Control, therefore, includes the detailed methods and actual practices by which corporations, government agencies, or individuals attempt to manipulate people (or enroll them to manipulate themselves and others) into making technical choices that serve power, rather than rationality, liberty, elegance, or any other geekly concern.

Consider the subject of evil. During my conversations with Sean Doyle in the late 1990s, as well as with a number of other geeks, the term *evil* was regularly used to refer to some kind of design or technical problem. I asked Sean what he meant.

> SD: [*Evil* is] just a term I use to say that something's wrong, but usually it means something is wrong on purpose, there was agency behind it. I can't remember [the example you gave] but I think it may have been some GE equipment, where it has this default where it likes to send things in its own private format rather than in DICOM [the radiology industry standard for digital images], if you give it a choice. I don't know why they would have done something like that,

it doesn't solve any backward compatibility problem, it's really just an exclusionary sort of thing. So I guess there's Evil like that. . . .

CK: one of the other examples that you had . . . was something with Internet Explorer 3.0?

SD: Yes, oh yes, there are so many things with IE3 that are completely Evil. Like here's one of them: in the http protocol there's a thing called the "user agent field" where a browser announces to the server who it is. If you look at IE, it announces that it is Mozilla, which is the [code-name for] Netscape. Why did they do this? Well because a lot of the web servers were sending out certain code that said, if it were Mozilla they would serve the stuff down, [if not] they would send out something very simple or stupid that would look very ugly. But it turned out that [IE3, or maybe IE2] didn't support things when it first came out. Like, I don't think they supported tables, and later on, their versions of Javascript were so different that there was no way it was compatible—it just added tremendous complexity. It was just a way of pissing on the Internet and saying there's no law that says we have to follow these Internet standards. We can do as we damn well please, and we're so big that you can't stop us. So I view it as Evil in that way. I mean they obviously have the talent to do it. They obviously have the resources to do it. They've obviously done the work, it's just that they'll have this little twitch where they won't support a certain MIME type or they'll support some things differently than others.

CK: But these kinds of incompatibility issues can happen as a result of a lack of communication or coordination, which might involve agency at some level, right?

SD: Well, I think of that more as Stupidity than Evil [laughter]. No, Evil is when there is an opportunity to do something, and an understanding that there is an opportunity to, and resources and all that—and then you do something just to spite the other person. You know I'm sure it's like in messy divorces, where you would rather sell the property at half its value rather than have it go to the other person.

Sean relates control to power by casting the decisions of a large corporation in a moral light. Although the specific allegory of the Protestant Reformation does not operate here, the details do. Microsoft's decision to manipulate Internet Explorer's behavior stems not from a lack of technical sophistication, nor is it an "accident" of complexity, according to Sean, but is a deliberate assertion of economic and political power to corrupt the very details by which software has been created and standardized and is expected to function. The clear goal of this activity is *conversion*, the expansion of Microsoft's flock through a detailed control of the beliefs and practices (browsers and functionality) of computer users. Calling Microsoft "Evil" in this way has much the same meaning as questioning the Catholic Church's use of ritual, ceremony, literacy, and history—the details of the "implementation" of religion, so to speak.

Or, in the terms of the Protestant Reformation itself, the practices of conversion as well as those of liberation, learning, and self-help are central to the story. It is not an accident that many historians of the Reformation themselves draw attention to the promises of liberation through reformation "information technologies."[12] Colloquial (and often academic) assertions that the printing press was technologically necessary or sufficient to bring the Reformation about appear constantly as a parable of this new age of information. Often the printing press is the only "technological" cause considered, but scholars of the real, historical Reformation also pay close attention to the fact of widespread literacy, to circulating devotional pamphlets, catechisms, and theological tracts, as well as to the range of transformations of political and legal relationships that occurred simultaneously with the introduction of the printing press.

One final way to demonstrate the effectiveness of these allegories—their ability to work on the minds of geeks—is to demonstrate how they have started to work on *me,* to demonstrate how much of a geek I have become—a form of participant allegorization, so to speak. The longer one considers the problems that make up the contemporary political economy of information technology that geeks inhabit, the more likely it is that these allegories will start to present themselves almost automatically—as, for instance, when I read *The Story of A,* a delightful book having nothing to do with geeks, a book about literacy in early America. The author, Patricia Crain, explains that the Christ's cross (see above) was often used in the creation of hornbooks or battledores, small leather-backed paddles inscribed with the Lord's Prayer and the alphabet, which were used to teach children their ABCs from as early as the fifteenth century until as late as the nineteenth: "In its early print manifestations, the pedagogical alphabet is headed not by the letter A but by the 'Christ's Cross': ✠. . . . Because the alphabet is associated with Catholic Iconography, as if the two sets of signs were really part of one semiological system, one of the struggles of the Reformation would be to wrest the alphabet away from the Catholic Church."[13]

Here, allegorically, the Catholic Church's control of the alphabet (like Microsoft's programming of Internet Explorer to blur public standards for the Internet) is not simply ideological; it is not just a fantasy of origin or ownership planted in the fallow mental soil of believers, but in fact a very specific, very nonsubjective, and very media-specific normative tool of control. Crain explains further: "Today ✠ represents the imprimatur of the Catholic Church on copyright pages. In its connection to the early modern alphabet as well, this cross carries an imprimatur or licensing effect. This 'let it be printed,' however, is directed not to the artisan printer but to the mind and memory of the young scholar. . . . Like modern copyright, the cross authorizes the existence of the alphabet and associates the letters with sacred authorship, especially since another long-lived function of ✠ in liturgical missals is to mark gospel passages. The symbol both conveys information and generates ritual behavior."[14]

The © today carries as much if not more power, both ideologically and legally, as the cross of the Catholic church. It is the very symbol of authorship, even though in origin and in function it governs only ownership and rights. Magical thinking about copyright abounds, but one important function of the symbol ©, if not its legal implications, is to achieve the same thing as the Christ's cross: to associate in the mind of the reader the ownership of a particular text (or in this case, piece of software) with a particular organization or person. Furthermore, even though the symbol is an artifact of national and international law, it creates an association not between a text and the state or government, but between a text and particular corporations, publishers, printers, or authors.

Like the Christ's cross, the copyright symbol carries both a licensing effect (exclusive, limited or nonexclusive) and an imprimatur on the minds of people: "let it be imprinted in memory" that this is the work of such and such an author *and* that this is the *property* of such and such a corporation.

Without the allegory of the Protestant Reformation, the only available narrative for such evil—whether it be the behavior of Microsoft or of some other corporation—is that corporations are "competing in the marketplace according to the rules of capitalism" and thus when geeks decry such behavior, it's just sour grapes. If corporations are not breaking any laws, why shouldn't they be allowed to achieve control in this manner? In this narrative there is no room for a *moral evaluation of competition*—anything goes, it would seem. Claiming for Microsoft that it is simply playing by the rules of capitalism puts everyone else into either the competitor box or the noncompetitor box (the state and other noncompetitive organizations). Using the allegory of the Protestant Reformation, on the other hand, gives geeks a way to make sense of an unequal distribution among competing powers—between large and small corporations, and between market power and the details of control. It provides an alternate imagination against which to judge the technically and legally specific actions that corporations and individuals take, and to imagine forms of justified action in return.

Without such an allegory, geeks who oppose Microsoft are generally forced into the position of being anticapitalist or are forced to adopt the stance that all standards should be publicly generated and controlled, a position few wish to take. Indeed, many geeks would prefer a different kind of imaginary altogether—a recursive public, perhaps. Instead of an infrastructure subject to unequal distributions of power and shot through with "evil" distortions of technical control, there is, as geeks see it, the possibility for a "self-leveling" level playing field, an autotelic system of rules, both technical and legal, by which all participants are expected to compete equally. Even if it remains an imaginary, the allegory of the Protestant Reformation makes sense of (gives order to) the political economy of the contemporary information-technology world and allows geeks to conceive of their interests and actions according to a narrative of reformation, rather than one of revolution or submission. In the Reformation the interpretation or truth of Christian teaching was not primarily in question: it was not a doctrinal revolution, but a bureaucratic one. Likewise, geeks do not question the rightness of networks, software, or protocols and standards, nor are they against capitalism or intellectual property, but they do wish to maintain a space for critique and the moral evaluation of contemporary capitalism and competition.

Polymaths and Transhumanists

Usable pasts articulate the conjunction of "operating systems and social systems," giving narrative form to imaginations of moral and technical order. To say that there are no ready-to-narrate stories about contemporary political economy means only that the standard colloquial explanations of the state of the modern world do not do justice to the kinds of moral and technical imaginations of order that geeks possess by virtue of their practices. Geeks live in, and build, one kind of world—a world of software, networks, and infrastructures—but they are often confronted with stories and explanations that simply don't match up with their experience, whether in newspapers and on television, or among nongeek friends. To many geeks, proselytization seems an obvious route: why not help friends and neighbors to understand the hidden world of networks and software, since, they are quite certain, it will come to structure their lives as well?

Geeks gather through the Internet and, like a self-governing people, possess nascent ideas of independence, contract, and constitution by which they wish to govern themselves and resist governance by others.[15] Conventional political philosophies like libertarianism, anarchism, and (neo)liberalism only partially capture these social imaginaries precisely because they make no reference to the operating systems, software, and networks within which geeks live, work, and in turn seek to build and extend.

Geeks live in specific ways in time and space. They are not just users of technology, or a "network society," or a "virtual community," but embodied and imagining actors whose affinity for one another is enabled in new ways by the tools and technologies they have such deep affective connections to. They live in *this-network-here*, a historically unique form grounded in particular social, moral, national, and historical specificities which nonetheless relates to generalities such as progress, technology, infrastructure, and liberty. Geeks are by no means of one mind about such generalities though, and they often have highly developed means of thinking about them.

Foucault's article "What Is Enlightenment?" captures part of this problematic. For Foucault, Kant's understanding of modernity was an attempt to rethink the relationship between the passage of historical time and the subjective relationship that individuals have toward it.

Thinking back on Kant's text, I wonder whether we may not envisage modernity as an attitude rather than as a period of history. And by "attitude," I mean a mode of relating to contemporary reality; a voluntary choice made by certain people; in the end, a way of thinking and feeling; a way, too, of acting and behaving that at one and the same time marks a relation of belonging and presents itself as a task. No doubt a bit like what the Greeks called an *ethos*. And consequently, rather than seeking to distinguish the "modern era" from the "premodern" or "postmodern," I think it would be more useful to try to find out how the attitude of modernity, ever since its formation, has found itself struggling with attitudes of "countermodernity."[16]

In thinking through how geeks understand the present, the past, and the future, I pose the question of whether they are "modern" in this sense. Foucault makes use of Baudelaire as his foil for explaining in what the attitude of modernity consists: "For [Baudelaire,] being modern . . . consists in recapturing something eternal that is not beyond the present, or behind it, but within it."[17] He suggests that Baudelaire's understanding of modernity is "an attitude that makes it possible to grasp the 'heroic' aspect of the present moment . . . the will to 'heroize' the present."[18] *Heroic* here means something like redescribing the seemingly fleeting events of the present in terms that conjure forth the universal or eternal character that animates them. In Foucault's channeling of Baudelaire such an attitude is incommensurable with one that sees in the passage of the present into the future some version of autonomous progress (whether absolute spirit or decadent degeneration), and the tag he uses for this is "you have no right to despise the present." To be modern is to confront the present as a problem that can be transformed by human action, not as an inevitable outcome of processes beyond the scope of individual or collective human control, that is, "attitudes of counter-modernity." When geeks tell stories of the past to make sense of the future, it is often precisely in order to "heroize" the present in this sense—but not all geeks do so. Within the spectrum from polymath to transhumanist, there are attitudes of both modernity and countermodernity.

The questions I raise here are also those of *politics* in a classical sense: Are the geeks I discuss bound by an attitude toward the present that concerns such things as the relationship of the public to the private and the social (à la Hannah Arendt), the relationship

REFORMERS, POLYMATHS, TRANSHUMANISTS

of economics to liberty (à la John Stuart Mill and John Dewey), or the possibilities for rational organization of society through the application of scientific knowledge (à la Friedrich Hayek or Foucault)? Are geeks "enlightened"? Are they Enlightenment rationalists? What might this mean so long after the Enlightenment and its vigorous, wide-ranging critiques? How is their enlightenment related to the technical and infrastructural commitments they have made? Or, to put it differently, what makes enlightenment newly necessary *now*, in the milieu of the Internet, Free Software, and recursive publics? What kinds of relationships become apparent when one asks how these geeks relate their own conscious appreciation of the history and politics of their time to their everyday practices and commitments? Do geeks despise the present?

Polymaths and transhumanists speak differently about concepts like technology, infrastructure, networks, and software, and they have different ideas about their temporality and relationship to progress and liberty. Some geeks see technology as one kind of intervention into a constituted field of organizations, money, politics, and people. Some see it as an autonomous force made up of humans and impersonal forces of evolution and complexity. Different geeks speak about the role of technology and its relationship to the present and future in different ways, and how they understand this relationship is related to their own rich understandings of the complex technical and political environment they live and work in.

Polymaths Polymathy is "avowed dilettantism," not extreme intelligence. It results from a curiosity that seems to grip a remarkable number of people who spend their time on the Internet and from the basic necessity of being able to evaluate and incorporate sometimes quite disparate fields of knowledge in order to build workable software. Polymathy inevitably emerges in the context of large software and networking projects; it is a creature of constraints, a process bootstrapped by the complex sediment of technologies, businesses, people, money, and plans. It might also be posed in the negative: bad software design is often the result of not enough avowed dilettantism. Polymaths *must* know a very large and wide range of things in order to intervene in an existing distribution of machines, people, practices, and places. They *must* have a detailed sense of the present, and the project of the present, in order to imagine how the future might be different.

My favorite polymath is Sean Doyle. Sean built the first versions of a piece of software that forms the centerpiece of the radiological-image-management company Amicas. In order to build it Sean learned the following: Java, to program it; the mathematics of wavelets, to encode the images; the workflow of hospital radiologists and the manner in which they make diagnoses from images, to make the interface usable; several incompatible databases and the SQL database language, to build the archive and repository; and manual after manual of technical standards, the largest and most frightening of which was the Digital Imaging and Communication (DICOM) standard for radiological images. Sean also read *Science* and *Nature* regularly, looking for inspiration about interface design; he read books and articles about imaging very small things (mosquito knees), very large things (galaxies and interstellar dust), very old things (fossils), and very pretty things (butterfly-wing patterns as a function of developmental pathways). Sean also introduced me to Tibetan food, to Jan Svankmeyer films, to Open Source Software, to cladistics and paleoherpetology, to Disney's scorched-earth policy with respect to culture, and to many other awesome things.

Sean is clearly an unusual character, but not that unusual. Over the years I have met many people with a similar range and depth of knowledge (though rarely with Sean's humility, which does set him apart). Polymathy is an occupational hazard for geeks. There is no sense in which a good programmer, software architect, or information architect simply specializes in code. Specialization is seen not as an end in itself, but rather as a kind of technical prerequisite before other work—the real work—can be accomplished. The real work is the *design*, the process of inserting usable software into a completely unfamiliar amalgamation of people, organizations, machines, and practices. Design is hard work, whereas the technical stuff—like choosing the right language or adhering to a standard or finding a ready-made piece of code to plug in somewhere—is not.

It is possible for Internet geeks and software architects to think this way in part due to the fact that so many of the technical issues they face are both extremely well defined and very easy to address with a quick search and download. It is easy to be an avowed dilettante in the age of mailing lists, newsgroups, and online scientific publishing. I myself have learned whole swaths of technical practices in this manner, but I have *designed* no technology of note.

Sean's partner in Amicas, Adrian Gropper, also fits the bill of polymath, though he is not a programmer. Adrian, a physician and a graduate of MIT's engineering program, might be called a "high-functioning polymath." He scans the horizon of technical and scientific accomplishments, looking for ways to incorporate them into his vision of medical technology qua intervention. Sean mockingly calls these "delusions," but both agree that Amicas would be nowhere without them. Adrian and Sean exemplify how the meanings of technology, intervention, design, and infrastructure are understood by polymaths as a particular form of pragmatic intervention, a progress achieved through deliberate, piecemeal re-formation of existing systems. As Adrian comments:

> I firmly believe that in the long run the only way you can save money and improve healthcare is to add technology. I believe that more strongly than I believe, for instance, that if people invent better pesticides they'll be able to grow more rice, and it's for the universal good of the world to be able to support more people. I have some doubt as to whether I support people doing genetic engineering of crops and pesticides as being "to the good." But I do, however, believe that healthcare is different in that in the long run you can impact both the cost and quality of healthcare by adding technology. And you can call that a religious belief if you want, it's not rational. But I guess what I'm willing to say is that traditional healthcare that's not technology-based has pretty much run out of steam.[19]

In this conversation, the "technological" is restricted to the novel things that can make healthcare less costly (i.e., cost-reducing, not cost-cutting), ease suffering, or extend life. Certain kinds of technological intervention are either superfluous or even pointless, and Adrian can't quite identify this "class"—it isn't "technology" in general, but it includes some kinds of things that are technological. What is more important is that technology does not solve anything by itself; it does not obviate the political problems of healthcare rationing: "Now, however, you get this other problem, which is that the way that healthcare is rationed is through the fear of pain, financial pain to some extent, but physical pain; so if you have a technology that, for instance, makes it relatively painless to fix . . . I guess, bluntly put, it's cheaper to let people die in most cases, and that's just undeniable. So what I find interesting in all of this, is that most people who are dealing with the politics of healthcare resource management don't want to have this discussion, nobody wants to talk about this, the doctors don't want to talk about it, because it's too depressing to talk about the value of. . . . And they don't really have a mandate to talk about technology."[20]

Adrian's self-defined role in this arena is as a nonpracticing physician who is also an engineer and an entrepreneur—hence, his polymathy has emerged from his attempts to translate between doctors, engineers, and businesspeople. His goal is twofold: first, create technologies that save money and improve the allocation of healthcare (and the great dream of telemedicine concerns precisely this goal: the reallocation of the most valuable asset, individuals and their expertise); second, to raise the level of discussion in the business-cum-medical world about the role of technology in managing healthcare resources. Polymathy is essential, since Adrian's twofold mission requires understanding the language and lives of at least three distinct groups who work elbow-to-elbow in healthcare: engineers and software architects; doctors and nurses; and businessmen.

Technology has two different meanings according to Adrian's two goals: in the first case *technology* refers to the intervention by means of new technologies (from software, to materials, to electronics, to pharmaceuticals) in specific healthcare situations wherein high costs or limited access to care can be affected. Sometimes technology is allocated, sometimes it does the allocating. Adrian's goal is to match his knowledge of state-of-the-art technology—in particular, Internet technology—with a specific healthcare situation and thereby effect a reorganization of practices, people, tools, and information. The tool Amicas created was distinguished by its clever use of compression, Internet standards, and cheap storage media to compete with much larger, more expensive, much more entrenched "legacy" and "turnkey" systems. Whether Amicas invented something "new" is less interesting than the nature of this intervention into an existing milieu. This intervention is what Adrian calls "technology." For Amicas, the relevant technology—the important intervention—was the Internet, which Amicas conceived as a tool for changing the nature of the way healthcare was organized. Their goal was to replace the infrastructure of the hospital radiology department (and potentially the other departments as well) with the Internet. Amicas was able to confront and reform the practices of powerful, entrenched entities, from the administration of large hospitals to their corporate bedfellows, like HBOC, Agfa, Siemens, and GE.

With regard to raising the level of discussion, however, *technology* refers to a kind of political-rhetorical argument: technology does not save the world (nor does it destroy it); it only saves lives—and it does this only when one makes particular decisions about its allocation. Or, put differently, the means is technology, but the ends are still where the action is at. Thus, the hype surrounding information technology in healthcare is horrifying to Adrian: promises precede technologies, and the promises suggest that the means can replace the ends. Large corporations that promise "technology," but offer no real hard interventions (Adrian's first meaning of *technology*) that can be concretely demonstrated to reduce costs or improve allocation are simply a waste of resources. Such companies are doubly frustrating because they use "technology" as a blinder that allows people to *not* think about the hard problems (the ends) of allocation, equity, management, and organization; that is, they treat "technology" (the means) as if it were a solution *as such*.

Adrian routinely analyzes the rhetorical and practical uses of technology in healthcare with this kind of subtlety; clearly, such subtlety of thought is rare, and it sets Adrian apart as someone who understands that intervention into, and reform of, modern organizations and styles of thought has to happen through reformation— through the clever use of technology by people who understand it intimately—not through revolution. Reformation through technical innovation is opposed here to control through the consolidation of money and power.

In my observations, Adrian always made a point of making the technology—the software tools and picture-archiving system— easily accessible, easily demonstrable to customers. When talking to hospital purchasers, he often said something like "I can show you the software, and I can tell you the price, and I can demonstrate the problem it will solve." In contrast, however, an array of enormous corporations with salesmen and women (usually called consultants) were probably saying something more like "Your hospital needs more technology, our corporation is big and stable—give us this much money and we will solve your problem." For Adrian, the decision to "hold hands," as he put it, with the comfortably large corporation was irrational if the hospital could instead purchase a specific technology that did a specific thing, for a real price.

Adrian's reflections on technology are also reflections on the nature of progress. Progress is limited intervention structured by goals that are not set by the technology itself, even if entrepreneurial activity is specifically focused on finding new uses and new ideas for new technologies. But discussions about healthcare allocation—which Adrian sees as a problem amenable to certain kinds of technical solutions—are instead structured as if technology did not matter to the nature of the ends. It is a point Adrian resists: "I firmly believe that in the long run the only way you can save money and improve healthcare is to add technology."

Sean is similarly frustrated by the homogenization of the concept of technology, especially when it is used to suggest, for instance, that hospitals "lag behind" other industries with regard to computerization, a complaint usually made in order to either instigate investment or explain failures. Sean first objects to such a homogenous notion of "technological."

> I actually have no idea what that means, that it's lagging behind. Because certainly in many ways in terms of image processing or some very high-tech things it's probably way ahead. And if that means what's on people's desktops, ever since 19-maybe-84 or so when I arrived at MGH [Massachusetts General Hospital] there's been a computer on pretty much everyone's desktop. . . . It seems like most hospitals that I have been to seem to have a serious commitment to networks and automation, etcetera. . . . I don't know about a lot of manufacturing industries—they might have computer consoles there, but it's a different sort of animal. Farms probably lag *really* far behind, I won't even *talk* about amusement parks. In some sense, hospitals are very complicated little communities, and so to say that this thing as a whole is lagging behind doesn't make much sense.[21]

He also objects to the notion that such a lag results in failures caused by technology, rather than by something like incompetence or bad management. In fact, it might be fair to say that, for the polymath, sometimes technology actually dissolves. Its boundaries are not easily drawn, nor are its uses, nor are its purported "unintended consequences." On one side there are rules, regulations, protocols, standards, norms, and forms of behavior; on the other there are organizational structures, business plans and logic, human skills, and other machines. This complex milieu requires reform from within: it cannot be replaced wholesale; it cannot leap-

REFORMERS, POLYMATHS, TRANSHUMANISTS

frog other industries in terms of computerization, as intervention is always local and strategic; and it involves a more complex relationship to the project of the present than simply "lagging behind" or "leaping ahead."

Polymathy—inasmuch as it is a polymathy of the lived experience of the necessity for multiple expertise to suit a situation—turns people into pragmatists. Technology is never simply a solution to a problem, but always part of a series of factors. The polymath, unlike the technophobe, can see when technology matters and when it doesn't. The polymath has a very this-worldly approach to technology: there is neither mystery nor promise, only human ingenuity and error. In this manner, polymaths might better be described as Feyerabendians than as pragmatists (and, indeed, Sean turned out to be an avid reader of Feyerabend). The polymath feels there is no single method by which technology works its magic: it is highly dependent on rules, on patterned actions, and on the observation of contingent and contextual factors. Intervention into this already instituted field of people, machines, tools, desires, and beliefs requires a kind of scientific-technical genius, but it is hardly single, or even autonomous. This version of pragmatism is, as Feyerabend sometimes refers to it, simply a kind of *awareness*: of standards, of rules, of history, of possibility.[22] The polymath thus does not allow himself or herself to despise the present, but insists on both reflecting on it and intervening in it.

Sean and Adrian are avowedly scientific and technical people; like Feyerabend, they assume that their interlocutors believe in good science and the benefits of progress. They have little patience for Luddites, for new-agers, for religious intolerance, or for any other non-Enlightenment-derived attitude. They do not despise the present, because they have a well-developed sense of how provisional the conventions of modern technology and business are. Very little is sacred, and rules, when they exist, are fragile. Breaking them pointlessly is immodest, but innovation is often itself seen as a way of transforming a set of accepted rules or practices to other ends. Progress is limited intervention.[23]

How ironic, and troubling, then, to realize that Sean's and Adrian's company would eventually become the kind of thing they started Amicas in order to reform. Outside of the limited intervention, certain kinds of momentum seem irresistible: the demand for investment and funding rounds, the need for "professional management," and the inertia of already streamlined and highly conservative purchasing practices in healthcare. For Sean and Adrian, Amicas became a failure in its success. Nonetheless, they remain resolutely modern polymaths: they do not despise the present. As described in Kant's "What Is Enlightenment?" the duty of the citizen is broken into public and private: on the one hand, a duty to carry out the responsibilities of an office; on the other, a duty to offer criticism where criticism is due, as a "scholar" in a reading public. Sean's and Adrian's endeavor, in the form of a private start-up company, might well be understood as the expression of the scholar's duty to offer criticism, through the creation of a particular kind of technical critique of an existing (and by their assessment) ethically suspect healthcare system. The mixture of private capital, public institutions, citizenship, and technology, however, is something Kant could not have known—and Sean and Adrian's technical pursuits must be understood as something more: a kind of modern civic duty, in the service of liberty and responding to the particularities of contemporary technical life.[24]

Transhumanists Polymathy is born of practical and pragmatic engagement with specific situations, and in some ways is demanded by such exigencies. Opposite polymathy, however, and leaning more toward a concern with the whole, with totality and the universal, are attitudes that I refer to by the label transhumanism, which concerns the mode of belief in the Timeline of Technical Progress.[25]

Transhumanism, the movement and the philosophy, focuses on the power of technology to transcend the limitations of the human body as currently evolved. Subscribers believe—but already this is the wrong word—in the possibility of downloading consciousness onto silicon, of cryobiological suspension, of the near emergence of strong artificial intelligence and of various other forms of technical augmentation of the human body for the purposes of achieving immortality—or at least, much more life.[26]

Various groups could be reasonably included under this label. There are the most ardent purveyors of the vision, the Extropians; there are a broad class of people who call themselves transhumanists; there is a French-Canadian subclass, the Raelians, who are more an alien-worshiping cult than a strictly scientific one and are bitterly denounced by the first two; there are also the variety of cosmologists and engineers who do not formally consider themselves transhumanist, but whose beliefs participate in some way or another: Stephen Hawking, Frank Tipler and John Barrow (famous for their anthropic cosmological principle), Hans Moravic, Ray Kurzweil, Danny Hillis, and down the line through those who embrace the cognitive sciences, the philosophy of artificial intelligence, the philosophy of mind, the philosophy of science, and so forth.

Historically speaking, the line of descent is diffuse. Teilhard de Chardin is broadly influential, sometimes acknowledged, sometimes not (depending on the amount of mysticism allowed). A more generally recognized starting point is Julian Huxley's article "Transhumanism" in *New Bottles for New Wine*.[27] Huxley's transhumanism, like Teilhard's, has a strange whiff of Nietzsche about it, though it tends much more strongly in the direction of the evolutionary emergence of the superman than in the more properly moral sense Nietzsche gave it. After Huxley, the notion of transhumanism is too easily identified with eugenics, and it has become one of a series of midcentury subcultural currents which finds expression largely in small, non-mainstream places, from the libertarians to Esalen.[28]

For many observers, transhumanists are a lunatic fringe, bounded on either side by alien abductees and Ayn Rand–spouting objectivists. However, like so much of the fringe, it merely represents in crystalline form attitudes that seem to permeate discussions more broadly, whether as beliefs professed or as beliefs attributed. Transhumanism, while probably anathema to most people, actually reveals a very specific attitude toward technical innovation, technical intervention, and political life that is widespread among technically adept individuals. It is a belief that has everything to do also with the timeline of progress and the role of technology in it.

The transhumanist understanding of technological progress can best be understood through the sometimes serious and sometimes playful concept of the "singularity," popularized by the science-fiction writer and mathematician Vernor Vinge.[29] The "singularity" is the point at which the speed of technical progress is faster than human comprehension of that progress (and, by implication, than human control over the course). It is a kind of cave-man parable, perhaps most beautifully rendered by Stanley Kubrik's film *2001: A Space Odyssey* (in particular, in the jump-cut early in the film that turns a hurled bone into a spinning space station, recapitulating the remarkable adventure of technology in two short seconds of an otherwise seemingly endless film).

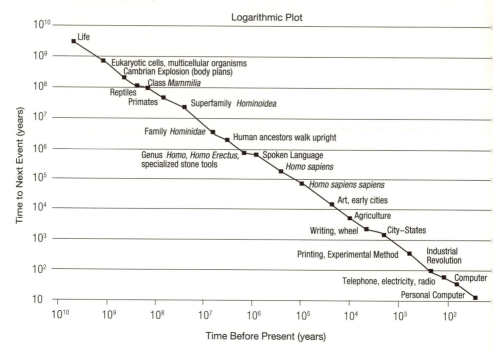

Countdown to Singularity

1. Illustration © 2005 Ray Kurzweil. Modifications © 2007 by C. Kelty. Original work licensed under a Creative Commons Attribution License: http://en.wikipedia.org/wiki/Image:PPTCountdowntoSingularityLog.jpg.

In figure 1, on the left hand of the timeline, there is history, or rather, there is a string of technological inventions (by which is implied that previous inventions set the stage for later ones) spaced such that they produce a logarithmic curve that can look very much like the doomsday population curves that started to appear in the 1960s. Each invention is associated with a name or sometimes a nation. Beyond the edge of the graph to the right side is the future: history changes here from a series of inventions to an autonomous self-inventing technology associated not with individual inventors but with a complex system of evolutionary adaptation that includes technological as well as biological forms. It is a future in which "humans" are no longer necessary to the progress of science and technology: technology-as-extension-of-humans on the left, a Borg-like autonomous technical intelligence on the right. The fundamental operation in constructing the "singularity" is the "reasoned extrapolation" familiar to the "hard science fiction" writer or the futurist. One takes present technology as the initial condition for future possibilities and extrapolates based on the (haphazardly handled) evidence of past technical speed-up and change.

The position of the observer is always a bit uncertain, since he or she is naturally projected at the highest (or lowest, depending on your orientation) point of this curve, but one implication is clear: that the function or necessity of human reflection on the present will disappear at the same time that humans do, rendering enlightenment a quaint, but necessary, step on the route to superrational, transhuman immortality.

Strangely, the notion that technical progress has *acceleration* seems to precede any sense of what the *velocity* of progress might mean in the first instance; technology is presumed to exist in absolute time—from the Big Bang to the heat death of the universe—and not in any relationship with human life or consciousness. The singularity is always described from the point of view of a god who is not God. The fact of technological speed-up is generally treated as the most obvious thing in the world, reinforced by the constant refrain in the media of the incredible pace of change in contemporary society.

Why is the singularity important? Because it always implies that the absolute fact of technical acceleration—this knowing glance into the future—should order the kinds of interventions that occur in the present. It is not mute waiting or eschatological certainty that governs this attitude; rather, it is a mode of historical consciousness that privileges the inevitability of technological progress over the inevitability of human power. Only by looking into the future can one manipulate the present in a way that will be widely meaningful, an attitude that could be expressed as something like "Those who do not learn from the future are condemned to suffer in it." Since it is a philosophy based on the success of human rationality and ingenuity, rationality and ingenuity are still clearly essential in the future. They lead, however, to a kind of posthuman state of constant technological becoming which is inconceivable to the individual human mind—and can only be comprehended by a transcendental intelligence that is not God.

Such is a fair description of some strands of transhumanism, and the reason I highlight them is to characterize the kinds of attitudes toward technology-as-intervention and the ideas of moral and technical order that geeks can evince. On the far side of polymathy, geeks are too close to the machine to see a big picture or to think about imponderable philosophical issues; on the transhuman side, by contrast, one is constantly reassessing the arcane details of everyday technical change with respect to a vision of the whole—a vision of the evolution of technology and its relationship to the humans that (for the time being) must create and attempt to channel it.

My favorite transhumanist is Eugen Leitl (who is, in fact, an authentic transhumanist and has been vice-chair of the World Transhumanist Association). Eugen is Russian-born, lives in Munich, and once worked in a cryobiology research lab. He is well versed in chemistry, nanotechnology, artificial-intelligence (AI) research, computational- and network-complexity research, artificial organs, cryobiology, materials engineering, and science fiction. He writes, for example,

> If you consider AI handcoded by humans, yes. However, given considerable computational resources (~cubic meter of computronium), and using suitable start population, you can coevolve machine intelligence on a time scale of much less than a year. After it achieves about a human level, it is potentially capable of entering an autofeedback loop. Given that even autoassembly-grade computronium is capable of running a human-grade intellect in a volume ranging from a sugar cube to an orange at a speed ranging from 10^4 . . . 10^6 it is easy to see that the autofeedback loop has explosive dynamics.
>
> (I hope above is intelligible, I've been exposed to weird memes for far too long).[30]

Eugen is also a polymath (and an autodidact to boot), but in the conventional sense. Eugen's polymathy is an *avocational* necessity: transhumanists need to keep up with all advances in technology and science in order to better assess what kinds of human-augmenting or human-obsolescing technologies are out there. It is not for work in this world that the transhumanist expands his or her knowledge, nor quite for the next, but for a "this world" yet to arrive.

Eugen and I were introduced during the Napster debates of 2001, which seemed at the time to be a knock-down, drag-out conflagration, but Eugen has been involved in so many online flame wars that he probably experienced it as a mere blip in an otherwise constant struggle with less-evolved intelligences like mine. Nonetheless, it was one of the more clarifying examples of how geeks think, and think differently, about technology, infrastructure, networks, and software. Transhumanism has no truck with old-fashioned humanism.

> > From: Ramu Narayan . . .
> > I don't like the
> > notion of technology as an unstoppable force with a will of its own that
> > has nothing to do with the needs of real people.
[Eugen Leitl:] Emergent large-scale behaviour is nothing new. How do you intend to control individual behaviour of a large population of only partially rational agents? They don't come with too many convenient behaviour-modifying hooks (pheromones as in social insects, but notice menarche-synch in females sharing quarters), and for a good reason. The few hooks we have (mob, war, politics, religion) have been notoriously abused, already. Analogous to apoptosis, metaindividuals may function using processes deleterious[*sic*] to its components (us).[31]

Eugen's understanding of what "technological progress" means is sufficiently complex to confound most of his interlocutors. For one surprising thing, it is not exactly inevitable. The manner in which Leitl argues with people is usually a kind of machine-gun prattle of coevolutionary, game-theoretic, cryptographic sorites. Eugen piles on the scientific and transhumanist reasoning, and his interlocutors slowly peel away from the discussion. But it isn't craziness, hype, or half-digested popular science—Eugen generally knows his stuff—it just fits together in a way that almost no one else can quite grasp. Eugen sees the large-scale adoption and proliferation of technologies (particularly self-replicating molecular devices and evolutionary software algorithms) as a danger that transcends all possibility of control at the individual or state level. Billions of individual decisions do not "average" into one will, but instead produce complex dynamics and hang perilously on initial conditions. In discussing the possibility of the singularity, Eugen suggests, "It could literally be a science-fair project [that causes the singularity]." If Francis Bacon's understanding of the relation between Man and Nature was that of master and possessor, Eugen's is its radicalization: Man is a powerful but ultimately arbitrary force in the progress of Life-Intelligence. Man is fully incorporated into Nature in this story, so much so that he dissolves into it. Eugen writes, when "life crosses over into this petri dish which is getting readied, things will become a lot more lively. . . . I hope we'll make it."

For Eugen, the arguments about technology that the polymaths involve themselves in couldn't be more parochial. They are important only insofar as they will set the "initial conditions" for the grand coevolutionary adventure of technology ahead of us. For the transhumanist, technology does *not* dissolve. Instead, it is the solution within which humans are dissolved. Suffering, allocation, decision making—all these are inessential to the ultimate outcome of technological progress; they are worldly affairs, even if they concern life and death, and as such, they can be either denounced or supported, but only with respect to fine-tuning the acceleration toward the singularity. For the transhumanist, one can't fight the inevitability of technical evolution, but one certainly can *contribute* to it. Technical progress is thus both law-like and subject to intelligent manipulation; technical progress is inevitable, but only because of the power of massively parallel human curiosity.

Considered as one of the modes of thought present in this-worldly political discussion, the transhumanist (like the polymath) turns technology into a rhetorical argument. Technology is the more powerful political argument because "it works." It is pointless to argue "about" technology, but not pointless to argue through and with it. It is pointless to talk about whether stopping technology is good or bad, because someone will simply build a technology that will invalidate your argument.

There is still a role for technical invention, but it is strongly distinguished from political, legal, cultural, or social interventions. For most transhumanists, there is no rhetoric here, no sophistry, just the pure truth of "it works": the pure, undeniable, unstoppable, and undeconstructable reality of technology. For the transhumanist attitude, the reality of "working code" has a reality that other assertions about the world do not. Extreme transhumanism replaces the life-world with the world of the computer, where bad (*ethically* bad) ideas *won't compile*. Less-staunch versions of transhumanism simply allow the confusion to operate opportunistically: the progress of technology is unquestionable (omniscient), and only its effects on humans are worth investigating.

The pure transhumanist, then, is a countermodern. The transhumanist despises the present for its intolerably slow descent into the future of immortality and superhuman self-improvement, and fears destruction because of too much turbulent (and ignorant) human resistance. One need have no individual conception of the present, no reflection on or synthetic understanding of it. One only need *contribute* to it correctly. One might even go so far as to suggest that forms of reflection on the present that do not contribute to technical progress endanger the very future of life-intelligence. Curiosity and technical innovation are not historical features of Western science, but natural features of a human animal that has created its own conditions for development. Thus, the transhumanists' historical consciousness consists largely of a timeline that makes ordered sense of our place on the progress toward the Singularity.

The moral of the story is not just that technology determines history, however. Transhumanism is a radically *anti*humanist position in which human agency or will—if it even exists—is not ontologically distinct from the agency of machines and animals and life itself. Even if it is necessary to organize, do things, make choices, participate, build, hack, *innovate*, this does not amount to a belief in the ability of humans to control their destiny, individually or collectively. In the end, the transhumanist cannot quite pinpoint exactly what part of this story is inevitable—except perhaps the story itself. Technology does not develop without millions of distributed humans contributing to it; humans cannot evolve without the explicit human adoption of life-altering and identity-altering technologies; evolution cannot become inevitable without the manipulation of environments and struggles for fitness. As in the dilemma of Calvinism (wherein one cannot know if one is saved by one's good works), the transhumanist must still create technology according to the particular and parochial demands of the day, but this by no means determines the eventual outcome of technological progress. It is a sentiment well articulated by Adam Ferguson and highlighted repeatedly by Friederich Hayek with respect to human society: "the result of human action, but not the execution of any human design."[32]

Conclusion

To many observers, geeks exhibit a perhaps bewildering mix of liberalism, libertarianism, anarchism, idealism, and pragmatism, yet tend to fall firmly into one or another constituted political category (liberal, conservative, socialist, capitalist, neoliberal, etc.). By showing how geeks make use of the Protestant Reformation as a usable past and how they occupy a spectrum of beliefs about progress, liberty, and intervention, I hope to resist this urge to classify. Geeks are an interesting case precisely because they are involved in the creation of *new things* that change the meaning of our constituted political categories. Their politics are mixed up and combined with the technical details of the Internet, Free Software, and the various and sundry organizations, laws, people, and practices that they deal with on a regular basis: operating systems *and* social systems. But such mixing does not make Geeks merely technoliberals or technoconservatives. Rather, it reveals how they think through the specific, historically unique situation of the Internet to the general problems of knowledge and power, liberty and enlightenment, progress and intervention.

Geeks are not a kind of *person*: geeks are geeks only insofar as they come together in new, technically mediated forms of their own creation and in ways that are not easy to identify (not language, not culture, not markets, not nations, not telephone books or databases). While their affinity is very clearly constituted through the Internet, the Internet is not the only reason for that affinity. It is this collective affinity that I refer to as a recursive public. Because it is impossible to understand this affinity by trying to identify particular types of people, it is necessary to turn to historically specific sets of practices that form the substance of their affinity. Free Software is an exemplary case—perhaps the exemplar—of a recursive public. To understand Free Software through its changing practices not only gives better access to the life-world of the geek but also reveals how the structure of a recursive public comes into being and manages to persist and transform, how it can become a powerful form of life that extends its affinities beyond technophile geeks into the realms of ordinary life.

PART II | FREE SOFTWARE

The Movement 3.

Part II of *Two Bits* describes what Free Software is and where it came from, with each of its five chapters detailing the historical narrative of a particular kind of practice: creating a movement, sharing source code, conceptualizing openness or open systems, writing copyright (and copyleft) licenses, and coordinating collaborations. Taken together, the stories describe Free Software. The stories have their endpoint (or starting point, genealogically speaking) in the years 1998–99, when Free Software burst onto the scene: on the cover of Forbes magazine, as part of the dotcom boom, and in the boardrooms of venture-capital firms and corporations like IBM and Netscape. While the chapters that make up part II can be read discretely to understand the practices that are the sine qua non of Free Software, they can also be read continuously, as a meandering story of the history of software and networks stretching from the late 1950s to the present.

Rather than define what makes Free Software free or Open Source open, *Two Bits* treats the five practices as parts of a *collective technical experimental system*: each component has its own history, development, and temporality, but they come together as a package and emerge as a recognizable thing around 1998–99. As with any experimental system, changing the components changes the operation and outcomes of the whole. Free Software so conceived is a kind of experimental system: its practices can be adopted, adapted, and modulated in new contexts and new places, but it is one whose rules are collectively determined and frequently modified. It is possible to see in each of the five practices where choices about *how* to do Free Software reached, or surpassed, certain limits, but nonetheless remained part of a system whose identity finally firmed up in the period 1998–99 and after.

The first of these practices—the making of Free Software into a movement—is both the most immediately obvious and the most difficult to grasp. By the term *movement* I refer to the practice, among geeks, of arguing about and discussing the structure and meaning of Free Software: what it consists of, what it is for, and whether or not it is a movement. Some geeks call Free Software a movement, and some don't; some talk about the ideology and goals of Free Software, and some don't; some call it Free Software, while others call it Open Source. Amid all this argument, however, Free Software geeks recognize that they are all *doing the same thing*: the practice of creating a movement is the practice of talking about the meaning and necessity of the other four practices. It was in 1998–99 that geeks came to recognize that they were all doing the same thing and, almost immediately, to argue about why.[1]

One way to understand the movement is through the story of Netscape and the Mozilla Web browser (now known as Firefox). Not only does this story provide some context for the stories of geeks presented in part I—and I move here from direct participant observation to historical and archival research on a phenomenon that was occurring at roughly the same time—but it also contains all the elements necessary to understand Free Software. It is full of discussion and argument about the practices that make up Free Software: sharing source code, conceiving of openness, writing licenses, and coordinating collaborations.

Forking Free Software, 1997–2000

Free Software forked in 1998 when the term *Open Source* suddenly appeared (a term previously used only by the CIA to refer to unclassified sources of intelligence). The two terms resulted in two separate kinds of narratives: the first, regarding Free Software, stretched back into the 1980s, promoting software freedom and resistance to proprietary software "hoarding," as Richard Stallman, the head of the Free Software Foundation, refers to it; the second, regarding Open Source, was associated with the dotcom boom and the evangelism of the libertarian pro-business hacker Eric Raymond, who focused on the economic value and cost savings that Open Source Software represented, including the pragmatic (and polymathic) approach that governed the everyday use of Free Software in some of the largest online start-ups (Amazon, Yahoo!, HotWired, and others all "promoted" Free Software by using it to run their shops).

A critical point in the emergence of Free Software occurred in 1998–99: new names, new narratives, but also new wealth and new stakes. "Open Source" was premised on dotcom promises of cost-cutting and "disintermediation" and various other schemes to make money on it (Cygnus Solutions, an early Free Software company, playfully tagged itself as "Making Free Software More Affordable"). VA Linux, for instance, which sold personal-computer systems pre-installed with Open Source operating systems, had the largest single initial public offering (IPO) of the stock-market bubble, seeing a 700 percent share-price increase in one day. "Free Software" by contrast fanned kindling flames of worry over intellectual-property expansionism and hitched itself to a nascent legal resistance to the 1998 Digital Millennium Copyright Act and Sonny Bono Copyright Term Extension Act. Prior to 1998, *Free Software* referred either to the Free Software Foundation (and the watchful, micromanaging eye of Stallman) or to one of thousands of different commercial, avocational, or university-research projects, processes, licenses, and ideologies that had a variety of names: sourceware, freeware, shareware, open software, public domain software, and so on. The term *Open Source*, by contrast, sought to encompass them all in one movement.

The event that precipitated this attempted semantic coup d'état was the release of the source code for Netscape's Communicator

Web browser. It's tough to overestimate the importance of Netscape to the fortunes of Free Software. Netscape is justly famous for its 1995 IPO and its decision to offer its core product, Netscape Navigator, for free (meaning a compiled, binary version could be downloaded and installed "for zero dollars"). But Netscape is far more famous among geeks for giving away something else, in 1998: the source code to Netscape Communicator (née Navigator). Giving away the Navigator *application* endeared Netscape to customers and confused investors. Giving away the Communicator *source code* in 1998 endeared Netscape to geeks and confused investors; it was ignored by customers.

Netscape is important from a number of perspectives. Businesspeople and investors knew Netscape as the pet project of the successful businessman Jim Clarke, who had founded the specialty computer manufacturer, Silicon Graphics Incorporated (SGI). To computer scientists and engineers, especially in the small university town of Champaign-Urbana, Illinois, Netscape was known as the highest bidder for the WWW team at the National Center for Supercomputing Applications (NCSA) at the University of Illinois. That team—Marc Andreessen, Rob McCool, Eric Bina, Jon Mittelhauser, Aleks Totic, and Chris Houck—had created Mosaic, the first and most fondly remembered "graphical browser" for surfing the World Wide Web. Netscape was thus first known as Mosaic Communications Corporation and switched its name only after legal threats from NCSA and a rival firm, Spyglass. Among geeks, Netscape was known as home to a number of Free Software hackers and advocates, most notably Jamie Zawinski, who had rather flamboyantly broken rank with the Free Software Foundation by forking the GNU EMACS code to create what was first known as Lucid Emacs and later as XEmacs. Zawinski would go on to lead the newly free Netscape browser project, now known as Mozilla.

Meanwhile, most regular computer users remember Netscape both as an emblem of the dotcom boom's venture-fed insanity and as yet another of Microsoft's victims. Although Netscape exploded onto the scene in 1995, offering a feature-rich browser that was an alternative to the bare-bones Mosaic browser, it soon began to lose ground to Microsoft, which relatively quickly adopted the strategy of giving away its browser, Internet Explorer, as if it were part of the Windows operating system; this was a practice that the U.S. Department of Justice eventually found to be in violation of antitrust laws and for which Microsoft was convicted, but never punished.

The nature of Netscape's decision to release the source code differs based on which perspective it is seen from. It could appear to be a business plan modeled on the original success: give away your product and make money in the stock market. It could appear to be a strategic, last-gasp effort to outcompete Microsoft. It could also appear, and did appear to many geeks, to be an attempt to regain some of that "hacker-cred" it once had acquired by poaching the NCSA team, or even to be an attempt to "do the right thing" by making one of the world's most useful tools into Free Software. But why would Netscape reach such a conclusion? By what reasoning would such a decision seem to be correct? The reasons for Netscape's decision to "free the source" recapitulate the five core practices of Free Software—and provided key momentum for the new movement.

Sharing Source Code Netscape's decision to share its source code could only seem surprising in the context of the widespread practice of keeping source code secret; secrecy was a practice followed largely in order to prevent competitors from copying a program and competing with it, but also as a means to control the market itself. The World Wide Web that Andreessen's team at NCSA had cut their teeth on was itself designed to be "platform independent" and accessible by any device on the network. In practice, however, this meant that someone needed to create "browsers" for each different computer or device. Mosaic was initially created for UNIX, using the Motif library of the X11 Window System—in short, a very specific kind of access. Netscape, by contrast, prided itself on "porting" Netscape Navigator to nearly all available computer architectures. Indeed, by 1997, plans were under way to create a version of the browser—written in Java, the programming language created by Sun Microsystems to "write once, run anywhere"—that would be completely platform independent.

The Java-based Navigator (called Javagator, of course) created a problem, however, with respect to the practice of keeping source code secret. Whenever a program in Java was run, it created a set of "bytecodes" that were easy to reverse-engineer because they had to be transmitted from the server to the machine that ran the program and were thus visible to anyone who might know how and where to look. Netscape engineers flirted with the idea of deliberately obfuscating these bytecodes to deter competitors from copying them. How can one compete, the logic goes, if anyone can copy your program and make their own ersatz version?

Zawinski, among others, suggested that this was a bad idea: why not just *share* the source code and get people to help make it better? As a longtime participant in Free Software, Zawinski understood the potential benefits of receiving help from a huge pool of potential contributors. He urged his peers at Netscape to see the light. However, although he told them stories and showed them successes, he could never make the case that this was an intelligent *business* plan, only that it was an efficient software-engineering plan. From the perspective of management and investors, such a move seemed tantamount to simply giving away the intellectual property of the company itself.

Frank Hecker, a sales manager, made the link between the developers and management: "It was obvious to [developers] why it was important. It wasn't really clear from a senior management level why releasing the source code could be of use because nobody ever made the business case."[2] Hecker penned a document called "Netscape Source Code as Netscape Product" and circulated it to various people, including Andreessen and Netscape CEO Jim Barksdale. As the title suggests, the business case was that the source code could also be a product, and in the context of Netscape, whose business model was "give it away and make it up on the stock market," such a proposal seemed less insane than it otherwise might have: "When Netscape first made Navigator available for unrestricted download over the Internet, many saw this as flying in the face of conventional wisdom for the commercial software business, and questioned how we could possibly make money 'giving our software away.' Now of course this strategy is seen in retrospect as a successful innovation that was a key factor in Netscape's rapid growth, and rare is the software company today that does not emulate our strategy in one way or another. Among other things, this provokes the following question: What if we were to repeat this scenario, only this time with source code?"[3]

Under the influence of Hecker, Zawinski, and CTO Eric Hahn (who had also written various internal "heresy documents" suggesting similar approaches), Netscape eventually made the decision to share their source code with the outside world, a decision that resulted in a famous January 1998 press release describing the aims and benefits of doing so. The decision, at that particular point in Netscape's life, and in the midst of the dotcom boom, was certainly momentous, but it did not lead either to a financial windfall or to a suddenly superior product.[4]

Conceptualizing Open Systems Releasing the source code was, in a way, an attempt to regain the trust of the people who had first imagined the www. Tim Berners-Lee, the initial architect of the www, was always adamant that the protocol and all its implementations should be freely available (meaning either "in the public domain" or "released as Free Software"). Indeed, Berners-Lee had done just that with his first bare-bones implementations of the www, proudly declaring them to be in the public domain.

Over the course of the 1990s, the "browser wars" caused both Netscape and Microsoft to stray far from this vision: each had implemented its own extensions and "features" to the browsers and servers, extensions not present in the protocol that Berners-Lee had created or in the subsequent standards created by the World Wide Web Consortium (W3C). Included in the implementations were various kinds of "evil" that could make browsers fail to work on certain operating systems or with certain kinds of servers. The "browser wars" repeated an open-systems battle from the 1980s, one in which the attempt to standardize a network operating system (UNIX) was stymied by competition and secrecy, at the same time that consortiums devoted to "openness" were forming in order to try to prevent the spread of evil. Despite the fact that both Microsoft and Netscape were members of the W3C, the noncompatibility of their browsers clearly represented the manipulation of the standards process in the name of competitive advantage.

Releasing the source code for Communicator was thus widely seen as perhaps the only way to bypass the poisoned well of competitively tangled, nonstandard browser implementations. An Open Source browser could be made to comply with the standards—if not by the immediate members involved with its creation, then by creating a "fork" of the program that was standards compliant—because of the rights of redistribution associated with an Open Source license. Open Source would be the solution to an open-systems problem that had never been solved because it had never confronted the issue of intellectual property directly. Free Software, by contrast, had a well-developed solution in the GNU General

Public License, also known as copyleft license, that would allow the software to remain free and revive hope for maintaining open standards.

Writing Licenses Herein lies the rub, however: Netscape was immediately embroiled in controversy among Free Software hackers because it chose to write its own bespoke licenses for distributing the source code. Rather than rely on one of the existing licenses, such as the GNU GPL or the Berkeley Systems Distribution (BSD) or MIT licenses, they created their own: the Netscape Public License (NPL) and the Mozilla Public License. The immediate concerns of Netscape had to do with their existing network of contracts and agreements with other, third-party developers—both those who had in the past contributed parts of the existing source code that Netscape might not have the rights to redistribute as Free Software, and those who were expecting in the future to buy and redistribute a commercial version. Existing Free Software licenses were either too permissive, giving to third parties rights that Netscape itself might not have, or too restrictive, binding Netscape to make source code freely available (the GPL) when it had already signed contracts with buyers of the nonfree code.

It was a complex and specific business situation—a network of existing contracts and licensed code—that created the need for Netscape to write its own license. The NPL thus contained a clause that allowed Netscape special permission to relicense any particular contribution to the source code as a proprietary product in order to appease its third-party contracts; it essentially gave Netscape special rights that no other licensee would have. While this did not necessarily undermine the Free Software licenses—and it was certainly Netscape's prerogative—it was contrary to the spirit of Free Software: it broke the "recursive public" into two halves. In order to appease Free Software geeks, Netscape wrote one license for existing code (the NPL) and a different license for *new* contributions: the Mozilla Public License.

Neither Stallman nor any other Free Software hacker was entirely happy with this situation. Stallman pointed out three flaws: "One flaw sends a bad philosophical message, another puts the free software community in a weak position, while the third creates a major practical problem within the free software community. Two of the flaws apply to the Mozilla Public License as well." He urged people not to use the NPL. Similarly, Bruce Perens suggested, "Many companies have adopted a variation of the MPL [*sic*] for their own programs. This is unfortunate, because the NPL was designed for the specific business situation that Netscape was in at the time it was written, and is not necessarily appropriate for others to use. It should remain the license of Netscape and Mozilla, and others should use the GPL or the BSD or X licenses."[5]

Arguments about the fine details of licenses may seem scholastic, but the decision had a huge impact on the structure of the new product. As Steven Weber has pointed out, the choice of license tracks the organization of a product and can determine who and what kinds of contributions can be made to a project.[6] It is not an idle choice; every new license is scrutinized with the same intensity or denounced with the same urgency.

Coordinating Collaborations   One of the selling points of Free Software, and especially of its marketing *as* Open Source, is that it leverages the work of thousands or hundreds of thousands of volunteer contributors across the Internet. Such a claim almost inevitably leads to spurious talk of "self-organizing" systems and emergent properties of distributed collaboration. The Netscape press release promised to "harness the creative power of thousands of programmers on the Internet by incorporating their best enhancements," and it quoted CEO Jim Barksdale as saying, "By giving away the source code for future versions, we can ignite the creative energies of the entire Net community and fuel unprecedented levels of innovation in the browser market."[7] But as anyone who has ever tried to start or run a Free Software project knows, it never works out that way.

Software engineering is a notoriously hard problem.[8] The halls of the software industry are lined with the warning corpses of dead software methodologies. Developing software in the dotcom boom was no different, except that the speed of release cycles and the velocity of funding (the "burn rate") was faster than ever before. Netscape's in-house development methodologies were designed to meet these pressures, and as many who work in this field will attest, that method is some version of a semistructured, deadline-driven, caffeine- and smart-drink–fueled race to "ship."[9]

Releasing the Mozilla code, therefore, required a system of coordination that would differ from the normal practice of in-house software development by paid programmers. It needed to incorporate the contributions of outsiders—developers who didn't work for Netscape. It also needed to *entice* people to contribute, since that was the bargain on which the decision to free the source was based, and to allow them to track their contributions, so they could verify that their contributions were included or rejected for legitimate reasons. In short, if any magical Open Source self-organization were to take place, it would require a thoroughly transparent, Internet-based coordination system.

At the outset, this meant practical things: obtaining the domain name mozilla.org; setting up (and in turn releasing the source code for) the version-control system (the Free Software standard cvs), the version-control interface (Bonsai), the "build system" that managed and displayed the various trees and (broken) branches of a complex software project (Tinderbox), and a bug-reporting system for tracking bugs submitted by users and developers (Bugzilla). It required an organizational system within the Mozilla project, in which paid developers would be assigned to check submissions from inside and outside, and maintainers or editors would be designated to look at and verify that these contributions should be used.

In the end, the release of the Mozilla source code was both a success and a failure. Its success was long in coming: by 2004, the Firefox Web browser, based on Mozilla, had started to creep up the charts of most popular browsers, and it has become one of the most visible and widely used Free Software applications. The failure, however, was more immediate: Mozilla failed to reap the massive benefits for Netscape that the 1995 give-away of Netscape Navigator had. Zawinski, in a public letter of resignation in April 1999 (one year after the release), expressed this sense of failure. He attributed Netscape's decline after 1996 to the fact that it had "stopped innovating" and become too large to be creative, and described the decision to free the Mozilla source code as a return to this innovation: "[The announcement] was a beacon of hope to me .... [I]t was so crazy, it just might work. I took my cue and ran with it, registering the domain that night, designing the structure of the organization, writing the first version of the web site, and, along with my co-conspirators, explaining to room after room of Netscape employees and managers how free software worked, and what we had to do to make it work."[10] For Zawinski, the decision was both a chance for Netscape to return to its glory and an opportunity to prove the power of Free Software: "I saw it as a chance for the code to actually prosper. By making it not be a Netscape project, but rather, be a public project to which Netscape was merely a contributor, the fact that Netscape was no longer capable of building products wouldn't matter: the outsiders would show Netscape how it's done. By putting control of the web browser into the hands of anyone who cared to step up to the task, we would ensure that those people would keep it going, out of their own self-interest."[11]

But this promise didn't come true—or, at least, it didn't come true at the speed that Zawinski and others in the software world were used to. Zawinski offered various reasons: the project was primarily made up of Netscape employees and thus still appeared to be a Netscape thing; it was too large a project for outsiders to dive into and make small changes to; the code was too "crufty," that is, too complicated, overwritten, and unclean. Perhaps most important, though, the source code was not actually *working*: "We never distributed the source code to a working web browser, more importantly, to *the web browser that people were actually using*."[12]

Netscape failed to entice. As Zawinski put it, "If someone were running a web browser, then stopped, added a simple new command to the source, recompiled, and had that same web browser *plus their addition*, they would be motivated to do this again, and possibly to tackle even larger projects."[13] For Zawinski, the failure to "ship" a working browser was the biggest failure, and he took pains to suggest that this failure was not an indictment of Free Software as such: "Let me assure you that whatever problems the Mozilla project is having are not because open source doesn't work. Open source does work, but it is most definitely not a panacea. If there's a cautionary tale here, it is that you can't take a dying project, sprinkle it with the magic pixie dust of 'open source,' and have everything magically work out. Software is hard. The issues aren't that simple."[14]

Fomenting Movements The period from 1 April 1998, when the Mozilla source code was first released, to 1 April 1999, when Zawinski announced its failure, couldn't have been a headier, more exciting time for participants in Free Software. Netscape's decision to release the source code was a tremendous opportunity for geeks involved in Free Software. It came in the midst of the rollicking dot-com bubble. It also came in the midst of the widespread adoption of key Free Software tools: the Linux operating system for servers, the Apache Web server for Web pages, the perl and python scripting languages for building quick Internet applications, and a number of other lower-level tools like Bind (an implementation of the DNS protocol) or sendmail for e-mail.

Perhaps most important, Netscape's decision came in a period of fevered and intense self-reflection among people who had been involved in Free Software in some way, stretching back to the mid-1980s. Eric Raymond's article "The Cathedral and The Bazaar," delivered at the Linux Kongress in 1997 and the O'Reilly Perl Conference the same year, had started a buzz among Free Software hackers. It was cited by Frank Hecker and Eric Hahn at Netscape as one of the sources for their thinking about the decision to free Mozilla; Raymond and Bruce Perens had both been asked to consult with Netscape on Free Software strategy. In April of the same year Tim O'Reilly, a publisher of handbooks for Free Software, organized a conference called the Freeware Summit.

The Freeware Summit's very name indicated some of the concern about definition and direction. Stallman, despite his obvious centrality, but also because of it, was not invited to the Freeware Summit, and the Free Software Foundation was not held up as the core philosophical guide of this event. Rather, according to the press release distributed after the meeting, "The meeting's purpose was to facilitate a high-level discussion of the successes and challenges facing the developers. While this type of software has often been called 'freeware' or 'free software' in the past, the developers agreed that commercial development of the software is part of the picture, and that the terms 'open source' or 'sourceware' best describe the development method they support."[15]

It was at this summit that Raymond's suggestion of "Open Source" as an alternative name was first publicly debated.[16] Shortly thereafter, Raymond and Perens created the Open Source Initiative and penned "The Open Source Definition." All of this self-reflection was intended to capitalize on the waves of attention being directed at Free Software in the wake of Netscape's announcement.

The motivations for these changes came from a variety of sources—ranging from a desire to be included in the dotcom boom to a powerful (ideological) resistance to being ideological. Linus Torvalds loudly proclaimed that the reason to do Free Software was because it was "fun"; others insisted that it made better busi-

THE MOVEMENT

ness sense or that the stability of infrastructures like the Internet depended on a robust ability to improve them from any direction. But none of them questioned how Free Software got done or proposed to change it.

Raymond's paper "The Cathedral and the Bazaar" quickly became the most widely told story of how Open Source works and why it is important; it emphasizes the centrality of novel forms of coordination over the role of novel copyright licenses or practices of sharing source code. "The Cathedral and the Bazaar" reports Raymond's experiments with Free Software (the bazaar model) and reflects on the difference between it and methodologies adopted by industry (the cathedral model). The paper does not truck with talk of freedom and has no denunciations of software hoarding à la Stallman. Significantly, it also has no discussion of issues of licensing. Being a hacker, however, Raymond did give his paper a "revision-history," which proudly displays revision 1.29, 9 February 1998: "Changed 'free software' to 'open source.'"[17]

Raymond was determined to reject the philosophy of liberty that Stallman and the Free Software Foundation represented, but not in order to create a political movement of his own. Rather, Raymond (and the others at the Freeware Summit) sought to cash in on the rising tide of the Internet economy by turning the creation of Free Software into something that made more sense to investors, venture capitalists, and the stock-buying public. To Raymond, Stallman and the Free Software Foundation represented not freedom or liberty, but a kind of dogmatic, impossible communism. As Raymond was a committed libertarian, one might expect his core beliefs in the necessity of strong property rights to conflict with the strange communalism of Free Software—and, indeed, his rhetoric was focused on pragmatic, business-minded, profit-driven, and market-oriented uses of Free Software. For Raymond, the essentially interesting component of Free Software was not its enhancement of human liberty, but the innovation in software production that it represented (the "development model"). It was clear that Free Software achieved something amazing through a clever inversion of strong property rights, an inversion which could be expected to bring massive revenue in some other form, either through cost-cutting or, Netscape-style, through the stock market.

Raymond wanted the business world and the mainstream industry to recognize Free Software's potential, but he felt that Stallman's rhetoric was getting in the way. Stallman's insistence, for example, on calling corporate intellectual-property protection of software "hoarding" was doing more damage than good in terms of Free Software's acceptance among businesses, as a practice, if not exactly a product.

Raymond's papers channeled the frustration of an entire generation of Free Software hackers who may or may not have shared Stallman's dogmatic philosophical stance, but who nonetheless wanted to participate in the creation of Free Software. Raymond's paper, the Netscape announcement, and the Freeware Summit all played into a palpable anxiety: that in the midst of the single largest creation of paper wealth in U.S. history, those being enriched through Free Software and the Internet were not those who built it, who maintained it, or who *got it.*

The Internet giveaway was a conflict of propriety: hackers and geeks who had built the software that made it work, under the sign of making it free for all, were seeing that software generate untold wealth for people who had not built it (and furthermore, who had no intention of *keeping* it free for all). Underlying the creation of wealth was a commitment to a kind of permanent technical freedom—a moral order—not shared by those who were reaping the most profit. This anxiety regarding the expropriation of work (even if it had been a labor of love) was ramified by Netscape's announcement.

All through 1998 and 1999, buzz around Open Source built. Little-known companies such as Red Hat, VA Linux, Cygnus, Slackware, and SuSe, which had been providing Free Software support and services to customers, suddenly entered media and business consciousness. Articles in the mainstream press circulated throughout the spring and summer of 1998, often attempting to make sense of the name change and whether it meant a corresponding change in practice. A front-cover article in *Forbes*, which featured photos of Stallman, Larry Wall, Brian Behlendorf, and Torvalds (figure 2), was noncommittal, cycling between Free Software, Open Source, and Freeware.[18]

By early 1999, O'Reilly Press published *Open Sources: Voices from the Open Source Revolution,* a hastily written but widely read book. It included a number of articles—this time including one by Stallman—that cobbled together the first widely available public history of Free Software, both the practice and the technologies

2. "Peace, Love and Software," cover of *Forbes*, 10 August 1998. Used with permission of *Forbes* and Nathaniel Welch.

involved. Kirk McKusick's article detailed the history of important technologies like the BSD version of UNIX, while an article by Brian Behlendorf, of Apache, detailed the practical challenges of running Free Software projects. Raymond provided a history of hackers and a self-aggrandizing article about his own importance in creating the movement, while Stallman's contribution told his own version of the rise of Free Software.

By December 1999, the buzz had reached a fever pitch. When VA Linux, a legitimate company which actually made something real—computers with Linux installed on them—went public, its shares' value gained 700 percent in one day and was the single most valuable initial public offering of the era. VA Linux took the unconventional step of allowing contributors to the Linux kernel to buy into the stock before the IPO, thus bringing at least a partial set of these contributors into the mainstream Ponzi scheme of the Internet dotcom economy. Those who managed to sell their stock ended up benefiting from the boom, whether or not their contributions to Free Software truly merited it. In a roundabout way, Raymond, O'Reilly, Perens, and others behind the name change had achieved recognition for the central role of Free Software in the success of the Internet—and now its true name could be known: Open Source.

Yet nothing much changed in terms of the way things actually got done. Sharing source code, conceiving openness, writing licenses, coordinating projects—all these continued as before with no significant differences between those flashing the heroic mantle of freedom and those donning the pragmatic tunic of methodology. Now, however, stories proliferated; definitions, distinctions, details, and detractions filled the ether of the Internet, ranging from the philosophical commitments of Free Software to the parables of science as the "original open source" software. Free Software proponents refined their message concerning rights, while Open Source advocates refined their claims of political agnosticism or nonideological commitments to "fun." All these stories served to create movements, to evangelize and advocate and, as Eugen Leitl would say, to "corrupt young minds" and convert them to the cause. The fact that there are different narratives for identical practices is an advantageous fact: regardless of *why* people think they are doing what they are doing, they are all nonetheless contributing to the same mysterious thing.

A Movement?

To most onlookers, Free Software and Open Source seem to be overwhelmed with frenzied argument; the flame wars and disputes, online and off, seem to dominate everything. To attend a conference where geeks—especially high-profile geeks like Raymond, Stallman, and Torvalds—are present, one might suspect that the very detailed practices of Free Software are overseen by the brow-beating, histrionic antics of a few charismatic leaders and that ideological commitments result in divergent, incompatible, and affect-laden opposition which must of necessity take specific and incompatible forms. Strangely, this is far from the case: all this sound and fury doesn't much change what people do, even if it is a requirement of apprenticeship. It truly is all over but for the shouting.

According to most of the scholarly literature, the function of a movement is to narrate the shared goals and to recruit new members. But is this what happens in Free Software or Open Source?[19] To begin with, *movement* is an awkward word; not all participants would define their participation this way. Richard Stallman suggests that Free Software is social movement, while Open Source is a development methodology. Similarly some Open Source proponents see it as a pragmatic methodology and Free Software as a dogmatic philosophy. While there are specific entities like the Free Software Foundation and the Open Source Initiative, they do not comprise all Free Software or Open Source. Free Software and Open Source are neither corporations nor organizations nor consortia (for there are no organizations to consort); they are neither national, subnational, nor international; they are not "collectives" because no membership is required or assumed—indeed to hear someone assert "I belong" to Free Software or Open Source would sound absurd to anyone who does. Neither are they shady bands of hackers, crackers, or thieves meeting in the dead of night, which is to say that they are not an "informal" organization, because there is no formal equivalent to mimic or annul. Nor are they quite a crowd, for a crowd can attract participants who have no idea what the goal of the crowd is; also, crowds are temporary, while movements extend over time. It may be that *movement* is the best term of the lot, but unlike social movements, whose organization and momentum are fueled by shared causes or broken by ideological dispute, Free Software and Open Source share practices first, and ideologies second. It is this fact that is the strongest confirmation that they are a recursive public, a form of public that is as concerned with the material practical means of becoming public as it is with any given public debate.

The movement, as a practice of argument and discussion, is thus centered around core agreements about the other four kinds of practices. The discussion and argument have a specific function: to tie together divergent practices according to a wide consensus which tries to capture the *why* of Free Software. Why is it different from normal software development? Why is it necessary? Why

now? Why do people do it? Why do people use it? Can it be preserved and enhanced? None of these questions address the *how*:
how should source code circulate? How should a license be written?
Who should be in charge? All of these "hows" change slowly and
experimentally through the careful modulation of the practices,
but the "whys" are turbulent and often distracting. Nonetheless,
people engaged in Free Software—users, developers, supporters,
and observers—could hardly remain silent on this point, despite the
frequent demand to just "shut up and show me the code." "Figuring
out" Free Software also requires a practice of reflecting on what is
central to it and what is outside of it.

The movement, as a practice of discussion and argument, is made
up of stories. It is a practice of storytelling: affect- and intellect-laden
lore that orients existing participants toward a particular problem,
contests other histories, parries attacks from outside, and draws
in new recruits.[20] This includes proselytism and evangelism (and
the usable pasts of protestant reformations, singularities, rebellion
and iconoclasm are often salient here), whether for the reform of
intellectual-property law or for the adoption of Linux in the trenches
of corporate America. It includes both heartfelt allegiance in the
name of social justice as well as political agnosticism stripped of
all ideology.[21] Every time Free Software is introduced to someone,
discussed in the media, analyzed in a scholarly work, or installed
in a workplace, a story of either Free Software or Open Source is
used to explain its purpose, its momentum, and its temporality. At
the extremes are the prophets and proselytes themselves: Eric Raymond describes Open Source as an evolutionarily necessary outcome of the natural tendency of human societies toward economies
of abundance, while Richard Stallman describes it as a defense of
the fundamental freedoms of creativity and speech, using a variety
of philosophical theories of liberty, justice, and the defense of freedom.[22] Even scholarly analyses must begin with a potted history
drawn from the self-narration of geeks who make or advocate free
software.[23] Indeed, as a methodological aside, one reason it is so easy
to track such stories and narratives is because geeks like to tell and,
more important, like to archive such stories—to create Web pages,
definitions, encyclopedia entries, dictionaries, and mini-histories
and to save every scrap of correspondence, every fight, and every
resolution related to their activities. This "archival hubris" yields
a very peculiar and specific kind of fieldsite: one in which a kind

of "as-it-happens" ethnographic observation is possible not only through "being there" in the moment but also by being there in the massive, proliferating archives of moments past. Understanding the movement as a changing entity requires constantly glancing back at its future promises and the conditions of their making.

Stories of the movement are also stories of a recursive public. The fact that *movement* isn't quite the right word is evidence of a kind of grasping, a figuring out of why these practices make sense to all these geeks, in this place and time; it is a practice that is not so different from my own ethnographic engagement with it. Note that *both* Free Software and Open Source tell stories of movement(s): they are not divided by a commercial-noncommercial line, even if they are divided by ill-defined and hazy notions of their ultimate goals. The problem of a recursive public (or, in an alternate language, a recursive *market*) as a social imaginary of moral and technical order is common to both of them as part of their practices. Thus, stories about "the movement" are detailed stories about the technical and moral order that geeks inhabit, and they are bound up with the functions and fates of the Internet. Often these stories are themselves practices of inclusion and exclusion (e.g., "this license is not a Free Software license" or "that software is not an open system"); sometimes the stories are normative definitions about how Free Software *should* look. But they are, always, stories that reveal the shared moral and technical imaginations that make up Free Software as a recursive public.

Conclusion

Before 1998, there was no movement. There was the Free Software Foundation, with its peculiar goals, and a very wide array of other projects, people, software, and ideas. Then, all of a sudden, in the heat of the dotcom boom, Free Software was a movement. Suddenly, it was a problem, a danger, a job, a calling, a dogma, a solution, a philosophy, a liberation, a methodology, a business plan, a success, and an alternative. Suddenly, it was Open Source *or* Free Software, and it became necessary to choose sides. After 1998, debates about definition exploded; denunciations and manifestos and journalistic hagiography proliferated. Ironically, the creation of two names allowed people to identify *one thing*, for these two names referred to identical practices, licenses, tools, and organizations. Free Software and Open Source shared everything "material," but differed vocally and at great length with respect to ideology. Stallman was denounced as a kook, a communist, an idealist, and a dogmatic holding back the successful adoption of Open Source by business; Raymond and users of "open source" were charged with selling out the ideals of freedom and autonomy, with the dilution of the principles and the promise of Free Software, as well as with being stooges of capitalist domination. Meanwhile, both groups proceeded to create objects—principally software— using tools that they agreed on, concepts of openness that they agreed on, licenses that they agreed on, and organizational schemes that they agreed on. Yet never was there fiercer debate about the definition of Free Software.

On the one hand, the Free Software Foundation privileges the liberty and creativity of individual geeks, geeks engaged in practices of self-fashioning through the creation of software. It gives precedence to the liberal claim that without freedom of expression, individuals are robbed of their ability to self-determine. On the other hand, Open Source privileges organizations and processes, that is, geeks who are engaged in building businesses, nonprofit organizations, or governmental and public organizations of some form or another. It gives precedence to the pragmatist (or polymathic) view that getting things done requires flexible principles and negotiation, and that the public practice of building and running things should be separate from the private practice of ethical and political beliefs. Both narratives give geeks ways of making sense of a practice that they share in almost all of its details; both narratives give geeks a way to understand how Free Software or Open Source Software is different from the mainstream, proprietary software development that dominates their horizons. The narratives turn the haphazard participation and sharing that existed before 1998 into meaningful, goal-directed practices in the present, turning a class-in-itself into a class-for-itself, to use a terminology for the most part unwelcome among geeks.

If two radically opposed ideologies can support people engaged in identical practices, then it seems obvious that the real space of politics and contestation is at the level of these practices and their emergence. These practices emerge as a response to a reorientation of power and knowledge, a reorientation somewhat impervious to conventional narratives of freedom and liberty, or to pragmatic claims of methodological necessity or market-driven innovation. Were these conventional narratives sufficient, the practices would be merely bureaucratic affairs, rather than the radical transformations they are.

4. Sharing Source Code

Free Software would be nothing without shared source code. The idea is built into the very name "Open Source," and it is a requirement of all Free Software licenses that source code be open to view, not "welded shut." Perhaps ironically, source code is the most material of the five components of Free Software; it is both an expressive medium, like writing or speech, and a tool that performs concrete actions. It is a mnemonic that translates between the illegible electron-speed doings of our machines and our lingering ability to partially understand and control them as human agents. Many Free Software programmers and advocates suggest that "information wants to be free" and that sharing is a natural condition of human life, but I argue something contrary: sharing produces its own kind of moral and technical order, that is, "information makes people want freedom" and how they want it is related to how that information is created and circulated. In this chapter I explore the twisted and contingent history of how source code and its sharing have come to take the technical, legal, and pedagogical forms they have today, and how the norms of sharing have come to seem so natural to geeks.

Source code is essential to Free Software because of the historically specific ways in which it has come to be shared, "ported," and "forked." Nothing about the nature of source code requires that it be shared, either by corporations for whom secrecy and jealous protection are the norm or by academics and geeks for whom source code is usually only one expression, or implementation, of a greater idea worth sharing. However, in the last thirty years, norms of sharing source code—technical, legal, and pedagogical norms—have developed into a seemingly natural practice. They emerged through attempts to make software into a product, such as IBM's 1968 "unbundling" of software and hardware, through attempts to define and control it legally through trade secret, copyright, and patent law, and through attempts to teach engineers how to understand and to create more software.

The story of the norms of sharing source code is, not by accident, also the history of the UNIX operating system.[1] The UNIX operating system is a monstrous academic-corporate hybrid, an experiment in portability and sharing whose impact is widely and reverently acknowledged by geeks, but underappreciated more generally. The story of UNIX demonstrates the details of how source code has come to be shared, technically, legally, and pedagogically. In technical terms UNIX and the programming language C in which it was written demonstrated several key ideas in operating-systems theory and practice, and they led to the widespread "porting" of UNIX to virtually every kind of hardware available in the 1970s, all around the world. In legal terms UNIX's owner, AT&T, licensed it widely and liberally, in both binary and source-code form; the legal definition of UNIX as a product, however, was not the same as the technical definition of UNIX as an evolving experiment in portable operating systems—a tension that has continued throughout its lifetime. In pedagogical terms UNIX became the very paradigm of an "operating system" and was thereby ported not only in the technical sense from one machine to another, but from machines to minds, as computer-science students learning the meaning of "operating system" studied the details of the quasi-legally shared UNIX source code.[2]

The proliferation of UNIX was also a hybrid commercial-academic undertaking: it was neither a "public domain" object shared solely among academics, nor was it a conventional commercial product. Proliferation occurred through novel forms of academic sharing as well as through licensing schemes constrained by the peculiar status of AT&T, a regulated monopoly forbidden to enter the computer and software industry before 1984. Thus proliferation was not mere replication: it was not the sale of copies of UNIX, but a complex web of shared and re-shared chunks of source code, and the re-implementation of an elegant and simple conceptual scheme. As UNIX proliferated, it was stabilized in multiple ways: by academics seeking to keep it whole and self-compatible through contributions of source code; by lawyers at AT&T seeking to define boundaries that mapped onto laws, licenses, versions, and regulations; and by professors seeking to define it as an exemplar of the core concepts of operating-system theory. In all these ways, UNIX was a kind of primal recursive public, drawing together people for whom the meaning of their affiliation was the use, modification, and stabilization of UNIX.

The obverse of proliferation is differentiation: forking. UNIX is admired for its integrity as a conceptual thing and despised (or marveled at) for its truly tangled genealogical tree of ports and forks: new versions of UNIX, some based directly on the source code, some not, some licensed directly from AT&T, some sublicensed or completely independent.

Forking, like random mutation, has had both good and bad effects; on the one hand, it ultimately created versions of UNIX that were not compatible with themselves (a kind of autoimmune response), but it also allowed the merger of UNIX and the Arpanet, creating a situation wherein UNIX operating systems came to be not only the paradigm of operating systems but also the paradigm of *networked* computers, through its intersection with the development of the TCP/IP protocols that are at the core of the Internet.[3] By the mid-1980s, UNIX was a kind of obligatory passage point for anyone interested in networking, operating systems, the Internet, and especially, modes of creating, sharing, and modifying source code—so much so that UNIX has become known among geeks not just as an operating system but as a philosophy, an answer to a very old question in new garb: how shall we live, among a new world of machines, software, and networks?

Before Source

In the early days of computing machinery, there was no such thing as source code. Alan Turing purportedly liked to talk to the machine in binary. Grace Hopper, who invented an early compiler, worked as close to the Harvard Mark I as she could get: flipping switches and plugging and unplugging relays that made up the "code" of what the machine would do. Such mechanical and meticulous work hardly merits the terms *reading* and *writing*; there were no GOTO statements, no line numbers, only calculations that had to be translated from the pseudo-mathematical writing of engineers and human computers to a physical or mechanical configuration.[4] Writing and reading source code and programming languages was a long, slow development that became relatively widespread only by the mid-1970s. So-called higher-level languages began to appear in the late 1950s: FORTRAN, COBOL, Algol, and the "compilers" which allowed for programs written in them to be transformed into the illegible mechanical and valvular representations of the machine. It was in this era that the terms *source language* and *target language* emerged to designate the activity of translating higher to lower level languages.[5]

There is a certain irony about the computer, not often noted: the unrivaled power of the computer, if the ubiquitous claims are believed, rests on its *general* programmability; it can be made to do any calculation, in principle. The so-called universal Turing machine provides the mathematical proof.[6] Despite the abstract power of such certainty, however, we do not live in the world of The Computer—we live in a world of *computers*. The hardware systems that manufacturers created from the 1950s onward were so specific and idiosyncratic that it was inconceivable that one might write a program for one machine and then simply run it on another. "Programming" became a bespoke practice, tailored to each new machine, and while programmers of a particular machine may well have shared programs with each other, they would not have seen much point in sharing with users of a different machine. Likewise, computer scientists shared mathematical descriptions of algorithms and ideas for automation with as much enthusiasm as corporations jealously guarded theirs, but this sharing, or secrecy, did not extend to the sharing of the program itself. The need to "rewrite" a program for each machine was not just a historical accident, but was determined by the needs of designers and engineers and the vicissitudes of the market for such expensive machines.[7]

In the good old days of computers-the-size-of-rooms, the languages that humans used to program computers were mnemonics; they did not exist in the computer, but on a piece of paper or a specially designed code sheet. The code sheet gave humans who were not Alan Turing a way to keep track of, to share with other humans, and to think systematically about the invisible light-speed calculations of a complicated device. Such mnemonics needed to be "coded" on punch cards or tape; if engineers conferred, they conferred over sheets of paper that matched up with wires, relays, and switches—or, later, printouts of the various machine-specific codes that represented program and data.

With the introduction of programming languages, the distinction between a "source" language and a "target" language entered the practice: source languages were "translated" into the illegible target language of the machine. Such higher-level source languages were still mnemonics of sorts—they were certainly easier for humans to read and write, mostly on yellowing tablets of paper or special code sheets—but they were also structured enough that a source language could be input into a computer and translated into a target language which the designers of the hardware had specified. Inputting commands and cards and source code required a series of actions specific to each machine: a particular card reader or, later, a keypunch with a particular "editor" for entering the commands. Properly input and translated source code provided the machine with an assembled binary program that would, in fact, run (calculate, operate, control). It was a separation, an abstraction that allowed for a certain division of labor between the ingenious human authors and the fast and mechanical translating machines.

Even after the invention of programming languages, programming "on" a computer—sitting at a glowing screen and hacking through the night—was still a long time in coming. For example, only by about 1969 was it possible to sit at a keyboard, write source code, instruct the computer to compile it, then run the program—all without leaving the keyboard—an activity that was all but unimaginable in the early days of "batch processing."[8] Very few programmers worked in such a fashion before the mid-1970s, when text editors that allowed programmers to see the text on a screen rather than on a piece of paper started to proliferate.[9] We are, by now, so familiar with the image of the man or woman sitting at a screen interacting with this device that it is nearly impossible to imagine how such a seemingly obvious practice was achieved in the first place—through the slow accumulation of the tools and techniques for working on a new kind of writing—and how that practice exploded into a Babel of languages and machines that betrayed the promise of the general-purpose computing machine.

The proliferation of different machines with different architectures drove a desire, among academics especially, for the standardization of programming languages, not so much because any single language was better than another, but because it seemed necessary to most engineers and computer users to share an emerging corpus of algorithms, solutions, and techniques of all kinds, necessary to avoid reinventing the wheel with each new machine. Algol, a streamlined language suited to algorithmic and algebraic representations, emerged in the early 1960s as a candidate for international standardization. Other languages competed on different strengths: FORTRAN and COBOL for general business use; LISP for symbolic processing. At the same time, the desire for a standard "higher-level" language necessitated a bestiary of translating programs: compilers, parsers, lexical analyzers, and other tools designed to transform the higher-level (human-readable) language into a machine-specific lower-level language, that is, machine language, assembly language, and ultimately the mystical zeroes and ones that course through our machines. The idea of a standard language and the necessity of devising specific tools for translation are the origin of the problem of *portability*: the ability to move software—not just good ideas, but actual programs, written in a standard language—from one machine to another.

A standard source language was seen as a way to counteract the proliferation of different machines with subtly different architectures. Portable source code would allow programmers to imagine their programs as ships, stopping in at ports of call, docking on different platforms, but remaining essentially mobile and unchanged by these port-calls. Portable source code became the Esperanto of humans who had wrought their own Babel of tribal hardware machines.

Meanwhile, for the computer industry in the 1960s, portable source code was largely a moot point. Software and hardware were two sides of single, extremely expensive coin—no one, except engineers, cared what language the code was in, so long as it performed the task at hand for the customer. Each new machine needed to be different, faster, and, at first, bigger, and then smaller, than the last. The urge to differentiate machines from each other was not driven by academic experiment or aesthetic purity, but by a demand for marketability, competitive advantage, and the transformation of machines and software into products. Each machine had to do something really well, and it needed to be developed in secret, in order to beat out the designs and innovations of competitors. In the 1950s and 1960s the software was a core component of this marketable object; it was not something that in itself was differentiated or separately distributed—until IBM's famous decision in 1968 to "unbundle" software and hardware.

Before the 1970s, employees of a computer corporation wrote software in-house. The machine was the product, and the software was just an extra line-item on the invoice. IBM was not the first to conceive of software as an independent product with its own market, however. Two companies, Informatics and Applied Data Research, had explored the possibilities of a separate market in software.[10] Informatics, in particular, developed the first commercially successful software product, a business-management system called Mark IV, which in 1967 cost $30,000. Informatics' president Walter Bauer "later recalled that potential buyers were 'astounded' by the price of Mark IV. In a world accustomed to free software the price of $30,000 was indeed high."[11]

IBM's unbundling decision marked a watershed, the point at which "portable" source code became a conceivable idea, if not a practical reality, to many in the industry.[12] Rather than providing a complete package of hardware and software, IBM decided to differentiate its products: to sell software and hardware separately to consumers.[13] But portability was not simply a technical issue; it was a political-economic one as well. IBM's decision was driven both by its desire to create IBM software that ran on all IBM machines (a central goal of the famous OS/360 project overseen and diagnosed by Frederick Brooks) and as response to an antitrust suit filed by the U.S. Department of Justice.[14] The antitrust suit included as part of its claims the suggestion that the close tying of software and hardware represented a form of monopolistic behavior, and it prompted IBM to consider strategies to "unbundle" its product.

Portability in the business world meant something specific, however. Even if software could be made portable at a technical level—transferable between two different IBM machines—this was certainly no guarantee that it would be portable between customers. One company's accounting program, for example, may not suit another's practices. Portability was therefore hindered both by the diversity of machine architectures and by the diversity of business practices and organization. IBM and other manufacturers therefore saw no benefit to standardizing source code, as it could only provide an advantage to competitors.[15]

Portability was thus not simply a technical problem—the problem of running one program on multiple architectures—but also a kind of political-economic problem. The meaning of *product* was not always the same as the meaning of *hardware* or *software*, but was usually some combination of the two. At that early stage, the outlines of a contest over the meaning of *portable* or *shareable* source code are visible, both in the technical challenges of creating high-level languages and in the political-economic challenges that corporations faced in creating distinctive proprietary products.

The UNIX Time-Sharing System

Set against this backdrop, the invention, success, and proliferation of the UNIX operating system seems quite monstrous, an aberration of both academic and commercial practice that should have failed in both realms, instead of becoming the most widely used portable operating system in history and the very paradigm of an "operating system" in general. The story of UNIX demonstrates how portability became a reality and how the particular practice of sharing UNIX source code became a kind of de facto standard in its wake.

UNIX was first written in 1969 by Ken Thompson and Dennis Ritchie at Bell Telephone Labs in Murray Hill, New Jersey. UNIX was the dénouement of the MIT project Multics, which Bell Labs had funded in part and to which Ken Thompson had been assigned. Multics was one of the earliest complete time-sharing operating systems, a demonstration platform for a number of early innovations in time-sharing (multiple simultaneous users on one computer).[16] By 1968, Bell Labs had pulled its support—including Ken Thompson—from the project and placed him back in Murray Hill, where he and

Dennis Ritchie were stuck without a machine, without any money, and without a project. They were specialists in operating systems, languages, and machine architecture in a research group that had no funding or mandate to pursue these areas. Through the creative use of some discarded equipment, and in relative isolation from the rest of the lab, Thompson and Ritchie created, in the space of about two years, a complete operating system, a programming language called C, and a host of tools that are still in extremely wide use today. The name UNIX (briefly, UNICS) was, among other things, a puerile pun: a castrated Multics.

The absence of an economic or corporate mandate for Thompson's and Ritchie's creativity and labor was not unusual for Bell Labs; researchers were free to work on just about anything, so long as it possessed some kind of vague relation to the interests of AT&T. However, the lack of funding for a more powerful machine did restrict the kind of work Thompson and Ritchie could accomplish. In particular, it influenced the design of the system, which was oriented toward a super-slim control unit (a kernel) that governed the basic operation of the machine and an expandable suite of small, independent tools, each of which did one thing well and which could be strung together to accomplish more complex and powerful tasks.[17] With the help of Joseph Ossana, Douglas McIlroy, and others, Thompson and Ritchie eventually managed to agitate for a new PDP-11/20 based not on the technical merits of the UNIX operating system itself, but on its potential applications, in particular, those of the text-preparation group, who were interested in developing tools for formatting, typesetting, and printing, primarily for the purpose of creating patent applications, which was, for Bell Labs, and for AT&T more generally, obviously a laudable goal.[18]

UNIX was unique for many technical reasons, but also for a specific economic reason: it was never quite academic and never quite commercial. Martin Campbell-Kelly notes that UNIX was a "non-proprietary operating system of major significance."[19] Kelly's use of "non-proprietary" is not surprising, but it is incorrect. Although business-speak regularly opposed *open* to *proprietary* throughout the 1980s and early 1990s (and UNIX was definitely the former), Kelly's slip marks clearly the confusion between software ownership and software distribution that permeates both popular and academic understandings. UNIX was indeed proprietary—it was copyrighted and wholly owned by Bell Labs and in turn by Western Electric

SHARING SOURCE CODE

and AT&T—but it was not exactly commercialized or marketed by them. Instead, AT&T allowed individuals and corporations to install UNIX and to create UNIX-like derivatives *for very low licensing fees*. Until about 1982, UNIX was licensed to academics very widely for a very small sum: usually royalty-free with a minimal service charge (from about $150 to $800).[20] The conditions of this license allowed researchers to do what they liked with the software so long as they kept it secret: they could not distribute or use it outside of their university labs (or use it to create any commercial product or process), nor publish any part of it. As a result, throughout the 1970s UNIX was developed both by Thompson and Ritchie inside Bell Labs and by users around the world in a relatively informal manner. Bell Labs followed such a liberal policy both because it was one of a small handful of industry-academic research and development centers and because AT&T was a government monopoly that provided phone service to the country and was therefore forbidden to directly enter the computer software market.[21]

Being on the border of business and academia meant that UNIX was, on the one hand, shielded from the demands of management and markets, allowing it to achieve the conceptual integrity that made it so appealing to designers and academics. On the other, it also meant that AT&T treated it as a potential product in the emerging software industry, which included new legal questions from a changing intellectual-property regime, novel forms of marketing and distribution, and new methods of developing, supporting, and distributing software.

Despite this borderline status, UNIX was a phenomenal success. The reasons why UNIX was so popular are manifold; it was widely admired aesthetically, for its size, and for its clever design and tools. But the fact that it spread so widely and quickly is testament also to the existing community of eager computer scientists and engineers (and a few amateurs) onto which it was bootstrapped, users for whom a powerful, flexible, low-cost, modifiable, and fast operating system was a revelation of sorts. It was an obvious alternative to the complex, poorly documented, buggy operating systems that routinely shipped standard with the machines that universities and research organizations purchased. "It worked," in other words.

A key feature of the popularity of UNIX was the inclusion of the source code. When Bell Labs licensed UNIX, they usually provided a tape that contained the documentation (i.e., documentation that was part of the system, not a paper technical manual external to it), a binary version of the software, and the source code for the software. The practice of distributing the source code encouraged people to maintain it, extend it, document it—and to contribute those changes to Thompson and Ritchie as well. By doing so, users developed an interest in maintaining and supporting the project precisely because it gave them an opportunity and the tools to use their computer creatively and flexibly. Such a globally distributed community of users organized primarily by their interest in maintaining an operating system is a precursor to the recursive public, albeit confined to the world of computer scientists and researchers with access to still relatively expensive machines. As such, UNIX was not only a widely shared piece of quasi-commercial software (i.e., distributed in some form other than through a price-based retail market), but also the first to systematically include the source code as part of that distribution as well, thus appealing more to academics and engineers.[22]

Throughout the 1970s, the low licensing fees, the inclusion of the source code, and its conceptual integrity meant that UNIX was ported to a remarkable number of other machines. In many ways, academics found it just as appealing, if not more, to be involved in the creation and improvement of a cutting-edge system by licensing and porting the software themselves, rather than by having it provided to them, without the source code, by a company. Peter Salus, for instance, suggests that people experienced the lack of support from Bell Labs as a kind of spur to develop and share their own fixes. The means by which source code was shared, and the norms and practices of sharing, porting, forking, and modifying source code were developed in this period as part of the development of UNIX itself—the technical design of the system facilitates and in some cases mirrors the norms and practices of sharing that developed: operating systems and social systems.[23]

Sharing UNIX

Over the course of 1974–77 the spread and porting of UNIX was phenomenal for an operating system that had no formal system of distribution and no official support from the company that owned it, and that evolved in a piecemeal way through the contributions of people from around the world. By 1975, a user's group had developed: USENIX.[24] UNIX had spread to Canada, Europe, Australia, and Japan, and a number of new tools and applications were being both independently circulated and, significantly, included in the frequent releases by Bell Labs itself. All during this time, AT&T's licensing department sought to find a balance between allowing this circulation and innovation to continue, and attempting to maintain trade-secret status for the software. UNIX was, by 1980, without a doubt the most widely and deeply understood trade secret in computing history.

The manner in which the circulation of and contribution to UNIX occurred is not well documented, but it includes both technical and pedagogical forms of sharing. On the technical side, distribution took a number of forms, both in resistance to AT&T's attempts to control it and facilitated by its unusually liberal licensing of the software. On the pedagogical side, UNIX quickly became a paradigmatic object for computer-science students precisely because it was a working operating system that included the source code and that was simple enough to explore in a semester or two.

In *A Quarter Century of UNIX* Salus provides a couple of key stories (from Ken Thompson and Lou Katz) about how exactly the technical sharing of UNIX worked, how sharing, porting, and forking can be distinguished, and how it was neither strictly legal nor deliberately illegal in this context. First, from Ken Thompson: "The first thing to realize is that the outside world ran on releases of UNIX (V4, V5, V6, V7) but we did not. Our view was a continuum. V5 was what we had at some point in time and was probably out of date simply by the activity required to put it in shape to export. After V6, I was preparing to go to Berkeley to teach for a year. I was putting together a system to take. Since it was almost a release, I made a diff with V6 [a tape containing only the differences between the last release and the one Ken was taking with him]. On the way to Berkeley I stopped by Urbana-Champaign to keep an eye on Greg Chesson. . . . I left the diff tape there and I told him that I wouldn't mind if it got around."[25]

The need for a magnetic tape to "get around" marks the difference between the 1970s and the present: the distribution of software involved both the material transport of media *and* the digital copying of information. The desire to distribute bug fixes (the "diff" tape) resonates with the future emergence of Free Software: the fact that others had fixed problems and contributed them back to Thompson and Ritchie produced an obligation to see that the fixes were shared as widely as possible, so that they in turn might be ported to new machines. Bell Labs, on the other hand, would have seen this through the lens of software development, requiring a new release, contract renegotiation, and a new license fee for a new version. Thompson's notion of a "continuum," rather than a series of releases also marks the difference between the idea of an evolving common set of objects stewarded by multiple people in far-flung locales and the idea of a shrink-wrapped "productized" software package that was gaining ascendance as an economic commodity at the same time. When Thompson says "the outside world," he is referring not only to people outside of Bell Labs but to the way the world was seen from within Bell Labs by the lawyers and marketers who would create a new version. For the lawyers, the circulation of source code was a problem because it needed to be stabilized, not so much for commercial reasons as for legal ones—one license for one piece of software. Distributing updates, fixes, and especially new tools and additions written by people who were not employed by Bell Labs scrambled the legal clarity even while it strengthened the technical quality. Lou Katz makes this explicit.

> A large number of bug fixes was collected, and rather than issue them one at a time, a collection tape ("the 50 fixes") was put together by Ken [the same "diff tape," presumably]. Some of the fixes were quite important, though I don't remember any in particular. *I suspect that a significant fraction of the fixes were actually done by non-Bell people.* Ken tried to send it out, but the lawyers kept stalling and stalling and stalling. Finally, in complete disgust, someone "found a tape on Mountain Avenue" [the location of Bell Labs] which had the fixes. When the lawyers found out about it, they called every licensee and threatened them with dire consequences if they didn't destroy the tape, after trying to find out how they got the tape. I would guess that no one would actually tell them how they came by the tape (I didn't).[26]

Distributing the fixes involved not just a power struggle between the engineers and management, but was in fact clearly motivated by the fact that, as Katz says, "a significant fraction of the fixes were done by non-Bell people." This meant two things: first, that there was an obvious incentive to return the updated system to these people and to others; second, that it was not obvious that AT&T actually owned or could claim rights over these fixes—or, if they did, they needed to cover their legal tracks, which perhaps in part explains the stalling and threatening of the lawyers, who may have been buying time to make a "legal" version, with the proper permissions.

The struggle should be seen not as one between the rebel forces of UNIX development and the evil empire of lawyers and managers, but as a struggle between two modes of stabilizing the object known as UNIX. For the lawyers, stability implied finding ways to make UNIX look like a product that would meet the existing legal framework and the peculiar demands of being a regulated monopoly unable to freely compete with other computer manufacturers; the ownership of bits and pieces, ideas and contributions had to be strictly accountable. For the programmers, stability came through redistributing the most up-to-date operating system and sharing all innovations with all users so that new innovations might also be portable. The lawyers saw urgency in making UNIX legally stable; the engineers saw urgency in making UNIX technically stable and compatible with itself, that is, to prevent the *forking* of UNIX, the death knell for portability. The tension between achieving legal stability of the object and promoting its technical portability and stability is one that has repeated throughout the life of UNIX and its derivatives—and that has ramifications in other areas as well.

The identity and boundaries of UNIX were thus intricately formed through its sharing and distribution. Sharing produced its own form of moral and technical order. Troubling questions emerged immediately: were the versions that had been fixed, extended, and expanded still UNIX, and hence still under the control of AT&T? Or were the differences great enough that something else (not-UNIX) was emerging? If a tape full of fixes, contributed by non-Bell employees, was circulated to people who had licensed UNIX, and those fixes changed the system, was it still UNIX? Was it still UNIX in a legal sense or in a technical sense or both? While these questions might seem relatively scholastic, the history of the development of UNIX suggests something far more interesting: just about every possible modification has been made, legally and technically, but the *concept* of UNIX has remained remarkably stable.

Porting UNIX

Technical portability accounts for only part of UNIX's success. As a pedagogical resource, UNIX quickly became an indispensable tool for academics around the world. As it was installed and improved, it was taught and learned. The fact that UNIX spread first to university computer-science departments, and not to businesses, government, or nongovernmental organizations, meant that it also became part of the core pedagogical practice of a generation of programmers and computer scientists; over the course of the 1970s and 1980s, UNIX came to exemplify the very concept of an operating system, especially time-shared, multi-user operating systems. Two stories describe the porting of UNIX from machines to minds and illustrate the practice as it developed and how it intersected with the technical and legal attempts to stabilize UNIX as an object: the story of John Lions's *Commentary on Unix 6th Edition* and the story of Andrew Tanenbaum's Minix.

The development of a pedagogical UNIX lent a new stability to the concept of UNIX as opposed to its stability as a body of source code or as a legal entity. The porting of UNIX was so successful that even in cases where a ported version of UNIX *shares none of the same source code as the original,* it is still considered UNIX. The monstrous and promiscuous nature of UNIX is most clear in the stories of Lions and Tanenbaum, especially when contrasted with the commercial, legal, and technical integrity of something like Microsoft Windows, which generally exists in only a small number of forms (NT, ME, XP, 95, 98, etc.), possessing carefully controlled source code, immured in legal protection, and distributed only through sales and service packs to customers or personal-computer manufacturers. While Windows is much more widely used than UNIX, it is far from having become a paradigmatic pedagogical object; its integrity is predominantly legal, not technical or pedagogical. Or, in pedagogical terms, Windows is to fish as UNIX is to fishing lessons.

Lions's *Commentary* is also known as "the most photocopied document in computer science." Lions was a researcher and senior lecturer at the University of New South Wales in the early 1970s; after reading the first paper by Ritchie and Thompson on UNIX, he convinced his colleagues to purchase a license from AT&T.[27] Lions, like many researchers, was impressed by the quality of the system, and he was, like all of the UNIX users of that period, intimately familiar with the UNIX source code—a necessity in order to install, run, or repair it. Lions began using the system to teach his classes on operating systems, and in the course of doing so he produced a textbook of sorts, which consisted of the entire source code of UNIX version 6 (V6), along with elaborate, line-by-line commentary and explanation. The value of this textbook can hardly be underestimated. Access to machines and software that could be used to understand how a real system worked was very limited: "Real computers with real operating systems were locked up in machine rooms and committed to processing twenty four hours a day. UNIX changed that."[28] Berny Goodheart, in an appreciation of Lions's *Commentary*, reiterated this sense of the practical usefulness of the source code and commentary: "It is important to understand the significance of John's work at that time: for students studying computer science in the 1970s, complex issues such as process scheduling, security, synchronization, file systems and other concepts were beyond normal comprehension and were extremely difficult to teach—there simply wasn't anything available with enough accessibility for students to use as a case study. Instead a student's discipline in computer science was earned by punching holes in cards, collecting fan-fold paper printouts, and so on. Basically, a computer operating system in that era was considered to be a huge chunk of inaccessible proprietary code."[29]

Lions's commentary was a unique document in the world of computer science, containing a kind of key to learning about a central component of the computer, one that very few people would have had access to in the 1970s. It shows how UNIX was ported not only to machines (which were scarce) but also to the minds of young researchers and student programmers (which were plentiful). Several generations of both academic computer scientists and students who went on to work for computer or software corporations were trained on photocopies of UNIX source code, with a whiff of toner and illicit circulation: a distributed operating system in the textual sense.

Unfortunately, *Commentary* was also legally restricted in its distribution. AT&T and Western Electric, in hopes that they could maintain trade-secret status for UNIX, allowed only very limited circulation of the book. At first, Lions was given permission to distribute single copies only to people who already possessed a license for UNIX V6; later Bell Labs itself would distribute *Commentary* briefly, but only to licensed users, and not for sale, distribution, or copying. Nonetheless, nearly everyone seems to have possessed a dog-eared, nth-generation copy. Peter Reintjes writes, "We soon came into possession of what looked like a fifth generation photocopy and someone who shall remain nameless spent all night in the copier room spawning a sixth, an act expressly forbidden by a carefully worded disclaimer on the first page. Four remarkable things were happening at the same time. One, we had discovered the first piece of software that would inspire rather than annoy us; two, we had acquired what amounted to a literary criticism of that computer software; three, we were making the single most significant advancement of our education in computer science by actually reading an entire operating system; and four, we were breaking the law."[30]

Thus, these generations of computer-science students and academics shared a secret—a trade secret become open secret. Every student who learned the essentials of the UNIX operating system from a photocopy of Lions's commentary, also learned about AT&T's attempt to control its legal distribution on the front cover of their textbook. The parallel development of photocopying has a nice resonance here; together with home cassette taping of music and the introduction of the video-cassette recorder, photocopying helped drive the changes to copyright law adopted in 1976.

Thirty years later, and long after the source code in it had been completely replaced, Lions's *Commentary* is still widely admired by geeks. Even though Free Software has come full circle in providing students with an actual operating system that can be legally studied, taught, copied, and implemented, the kind of "literary criticism" that Lions's work represents is still extremely rare; even reading obsolete code with clear commentary is one of the few ways to truly understand the design elements and clever implementations that made the UNIX operating system so different from its predecessors and even many of its successors, few, if any of which have been so successfully ported to the minds of so many students.

Lions's *Commentary* contributed to the creation of a worldwide community of people whose connection to each other was formed by a body of source code, both in its implemented form and in its textual, photocopied form. This nascent recursive public not only understood itself as belonging to a technical elite which was constituted by its creation, understanding, and promotion of a particular technical tool, but also recognized itself as "breaking the law," a community constituted in opposition to forms of power that governed the circulation, distribution, modification, and creation of the very tools they were learning to make as part of their vocation. The material connection shared around the world by UNIX-loving geeks to their source code is not a mere technical experience, but a social and legal one as well.

Lions was not the only researcher to recognize that teaching the source code was the swiftest route to comprehension. The other story of the circulation of source code concerns Andrew Tanenbaum, a well-respected computer scientist and an author of standard textbooks on computer architecture, operating systems, and networking.[31] In the 1970s Tanenbaum had also used UNIX as a teaching tool in classes at the Vrije Universiteit, in Amsterdam. Because the source code was distributed with the binary code, he could have his students explore directly the implementations of the system, and he often used the source code and the Lions book in his classes. But, according to his *Operating Systems: Design and Implementation* (1987), "When AT&T released Version 7 [ca. 1979], it began to realize that UNIX was a valuable commercial product, so it issued Version 7 with a license that prohibited the source code from being studied in courses, in order to avoid endangering its status as a trade secret. Many universities complied by simply dropping the study of UNIX, and teaching only theory" (13). For Tanenbaum, this was an unacceptable alternative—but so, apparently, was continuing to break the law by teaching UNIX in his courses. And so he proceeded to create a completely new UNIX-like operating system that used not a single line of AT&T source code. He called his creation Minix. It was a stripped-down version intended to run on personal computers (IBM PCs), and to be distributed along with the textbook *Operating Systems*, published by Prentice Hall.[32]

Minix became as widely used in the 1980s as a teaching tool as Lions's source code had been in the 1970s. According to Tanenbaum, the Usenet group comp.os.minix had reached 40,000 members by the late 1980s, and he was receiving constant suggestions for changes and improvements to the operating system. His own commitment to teaching meant that he incorporated few of these suggestions, an effort to keep the system simple enough to be printed in a textbook and understood by undergraduates. Minix was freely available as source code, and it was a fully functioning operating system, even a potential alternative to UNIX that would run on a personal computer. Here was a clear example of the conceptual integrity of UNIX being communicated to another generation of computer-science students: Tanenbaum's textbook is not called "UNIX Operating Systems"—it is called *Operating Systems.* The clear implication is that UNIX represented the clearest example of the principles that should guide the creation of any operating system: it was, for all intents and purposes, state of the art even twenty years after it was first conceived.

Minix was not commercial software, but nor was it Free Software. It was copyrighted and controlled by Tanenbaum's publisher, Prentice Hall. Because it used no AT&T source code, Minix was also legally independent, a legal object of its own. The fact that it was intended to be legally distinct from, yet conceptually true to UNIX is a clear indication of the kinds of tensions that govern the creation and sharing of source code. The ironic apotheosis of Minix as the pedagogical gold standard for studying UNIX came in 1991–92, when a young Linus Torvalds created a "fork" of Minix, also rewritten from scratch, that would go on to become the paradigmatic piece of Free Software: Linux. Tanenbaum's purpose for Minix was that it remain a pedagogically useful operating system—small, concise, and illustrative—whereas Torvalds wanted to extend and expand his version of Minix to take full advantage of the kinds of hardware being produced in the 1990s. Both, however, were committed to source-code visibility and sharing as the swiftest route to complete comprehension of operating-systems principles.

Forking UNIX

Tanenbaum's need to produce Minix was driven by a desire to share the source code of UNIX with students, a desire AT&T was manifestly uncomfortable with and which threatened the trade-secret status of their property. The fact that Minix might be called a fork of UNIX is a key aspect of the political economy of operating systems and social systems. *Forking* generally refers to the creation of new, modified source code from an original base of source code, resulting in two distinct programs with the same parent. Whereas the modification of an engine results only in a modified engine, the modification of source code implies differentiation *and* reproduction, because of the ease with which it can be copied.

How could Minix—a complete rewrite—still be considered the same object? Considered solely from the perspective of trade-secret law, the two objects were distinct, though from the perspective of copyright there was perhaps a case for infringement, although AT&T did not rely on copyright as much as on trade secret. From a technical perspective, the functions and processes that the software accomplishes are the same, but the means by which they are coded to do so are different. And from a pedagogical standpoint, the two are identical—they exemplify certain core features of an operating system (file-system structure, memory paging, process management)—all the rest is optimization, or bells and whistles. Understanding the nature of forking requires also that UNIX be understood from a social perspective, that is, from the perspective of an operating system created and modified by user-developers around the world according to particular and partial demands. It forms the basis for the emergence of a robust recursive public.

One of the more important instances of the forking of UNIX's perambulatory source code and the developing community of UNIX co-developers is the story of the Berkeley Software Distribution and its incorporation of the TCP/IP protocols. In 1975 Ken Thompson took a sabbatical in his hometown of Berkeley, California, where he helped members of the computer-science department with their installations of UNIX, arriving with V6 and the "50 bug fixes" diff tape. Ken had begun work on a compiler for the Pascal programming language that would run on UNIX, and this work was taken up by two young graduate students: Bill Joy and Chuck Hartley. (Joy would later co-found Sun Microsystems, one of the most successful UNIX-based workstation companies in the history of the industry.)

Joy, above nearly all others, enthusiastically participated in the informal distribution of source code. With a popular and well-built Pascal system, and a new text editor called ex (later vi), he created the Berkeley Software Distribution (BSD), a set of tools that could be used in combination with the UNIX operating system. They were extensions to the original UNIX operating system, but not a complete, rewritten version that might replace it. By all accounts, Joy served as a kind of one-man software-distribution house, making tapes and posting them, taking orders and cashing checks—all in addition to creating software.[33] UNIX users around the world soon learned of this valuable set of extensions to the system, and before long, many were differentiating between AT&T UNIX and BSD UNIX.

According to Don Libes, Bell Labs allowed Berkeley to distribute its extensions to UNIX so long as the recipients also had a license from Bell Labs for the original UNIX (an arrangement similar to the one that governed Lions's *Commentary*).[34] From about 1976 until about 1981, BSD slowly became an independent distribution—indeed, a complete version of UNIX—well-known for the vi editor and the Pascal compiler, but also for the addition of virtual memory and its implementation on DEC's VAX machines.[35] It should be clear that the unusual quasi-commercial status of AT&T's UNIX allowed for this situation in a way that a fully commercial computer corporation would never have allowed. Consider, for instance, the fact that many UNIX users—students at a university, for instance—could not essentially know whether they were using an AT&T product or something called BSD UNIX created at Berkeley. The operating system functioned in the same way and, except for the presence of copyright notices that occasionally flashed on the screen, did not make any show of asserting its brand identity (that would come later, in the 1980s). Whereas a commercial computer manufacturer would have allowed something like BSD only if it were incorporated into and distributed as a single, marketable, and identifiable product with a clever name, AT&T turned something of a blind eye to the proliferation and spread of AT&T UNIX and the result were forks in the project: distinct bodies of source code, each an instance of something called UNIX.

As BSD developed, it gained different kinds of functionality than the UNIX from which it was spawned. The most significant development was the inclusion of code that allowed it to connect computers to the Arpanet, using the TCP/IP protocols designed by Vinton Cerf and Robert Kahn. The TCP/IP protocols were a key feature of the Arpanet, overseen by the Information Processing and Techniques Office (IPTO) of the Defense Advanced Research Projects Agency (DARPA) from its inception in 1967 until about 1977. The goal of the protocols was to allow different networks, each with its own machines and administrative boundaries, to be connected to each other.[36] Although there is a common heritage—in the form of J. C. R. Licklider—which ties the imagination of the time-sharing operating system to the creation of the "galactic network," the Arpanet initially developed completely independent of UNIX.[37] As a time-sharing operating system, UNIX was meant to allow the sharing of resources on a *single* computer, whether mainframe or minicomputer, but it was not initially intended to be connected to a network of other computers running UNIX, as is the case today.[38] The goal of Arpanet, by contrast, was explicitly to achieve the sharing of resources located on diverse machines across diverse networks.

To achieve the benefits of TCP/IP, the resources needed to be implemented in all of the different operating systems that were connected to the Arpanet—whatever operating system and machine happened to be in use at each of the nodes. However, by 1977, the original machines used on the network were outdated and increasingly difficult to maintain and, according to Kirk McKusick, the greatest expense was that of porting the old protocol software to new machines. Hence, IPTO decided to pursue in part a strategy of achieving coordination at the operating-system level, and they chose UNIX as one of the core platforms on which to standardize. In short, they had seen the light of portability. In about 1978 IPTO granted a contract to Bolt, Beranek, and Newman (BBN), one of the original Arpanet contractors, to integrate the TCP/IP protocols into the UNIX operating system.

But then something odd happened, according to Salus: "An initial prototype was done by BBN and given to Berkeley. Bill [Joy] immediately started hacking on it because it would only run an Ethernet at about 56K/sec utilizing 100% of the CPU on a 750. . . . Bill lobotomized the code and increased its performance to on the order of 700KB/sec. This caused some consternation with BBN when they came in with their 'finished' version, and Bill wouldn't accept it. There were battles for years after, about which version would be in the system. The Berkeley version ultimately won."[39]

Although it is not clear, it appears BBN intended to give Joy the code in order to include it in his BSD version of UNIX for distribution, and that Joy and collaborators intended to cooperate with Rob Gurwitz of BBN on a final implementation, but Berkeley insisted on "improving" the code to make it perform more to their needs, and BBN apparently dissented from this.[40] One result of this scuffle between BSD and BBN was a genuine fork: two bodies of code that did the same thing, competing with each other to become the standard UNIX implementation of TCP/IP. Here, then, was a case of sharing source code that led to the creation of different versions of software—sharing without collaboration. Some sites used the BBN code, some used the Berkeley code.

Forking, however, does not imply permanent divergence, and the continual improvement, porting, and sharing of software can have odd consequences when forks occur. On the one hand, there are *particular pieces of source code*: they must be identifiable and exact, and prepended with a copyright notice, as was the case of the Berkeley code, which was famously and vigorously policed by the University of California regents, who allowed for a very liberal distribution of BSD code on the condition that the copyright notice was retained. On the other hand, there are particular named *collections of code* that work together (e.g., UNIX™, or DARPA-approved UNIX, or later, Certified Open Source [sm]) and are often identified by a trademark symbol intended, legally speaking, to differentiate products, not to assert ownership of particular instances of a product.

The odd consequence is this: Bill Joy's specific TCP/IP code was incorporated not only into BSD UNIX, but also into other versions of UNIX, including the UNIX distributed by AT&T (which had originally licensed UNIX to Berkeley) with the Berkeley copyright notice removed. This bizarre, tangled bank of licenses and code resulted in a famous suit and countersuit between AT&T and Berkeley, in which the intricacies of this situation were sorted out.[41] An innocent bystander, expecting UNIX to be a single thing, might be surprised to find that it takes different forms for reasons that are all but impossible to identify, but the cause of which is clear: different versions of sharing in conflict with one another; different moral and technical imaginations of order that result in complex entanglements of value and code.

The BSD fork of UNIX (and the subfork of TCP/IP) was only one of many to come. By the early 1980s, a proliferation of UNIX forks had emerged and would be followed shortly by a very robust commercialization. At the same time, the circulation of source code started to slow, as corporations began to compete by adding features and creating hardware specifically designed to run UNIX (such as the Sun Sparc workstation and the Solaris operating system, the result of Joy's commercialization of BSD in the 1980s). The question of how to make all of these versions work together eventually became the subject of the open-systems discussions that would dominate the workstation and networking sectors of the computer market from the early 1980s to 1993, when the dual success of Windows NT and the arrival of the Internet into public consciousness changed the fortunes of the UNIX industry.

A second, and more important, effect of the struggle between BBN and BSD was simply the widespread adoption of the TCP/IP protocols. An estimated 98 percent of computer-science departments in the United States and many such departments around the world incorporated the TCP/IP protocols into their UNIX systems and gained instant access to Arpanet.[42] The fact that this occurred when it did is important: a few years later, during the era of the commercialization of UNIX, these protocols might very well not have been widely implemented (or more likely implemented in incompatible, nonstandard forms) by manufacturers, whereas before 1983, university computer scientists saw every benefit in doing so if it meant they could easily connect to the largest single computer network on the planet. The large, already functioning, relatively standard implementation of TCP/IP on UNIX (and the ability to look at the source code) gave these protocols a tremendous advantage in terms of their survival and success as the basis of a global and singular network.

Conclusion

The UNIX operating system is not just a technical achievement; it is the creation of a set of norms for sharing source code in an unusual environment: quasi-commercial, quasi-academic, networked, and planetwide. Sharing UNIX source code has taken three basic forms: porting source code (transferring it from one machine to another); teaching source code, or "porting" it to students in a pedagogical setting where the use of an actual working operating system vastly facilitates the teaching of theory and concepts; and forking source code (modifying the existing source code to do something new or different). This play of proliferation and differentiation is essential to the remarkably stable identity of UNIX, but that identity exists in multiple forms: technical (as a functioning, self-compatible operating system), legal (as a license-circumscribed version subject to intellectual property and commercial law), and pedagogical (as a conceptual exemplar, the paradigm of an operating system). Source code shared in this manner is essentially unlike any other kind of source code in the world of computers, whether academic or commercial. It raises troubling questions about standardization, about control and audit, and about legitimacy that haunts not only UNIX but the Internet and its various "open" protocols as well.

Sharing source code in Free Software looks the way it does today because of UNIX. But UNIX looks the way it does not because of the inventive genius of Thompson and Ritchie, or the marketing and management brilliance of AT&T, but because *sharing produces its own kind of order*: operating systems and social systems. The fact that geeks are wont to speak of "the UNIX philosophy" means that UNIX is not just an operating system but a way of organizing the complex relations of life and work through technical means; a way of charting and breaching the boundaries between the academic, the aesthetic, and the commercial; a way of implementing ideas of a moral and technical order. What's more, as source code comes to include more and more of the activities of everyday communication and creation—as it comes to replace writing and supplement thinking—the genealogy of its portability and the history of its forking will illuminate the kinds of order emerging in practices and technologies far removed from operating systems—but tied intimately to the UNIX philosophy.

Conceiving Open Systems 5.

The great thing about standards is that there are
so many to choose from.[1]

Openness is an unruly concept. While *free* tends toward ambiguity (free as in speech, or free as in beer?), *open* tends toward obfuscation. Everyone claims to be open, everyone has something to share, everyone agrees that being open is the obvious thing to do—after all, openness is the other half of "open source"—but for all its obviousness, being "open" is perhaps the most complex component of Free Software. It is never quite clear whether being open is a means or an end. Worse, the opposite of open in this case (specifically, "open systems") is not closed, but "proprietary"—signaling the complicated imbrication of the technical, the legal, and the commercial.

In this chapter I tell the story of the contest over the meaning of "open systems" from 1980 to 1993, a contest to create a simultaneously moral and technical infrastructure within the computer industry.[2] The infrastructure in question includes technical components—the UNIX operating system and the TCP/IP protocols of the Internet as open systems—but it also includes "moral" components, including the demand for structures of fair and open competition, antimonopoly and open markets, and open-standards processes for high-tech networked computers and software in the 1980s.[3] By *moral,* I mean imaginations of the proper order of collective political and commercial action; referring to much more than simply how individuals should act, *moral* signifies a vision of how economy and society should be ordered collectively.

The open-systems story is also a story of the blind spot of open systems—in that blind spot is intellectual property. The story reveals a tension between incompatible moral-technical orders: on the one hand, the promise of multiple manufacturers and corporations creating interoperable components and selling them in an open, heterogeneous market; on the other, an intellectual-property system that encouraged jealous guarding and secrecy, and granted monopoly status to source code, designs, and ideas in order to differentiate products and promote competition. The tension proved irresolvable: without shared source code, for instance, interoperable operating systems are impossible. Without interoperable operating systems, internetworking and portable applications are impossible. Without portable applications that can run on any system, open markets are impossible. Without open markets, monopoly power reigns.

Standardization was at the heart of the contest, but by whom and by what means was never resolved. The dream of open systems, pursued in an entirely unregulated industry, resulted in a complicated experiment in novel forms of standardization and cooperation. The creation of a "standard" operating system based on UNIX is the story of a failure, a kind of "figuring out" gone haywire, which resulted in huge consortia of computer manufacturers attempting to work together and compete with each other at the same time. Meanwhile, the successful creation of a "standard" networking protocol—known as the Open Systems Interconnection Reference Model (OSI)—is a story of failure that hides a larger success; OSI was eclipsed in the same period by the rapid and ad hoc adoption of the Transmission Control Protocol/Internet Protocol (TCP/IP), which used a radically different standardization process and which succeeded for a number of surprising reasons, allowing the Internet to take the form it did in the 1990s and ultimately exemplifying the moral-technical imaginary of a recursive public—and one at the heart of the practices of Free Software.

The *conceiving* of openness, which is the central plot of these two stories, has become an essential component of the contemporary practice and power of Free Software. These early battles created a kind of widespread readiness for Free Software in the 1990s, a recognition of Free Software as a removal of open systems' blind spot, as much as an exploitation of its power. The geek ideal of openness and a moral-technical order (the one that made Napster so significant an event) was forged in the era of open systems; without this concrete historical conception of how to maintain openness in technical and moral terms, the recursive public of geeks would be just another hierarchical closed organization—a corporation manqué—and not an independent public serving as a check on the kinds of destructive power that dominated the open-systems contest.

Hopelessly Plural

Big iron, silos, legacy systems, turnkey systems, dinosaurs, mainframes: with the benefit of hindsight, the computer industry of the 1960s to the 1980s appears to be backward and closed, to have literally painted itself into a corner, as an early Intel advertisement suggests (figure 3). Contemporary observers who show disgust and impatience with the form that computers took in this era are without fail supporters of open systems and opponents of proprietary systems that "lock in" customers to specific vendors and create artificial demands for support, integration, and management of resources. Open systems (were it allowed to flourish) would solve all these problems.

Given the promise of a "general-purpose computer," it should seem ironic at best that open systems needed to be created. But the general-purpose computer never came into being. We do not live in the world of The Computer, but in a world of computers: myriad, incompatible, specific machines. The design of specialized machines (or "architectures") was, and still is, key to a competitive industry in computers. It required CPUs and components and associated software that could be clearly qualified and marketed

THE DIFFERENCE BETWEEN AN OPEN SYSTEM AND EVERYTHING ELSE.

One day you have a corner on the market. Next day, you're stuck in that corner. That's the technology trap so many computer systems back you into.

Not ours. Our systems are open systems. They're based on industry standards. Many of which we created. And those standards are just one of the things that keep you from getting trapped, slowed down or otherwise beat out by your competition.

Take for example, our new multi-user supermicro systems for OEMs. They're based on the world's most powerful (and popular) multitasking microprocessor, our iAPX 286. So you can look forward to exceptional price/performance in either real time or commercial applications.

And, because the super-micro is an open system, you get configurability to spare.

Which means you can take advantage of the products and applications already available from hundreds of MUL-TIBUS and independent software vendors (including ourselves). To add speech, graphics, Ethernet® net-working, or whatever you need to make the sale. Easily. But that's just today.

With our open systems, you're also never shut out of a market, even if you're a bit late getting in. With Intel, you'll always have the latest VLSI technology at hand. To make certain you always get a foot or two

in the door.

What's more, we can offer our Open Systems Maintenance program. Which covers your Intel system—from 82 worldwide service locations—even if the problem wasn't caused by one of our parts. Just what you'd expect from a billion-dollar-plus company.

To find out more about the difference our systems can make for your business, ask for our open systems brochure. Just call 800-538-1876; in California, 800-672-1833. Or write Intel, Lit. Dept. Z-18, 3065 Bowers Ave., Santa Clara, CA 95051.

You'll find that our open systems give you a power-ful advantage. Especially when it comes time to close the sale.

intel

3. Open systems is the solution to painting yourself into a corner. Intel advertisement, *Wall Street Journal*, 30 May 1984.

as distinct products: the DEC PDP-11 or the IBM 360 or the CDC 6600. On the Fordist model of automobile production, the computer industry's mission was to render desired functions (scientific calculation, bookkeeping, reservations management) in a large box with a button on it (or a very large number of buttons on increasingly smaller boxes). Despite the theoretical possibility, such computers were not designed to do anything, but, rather, to do specific kinds of calculations exceedingly well. They were objects customized to particular markets.

The marketing strategy was therefore extremely stable from about 1955 to about 1980: identify customers with computing needs, build a computer to serve them, provide them with all of the equipment, software, support, or peripherals they need to do the job—and charge a large amount. Organizationally speaking, it was an industry dominated by "IBM and the seven dwarfs": Hewlett-Packard, Honeywell, Control Data, General Electric, NCR, RCA, Univac, and Burroughs, with a few upstarts like DEC in the wings.

By the 1980s, however, a certain inversion had happened. Computers had become smaller and faster; there were more and more of them, and it was becoming increasingly clear to the "big iron" manufacturers that what was most valuable to users was the information they generated, not the machines that did the generating. Such a realization, so the story goes, leads to a demand for interchangeability, interoperability, information sharing, and networking. It also presents the nightmarish problems of conversion between a bewildering, heterogeneous, and rapidly growing array of hardware, software, protocols, and systems. As one conference paper on the subject of evaluating open systems put it, "At some point a large enterprise will look around and see a huge amount of equipment and software that will not work together. Most importantly, the information stored on these diverse platforms is not being shared, leading to unnecessary duplication and lost profit."⁴

Open systems emerged in the 1980s as the name of the solution to this problem: an approach to the design of systems that, if all participants were to adopt it, would lead to widely interoperable, integrated machines that could send, store, process, and receive the user's information. In marketing and public-relations terms, it would provide "seamless integration."

In theory, open systems was simply a question of standards adoption. For instance, if all the manufacturers of UNIX systems could be convinced to adopt the same basic standard for the operating system, then seamless integration would naturally follow as all the various applications could be written once to run on any variant UNIX system, regardless of which company made it. In reality, such a standard was far from obvious, difficult to create, and even more difficult to enforce. As such, the meaning of *open systems* was "hopelessly plural," and the term came to mean an incredibly diverse array of things.

"Openness" is precisely the kind of concept that wavers between end and means. Is openness good in itself, or is openness a means to achieve something else—and if so what? Who wants to achieve openness, and for what purpose? Is openness a goal? Or is it a means by which a different goal—say, "interoperability" or "integration"—is achieved? Whose goals are these, and who sets them? Are the goals of corporations different from or at odds with the goals of university researchers or government officials? Are there large central visions to which the activities of all are ultimately subordinate?

Between 1980 and 1993, no person or company or computer industry consortium explicitly set openness as the goal that organizations, corporations, or programmers should aim at, but, by the same token, hardly anyone dissented from the demand for openness. As such, it appears clearly as a kind of cultural imperative, reflecting a longstanding social imaginary with roots in liberal democratic notions, versions of a free market and ideals of the free exchange of knowledge, but confronting changed technical conditions that bring the moral ideas of order into relief, and into question.

In the 1980s everyone seemed to want some kind of openness, whether among manufacturers or customers, from General Motors to the armed forces.[5] The debates, both rhetorical and technical, about the meaning of open systems have produced a slough of writings, largely directed at corporate IT managers and CIOs. For instance, Terry A. Critchley and K. C. Batty, the authors of *Open Systems: The Reality* (1993), claim to have collected over a hundred definitions of *open systems*. The definitions stress different aspects— from interoperability of heterogeneous machines, to compatibility of different applications, to portability of operating systems, to legitimate standards with open-interface definitions—including those that privilege ideologies of a free market, as does Bill Gates's definition: "There's nothing more open than the PC market. . . . [U]sers can choose the latest and greatest software." The range

CONCEIVING OPEN SYSTEMS

of meanings was huge and oriented along multiple axes: what, to whom, how, and so on. *Open systems* could mean that source code was open to view or that only the specifications or interfaces were; it could mean "available to certain third parties" or "available to everyone, including competitors"; it could mean self-publishing, well-defined interfaces and application programming interfaces (APIs), or it could mean sticking to standards set by governments and professional societies. To cynics, it simply meant that the marketing department liked the word *open* and used it a lot.

One part of the definition, however, was both consistent and extremely important: the opposite of an "open system" was not a "closed system" but a "proprietary system." In industries other than networking and computing the word *proprietary* will most likely have a positive valence, as in "our exclusive proprietary technology." But in the context of computers and networks such a usage became anathema in the 1980s and 1990s; what customers reportedly wanted was a system that worked nicely with other systems, and that system had to be by definition open since no single company could provide all of the possible needs of a modern business or government agency. And even if it could, it shouldn't be allowed to. For instance, "In the beginning was the word and the word was 'proprietary.' IBM showed the way, purveying machines that existed in splendid isolation. They could not be operated using programs written for any other computer; they could not communicate with the machines of competitors. If your company started out buying computers of various sizes from the International Business Machines Corporation because it was the biggest and best, you soon found yourself locked as securely to Big Blue as a manacled wretch in a medieval dungeon. When an IBM rival unveiled a technologically advanced product, you could only sigh; it might be years before the new technology showed up in the IBM line."[6]

With the exception of IBM (and to some extent its closest competitors: Hewlett-Packard, Burroughs, and Unisys), computer corporations in the 1980s sought to distance themselves from such "medieval" proprietary solutions (such talk also echoes that of usable pasts of the Protestant Reformation often used by geeks). New firms like Sun and Apollo deliberately berated the IBM model. Bill Joy reportedly called one of IBM's new releases in the 1980s a "grazing dinosaur 'with a truck outside pumping its bodily fluids through it.'"[7]

Open systems was never a simple solution though: all that complexity in hardware, software, components, and peripherals could only be solved by pushing hard for standards—even for a single standard. Or, to put it differently, during the 1980s, everyone agreed that open systems was a great idea, but no one agreed on *which* open systems. As one of the anonymous speakers in *Open Systems: The Reality* puts it, "It took me a long time to understand what (the industry) meant by open vs. proprietary, but I finally figured it out. From the perspective of any one supplier, open meant 'our products.' Proprietary meant 'everyone else's products.'"[8]

For most supporters of open systems, the opposition between *open* and *proprietary* had a certain moral force: it indicated that corporations providing the latter were dangerously close to being evil, immoral, perhaps even criminal monopolists. Adrian Gropper and Sean Doyle, the principals in Amicas, an Internet teleradiology company, for instance, routinely referred to the large proprietary healthcare-information systems they confronted in these terms: open systems are the way of light, not dark. Although there are no doubt arguments for closed systems—security, privacy, robustness, control—the demand for interoperability does not mean that such closure will be sacrificed.[9] Closure was also a choice. That is, open systems was an issue of sovereignty, involving the right, in a moral sense, of a customer to control a technical order hemmed in by firm standards that allowed customers to combine a number of different pieces of hardware and software purchased in an open market and to control the configuration themselves—not enforced openness, but the right to decide oneself on whether and how to be open or closed.

The open-systems idea of moral order conflicts, however, with an idea of moral order represented by intellectual property: the right, encoded in law, to assert ownership over and control particular bits of source code, software, and hardware. The call for and the market in open systems were never imagined as being opposed to intellectual property as such, even if the opposition between open and proprietary seemed to indicate a kind of subterranean recognition of the role of intellectual property. The issue was never explicitly broached. Of the hundred definitions in *Open Systems*, only one definition comes close to including legal issues: "Speaker at Interop '90 (paraphrased and maybe apocryphal): 'If you ask to gain access to a technology and the response you get back is a price list, then that technology is "open." If what you get back is a letter from a lawyer, then it's not "open." ' "[10]

Openness here is not equated with freedom to copy and modify, but with the freedom to *buy* access to any aspect of a system without signing a contract, a nondisclosure agreement, or any other legal document besides a check. The ground rules of competition are unchallenged: the existing system of intellectual property—a system that was expanded and strengthened in this period—was a sine qua non of competition.

Openness understood in this manner means an open market in which it is possible to buy standardized things which are neither obscure nor secret, but can be examined and judged—a "commodity" market, where products have functions, where quality is comparable and forms the basis for vigorous competition. What this notion implies is *freedom from monopoly control* by corporations over products, a freedom that is nearly impossible to maintain when the entire industry is structured around the monopoly control of intellectual property through trade secret, patent, or copyright. The blind spot hides the contradiction between an industry imagined on the model of manufacturing distinct and tangible products, and the reality of an industry that wavers somewhere between service and product, dealing in intangible intellectual property whose boundaries and identity are in fact defined by how they are exchanged, circulated, and shared, as in the case of the proliferation and differentiation of the UNIX operating system.

There was no disagreement about the necessity of intellectual property in the computer industry of the 1980s, and there was no perceived contradiction in the demands for openness. Indeed, openness could only make sense if it were built on top of a stable system of intellectual property that allowed competitors to maintain clear definitions of the boundaries of their products. But the creation of interoperable components seemed to demand a relaxation of the secrecy and guardedness necessary to "protect" intellectual property. Indeed, for some observers, the problem of openness created the opportunity for the worst kinds of cynical logic, as in this example from Regis McKenna's *Who's Afraid of Big Blue?*

> Users want open environments, so the vendors had better comply. In fact, it is a good idea to support new standards early. That way, you can help control the development of standards. Moreover, you can

take credit for driving the standard. Supporting standards is a way to demonstrate that you're on the side of users. On the other hand, companies cannot compete on the basis of standards alone. Companies that live by standards can die by standards. Other companies, adhering to the same standards, *could win on the basis of superior manufacturing technology.* If companies do nothing but adhere to standards, then all computers will become commodities, and nobody will be able to make any money. Thus, companies must keep something proprietary, something to differentiate their products.[11]

By such an account, open systems would be tantamount to economic regression, a state of pure competition on the basis of manufacturing superiority, and not on the basis of the competitive advantage granted by the monopoly of intellectual property, the clear hallmark of a high-tech industry.[12] It was an irresolvable tension between the desire for a cooperative, market-based infrastructure and the structure of an intellectual-property system ill-suited to the technical realities within which companies and customers operated—a tension revealing the reorientation of knowledge and power with respect to creation, dissemination, and modification of knowledge.

From the perspective of intellectual property, ideas, designs, and source code are everything—if a company were to release the source code, and allow other vendors to build on it, then what exactly would they be left to sell? Open systems did not mean anything like free, open-source, or public-domain computing. But the fact that competition required some form of collaboration was obvious as well: standard software and network systems were needed; standard markets were needed; standard norms of innovation within the constraints of standards were needed. In short, the challenge was not just the creation of competitive products but the creation of a standard *infrastructure*, dealing with the technical questions of availability, modifiability, and reusability of components, and the moral questions of the proper organization of competition and collaboration across diverse domains: engineers, academics, the computer industry, and the industries it computerized. What follows is the story of how UNIX entered the open-systems fray, a story in which the tension between the conceiving of openness and the demands of intellectual property is revealed.

Open Systems One: Operating Systems

In 1980 UNIX was by all accounts the most obvious choice for a standard operating system for a reason that seemed simple at the outset: it ran on more than one kind of hardware. It had been installed on DEC machines and IBM machines and Intel processors and Motorola processors—a fact exciting to many professional programmers, university computer scientists, and system administrators, many of whom also considered UNIX to be the best designed of the available operating systems.

There was a problem, however (there always is): UNIX belonged to AT&T, and AT&T had licensed it to multiple manufacturers over the years, in addition to allowing the source code to circulate more or less with abandon throughout the world and to be ported to a wide variety of different machine architectures. Such proliferation, albeit haphazard, was a dream come true: a single, interoperable operating system running on all kinds of hardware. Unfortunately, proliferation would also undo that dream, because it meant that as the markets for workstations and operating systems heated up, the existing versions of UNIX hardened into distinct and incompatible versions with different features and interfaces. By the mid 1980s, there were multiple competing efforts to standardize UNIX, an endeavour that eventually went haywire, resulting in the so-called UNIX wars, in which "gangs" of vendors (some on both sides of the battle) teamed up to promote competing standards. The story of how this happened is instructive, for it is a story that has been reiterated several times in the computer industry.[13]

As a hybrid commercial-academic system, UNIX never entered the market as a single thing. It was licensed in various ways to different people, both academic and commercial, and contained additions and tools and other features that may or may not have originated at (or been returned to) Bell Labs. By the early 1980s, the Berkeley Software Distribution was in fact competing with the AT&T version, even though BSD was a sublicensee—and it was not the only one. By the late 1970s and early 1980s, a number of corporations had licensed UNIX from AT&T for use on new machines. Microsoft licensed it (and called it Xenix, rather than licensing the name UNIX as well) to be installed on Intel-based machines. IBM, Unisys, Amdahl, Sun, DEC, and Hewlett-Packard all followed suit and created their own versions and names: HP-UX, A/UX, AIX, Ultrix, and so on. Given the ground rules of trade secrecy and intellectual property, each of these licensed versions needed to be made legally distinct—if they were to compete with each other. Even if "UNIX" remained conceptually pure in an academic or pedagogical sense, every manufacturer would nonetheless have to tweak, to extend, to optimize in order to differentiate. After all, "if companies do nothing but adhere to standards, then all computers will become commodities, and nobody will be able to make any money."[14]

It was thus unlikely that any of these corporations would contribute the changes they made to UNIX back into a common pool, and certainly not back to AT&T which subsequent to the 1984 divestiture finally released their own commercial version of UNIX, called UNIX System V. Very quickly, the promising "open" UNIX of the 1970s became a slough of alternative operating systems, each incompatible with the next thanks to the addition of market-differentiating features and hardware-specific tweaks. According to Pamela Gray, "By the mid-1980s, there were more than 100 versions in active use" centered around the three market leaders, AT&T's System V, Microsoft/SCO Xenix, and the BSD.[15] By 1984, the differences in systems had become significant—as in the case of the BSD additions of the TCP/IP protocols, the vi editor, and the Pascal compiler—and created not only differentiation in terms of quality but also incompatibility at both the software and networking levels.

Different systems of course had different user communities, based on who was the customer of whom. Eric Raymond suggests that in the mid-1980s, independent hackers, programmers, and computer scientists largely followed the fortunes of BSD: "The divide was roughly between longhairs and shorthairs; programmers and technical people tended to line up with Berkeley and BSD, more business-oriented types with AT&T and System V. The longhairs, repeating a theme from Unix's early days ten years before, liked to see themselves as rebels against a corporate empire; one of the small companies put out a poster showing an X-wing-like space fighter marked "BSD" speeding away from a huge AT&T 'death star' logo left broken and in flames."[16]

So even though UNIX had become the standard operating system of choice for time-sharing, multi-user, high-performance computers by the mid-1980s, there was no such thing as UNIX. Competitors

CONCEIVING OPEN SYSTEMS

in the UNIX market could hardly expect the owner of the system, AT&T, to standardize it and compete with them at the same time, and the rest of the systems were in some legal sense still derivations from the original AT&T system. Indeed, in its licensing pamphlets, AT&T even insisted that UNIX was not a noun, but an adjective, as in "the UNIX system."[17]

The dawning realization that the proliferation of systems was not only spreading UNIX around the world but also spreading it thin and breaking it apart led to a series of increasingly startling and high-profile attempts to "standardize" UNIX. Given that the three major branches (BSD, which would become the industry darling as Sun's Solaris operating system; Microsoft, and later SCO Xenix; and AT&T's System V) all emerged from the same AT&T and Berkeley work done largely by Thompson, Ritchie, and Joy, one would think that standardization would be a snap. It was anything but.

Figuring Out Goes Haywire

Figuring out the moral and technical order of open systems went haywire around 1986–88, when there were no fewer than four competing international standards, represented by huge consortia of computer manufacturers (many of whom belonged to multiple consortia): POSIX, the X/Open consortium, the Open Software Foundation, and UNIX International. The blind spot of open systems had much to do with this crazy outcome: academics, industry, and government could not find ways to agree on standardization. One goal of standardization was to afford customers choice; another was to allow competition unconstrained by "artificial" means. A standard body of source code was impossible; a standard "interface definition" was open to too much interpretation; government and academic standards were too complex and expensive; no particular corporation's standard could be trusted (because they could not be trusted to reveal it in advance of their own innovations); and worst of all, customers kept buying, and vendors kept shipping, and the world was increasingly filled with diversity, not standardization.

UNIX *proliferated* quickly because of porting, leading to multiple instances of an operating system with substantially similar source code shared by academics and licensed by AT&T. But it *differentiated* just as quickly because of forking, as particular features were added to different ports. Some features were reincorporated into the "main" branch—the one Thompson and Ritchie worked on—but the bulk of these mutations spread in a haphazard way, shared through users directly or implemented in newly formed commercial versions. Some features were just that, features, but others could extend the system in ways that might make an application possible on one version, but not on another.

The proliferation and differentiation of UNIX, the operating system, had peculiar effects on the emerging market for UNIX, the product: technical issues entailed design and organizational issues. The original UNIX looked the way it did because of the very peculiar structure of the organization that created and sustained UNIX: Bell Labs and the worldwide community of users and developers. The newly formed competitors, conceiving of UNIX as a product distinct from the original UNIX, adopted it precisely because of its portability and because of the promise of open systems as an alternative to "big iron" mainframes. But as UNIX was funneled into existing corporations with their own design and organizational structures, it started to become incompatible with itself, and the desire for competition in open systems necessitated efforts at UNIX standardization.

The first step in the standardization of open systems and UNIX was the creation of what was called an "interface definition," a standard that enumerated the minimum set of functions that any version of UNIX should support *at the interface level*, meaning that any programmer who wrote an application could expect to interact with any version of UNIX on any machine in the same way and get the same response from the machine (regardless of the specific implementation of the operating system or the source code that was used). Interface definitions, and extensions to them, were ideally to be published and freely available.

The interface definition was a standard that emphasized portability, not at the source-code or operating-system level, but at the application level, allowing applications built on any version of UNIX to be installed and run on any other. The push for such a standard came first from a UNIX user group founded in 1980 by Bob Marsh and called, after the convention of file hierarchies in the UNIX interface, "/usr/group" (later renamed Uniforum). The 1984 /usr/ group standard defined a set of system calls, which, however, "was

CONCEIVING OPEN SYSTEMS

immediately ignored and, for all practical purposes, useless."[18] It seemed the field was changing too fast and UNIX proliferating and innovating too widely for such a standard to work.

The /usr/group standard nevertheless provided a starting point for more traditional standards organizations—the Institute of Electrical and Electronics Engineers (IEEE) and the American National Standards Institute (ANSI)—to take on the task. Both institutions took the /usr/group standard as a basis for what would be called IEEE P1003 Portable Operating System Interface for Computer Environments (POSIX). Over the next three years, from 1984 to 1987, POSIX would work diligently at providing a standard interface definition for UNIX.

Alongside this development, the AT&T version of UNIX became the basis for a different standard, the System V Interface Definition (SVID), which attempted to standardize a set of functions similar but not identical to the /usr/group and POSIX standards. Thus emerged two competing definitions for a standard interface to a system that was rapidly proliferating into hundreds of tiny operating-system fiefdoms.[19] The danger of AT&T setting the standard was not lost on any of the competing manufacturers. Even if they created a thoroughly open standard-interface definition, AT&T's version of UNIX would be the first to implement it, and they would continually have privileged knowledge of any changes: if they sought to change the implementation, they could change the standard; if they received demands that the standard be changed, they could change their implementation before releasing the new standard.

In response to this threat, a third entrant into the standards race emerged: X/Open, which comprised a variety of European computer manufacturers (including AT&T!) and sought to develop a standard that encompassed both SVID and POSIX. The X/Open initiative grew out of European concern about the dominance of IBM and originally included Bull, Ericsson, ICL, Nixdorf, Olivetti, Philips, and Siemens. In keeping with a certain 1980s taste for the integration of European economic activity vis-à-vis the United States and Japan, these manufacturers banded together both to distribute a unified UNIX operating system in Europe (based initially on the BSD and Sun versions of UNIX) and to attempt to standardize it at the same time.

X/Open represented a subtle transformation of standardization efforts and of the organizational definition of open systems. While the /usr/group standard was developed by individuals who used UNIX, and the POSIX standard by an acknowledged professional society (IEEE), the X/Open group was a collective of computer corporations that had banded together to fund an independent entity to help further the cause of a standard UNIX. This paradoxical situation—of a need to share a standard among all the competitors and the need to keep the details of that standardized product secret to maintain an advantage—was one that many manufacturers, especially the Europeans with their long experience of IBM's monopoly, understood as mutually destructive. Hence, the solution was to engage in a kind of *organizational* innovation, to create a new form of metacorporate structure that could strategically position itself as at least temporarily interested in *collaboration* with other firms, rather than in competition. Thus did stories and promises of open systems wend their way from the details of technical design to those of organizational design to the moral order of competition and collaboration, power and strategy. "Standards" became products that corporations sought to "sell" to their own industry through the intermediary of the consortium.

In 1985 and 1986 the disarrayed state of UNIX was also frustrating to the major U.S. manufacturers, especially to Sun Microsystems, which had been founded on the creation of a market for UNIX-based "workstations," high-powered networked computers that could compete with mainframes and personal computers at the same time. Founded by Bill Joy, Vinod Khosla, and Andreas Bechtolsheim, Sun had very quickly become an extraordinarily successful computer company. The business pages and magazines were keen to understand whether workstations were viable competitors to PCs, in particular to those of IBM and Microsoft, and the de facto standard DOS operating system, for which a variety of extremely successful business-, personal-, and home-computer applications were written.

Sun seized on the anxiety around open systems, as is evident in the ad it ran during the summer of 1987 (figure 4). The ad plays subtly on two anxieties: the first is directed at the consumer and suggests that only with Sun can one actually achieve interoperability among all of one business' computers, much less across a network or industry; the second is more subtle and plays to fears within the computer industry itself, the anxiety that Sun might merge with one

Announcing the biggest merger in the computer business.

The merger of every computer in your business.

4a and 4b. Open systems anxiety around mergers and compatibility. Sun Microsystems advertisement, *Wall Street Journal*, 9 July 1987.

of the big corporations, AT&T or Unisys, and corner the market in open systems by producing the de facto standard.

In fact, in October 1987 Sun announced that it had made a deal with AT&T. AT&T would distribute a workstation based on Sun's SPARC line of workstations and would acquire 20 percent of Sun.[20] As part of this announcement, Sun and AT&T made clear that they intended to merge two of the dominant versions of UNIX on the market: AT&T's System V and the BSD-derived Solaris. This move clearly frightened the rest of the manufacturers interested in UNIX and open systems, as it suggested a kind of super-power alignment that would restructure (and potentially dominate) the market. A 1988 article in the *New York Times* quotes an industry analyst who characterizes the merger as "a matter of concern at the highest levels of every major computer company in the United States, and possibly the world," and it suggests that competing manufacturers "also fear that AT&T will gradually make Unix a proprietary product, usable only on AT&T or Sun machines."[21] The industry anxiety was great enough that in March Unisys (a computer manufacturer, formerly Burroughs-Sperry) announced that it would work with AT&T and Sun to bring UNIX to its mainframes and to make its business applications run on UNIX. Such a move was tantamount to Unisys admitting that there would be no future in *proprietary* high-end computing—the business on which it had hitherto built its reputation—unless it could be part of the consortium that could own the standard.[22]

In response to this perceived collusion a group of U.S. and European companies banded together to form another rival organization— one that partially overlapped with X/Open but now *included* IBM—this one called the Open Software Foundation. A nonprofit corporation, the foundation included IBM, Digital Equipment, Hewlett-Packard, Bull, Nixdorf, Siemens, and Apollo Computer (Sun's most direct competitor in the workstation market). Their goal was explicitly to create a "competing standard" for UNIX that would be available on the hardware they manufactured (and based, according to some newspaper reports, on IBM's AIX, which was to be called OSF/1). AT&T appeared at first to support the foundation, suggesting that if the Open Software Foundation could come up with a standard, then AT&T would make System V compatible with it. Thus, 1988 was the summer of open love. Every major computer manufacturer in the world was now part of some consortium or another, and some were part of two—each promoting a separate standard.

Of all the corporations, Sun did the most to brand itself as the originator of the open-systems concept. They made very broad claims for the success of open-systems standardization, as for instance in an ad from August 1988 (figure 5), which stated in part:

> But what's more, those sales confirm a broad acceptance of the whole idea behind Sun.
>
> The Open Systems idea.
>
> Systems based on standards so universally accepted that they allow combinations of hardware and software from literally thousands of independent vendors. . . .
>
> So for the first time, you're no longer locked into the company who made your computers. Even if it's us.

The ad goes on to suggest that "in a free market, the best products win out," even as Sun played both sides of every standardization battle, cooperating with both AT&T and with the Open Software Foundation. But by October of that year, it was clear to Sun that

It pays to be open.

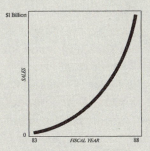

During the past year we delivered $1 billion worth of Sun computers and services.

Gratifying enough for a company that's just six years old.

But what's more, those sales confirm a broad acceptance of the whole idea behind Sun.

The Open Systems idea.

Systems based on standards so universally accepted that they allow combinations of hardware and software from literally thousands of independent vendors. Standards such as the UNIX® Operating System, SPARC™ processors, NFS™ networking software, and the OPEN LOOK™ user interface.

So for the first time, you're no longer locked into the company who made your computers. Even if it's us.

Not that we're concerned.

In a free market, the best products win out. And nobody but Sun offers such powerful and cost-effective distributed computing solutions. Solutions based on state-of-the-art workstations, servers, and networking software.

A Sun workstation moves data many times faster than a PC, and its graphics are stunningly better. A Sun server is as powerful as a supermini, but at a fraction of the cost. And Sun's networking software provides the entire system with an unmatched ability to talk to any kind of computer. Transparently. Effortlessly. Desk to desk. Desk to mini. Desk to mainframe.

Even continent to continent.

It was our vision that this kind of distributed computing could bring unprecedented power not only to the individual, but to the workgroup, and the entire company—in ways that PCs, minis and mainframes never could. Power to access, share, and work with information to a greater degree than ever before.

That is the very reason Sun's Open Systems are at work today throughout all industries. From finance to semiconductors, automotive to aviation. At the largest and smallest corporations in the world, all over the world.

These companies saw the future the same way we did. Wide open.

sun
microsystems

The Network *Is* The Computer™

---

5. It pays to be open: Sun's version of profitable and successful open systems. Sun Microsystems advertisement, *New York Times*, 2 August 1988.

the idea hadn't really become "so universal" just yet. In that month AT&T and Sun banded together with seventeen other manufacturers and formed a rival consortium: Unix International, a coalition of the willing that would back the AT&T UNIX System V version as the one true open standard. In a full-page advertisement from Halloween of 1988 (figure 6), run simultaneously in the *New York Times*, the *Washington Post*, and the *Wall Street Journal*, the rhetoric of achieved success remained, but now instead of "the Open Systems idea," it was "your demand for UNIX System V-based solutions that ushered in the era of open architecture." Instead of a standard for all open systems, it was a war of all against all, a war to assure customers that they had made, not the right choice of hardware or software, but the right choice of *standard.*

The proliferation of standards and standards consortia is often referred to as the UNIX wars of the late 1980s, but the creation of such consortia did not indicate clearly drawn lines. Another metaphor that seems to have been very popular in the press at the time was that of "gang" warfare (no doubt helped along by the creation of another industry consortia informally called the Gang of Nine, which were involved in a dispute over whether MicroChannel or EISA buses should be installed in PCs). The idea of a number of companies forming gangs to fight with each other, Bloods-and-Crips style—or perhaps more Jets-and-Sharks style, minus the singing —was no doubt an appealing metaphor at the height of Los Angeles's very real and high-profile gang warfare. But as one article in the *New York Times* pointed out, these were strange gangs: "Since 'openness' and 'cooperation' are the buzzwords behind these alliances, the gang often asks its enemy to join. Often the enemy does so, either so that it will not seem to be opposed to openness or to keep tabs on the group. IBM was invited to join the corporation for Open Systems, even though the clear if unstated motive of the group was to dilute IBM's influence in the market. AT&T negotiated to join the Open Software Foundation, but the talks collapsed recently. Some companies find it completely consistent to be members of rival gangs. . . . About 10 companies are members of both the Open Software Foundation and its archrival Unix International."[23]

The proliferation of these consortia can be understood in various ways. One could argue that they emerged at a time—during the Reagan administration—when antitrust policing had diminished to

CONCEIVING OPEN SYSTEMS

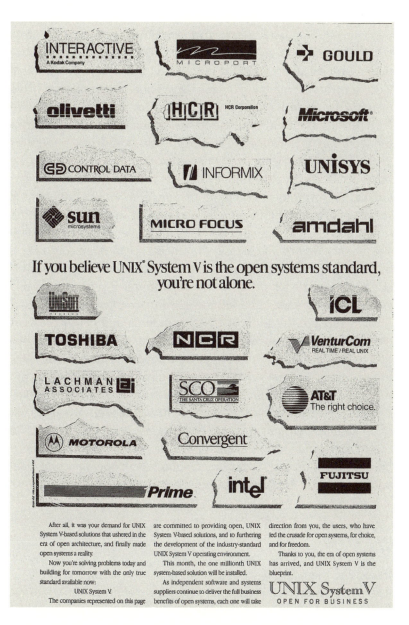

6. The UNIX Wars, Halloween 1988. UNIX International advertisement, *Wall Street Journal* and *New York Times*, 31 October 1988.

the point where computer corporations did not see such collusion as a risky activity vis-à-vis antitrust policing. One could also argue that these consortia represented a recognition that the focus on hardware control (the meaning of *proprietary*) had been replaced with a focus on the control of the "open standard" by one or several manufacturers, that is, that competition was no longer based on superior products, but on "owning the standard." It is significant that the industry consortia quickly overwhelmed national efforts, such as the IEEE POSIX standard, in the media, an indication that no one was looking to government or nonprofits, or to university professional societies, to settle the dispute by declaring a standard, but rather to industry itself to hammer out a standard, de facto or otherwise. Yet another way to understand the emergence of these consortia is as a kind of mutual policing of the market, a kind of paranoid strategy of showing each other just enough to make sure that no one would leapfrog ahead and kill the existing, fragile competition.

What this proliferation of UNIX standards and consortia most clearly represents, however, is the blind spot of open systems: the difficulty of having collaboration and competition at the same time in the context of intellectual-property rules that incompletely capture the specific and unusual characteristics of software. For participants in this market, the structure of intellectual property was unassailable—without it, most participants assumed, innovation would cease and incentives disappear. Despite the fact that secrecy haunted the industry, its customers sought both openness and compatibility. These conflicting demands proved irresolvable.

Denouement

Ironically, the UNIX wars ended not with the emergence of a winner, but with the reassertion of proprietary computing: Microsoft Windows and Windows NT. Rather than open systems emerging victorious, ushering in the era of seamless integration of diverse components, the reverse occurred: Microsoft managed to grab a huge share of computer markets, both desktop and high-performance, by leveraging its brand, the ubiquity of DOS, and application-software developers' dependence on the "Wintel" monster (Windows plus Intel chips). Microsoft triumphed, largely for the same reasons the open-systems dream failed: the legal structure of intel-

CONCEIVING OPEN SYSTEMS

lectual property favored a strong corporate monopoly on a single, branded product over a weak array of "open" and competing components. There was no large gain to investors, or to corporations, from an industry of nice guys sharing the source code and making the components work together. Microsoft, on the other hand, had decided to do so internal to itself; it did not necessarily need to form consortia or standardize its operating systems, if it could leverage its dominance in the market to spread the operating system far and wide. It was, as standards observers like to say, the triumph of de facto standardization over de jure. It was a return to the manacled wretches of IBM's monopoly—but with a new dungeon master.

The denouement of the UNIX standards story was swift: AT&T sold its UNIX System Labs (including all of the original source and rights) to Novell in 1993, who sold it in turn to SCO two years later. Novell sold (or transferred) the trademark name UNIX™ to the X/Open group, which continued to fight for standardization, including a single universal UNIX specification. In 1996 X/Open and the Open Software Foundation merged to form the Open Group.[24] The Open Group eventually joined forces with IEEE to turn POSIX into a single UNIX specification in 2001. They continue to push the original vision of open systems, though they carefully avoid using the name or concept, referring instead to the trademarked mouthful "Boundaryless Information Flow" and employing an updated and newly inscrutable rhetoric: "Boundaryless Information Flow, a shorthand representation of 'access to integrated information to support business process improvements' represents a desired state of an enterprise's infrastructure and is specific to the business needs of the organization."[25]

The Open Group, as well as many other participants in the history of open systems, recognize the emergence of "open source" as a return to the now one true path of boundaryless information flow. Eric Raymond, of course, sees continuity and renewal (not least of which in his own participation in the Open Source movement) and in his *Art of UNIX Programming* says, "The Open Source movement is building on this stable foundation and is creating a resurgence of enthusiasm for the UNIX philosophy. In many ways Open Source can be seen as the true delivery of Open Systems that will ensure it continues to go from strength to strength."[26]

This continuity, of course, deliberately disavows the centrality of the legal component, just as Raymond and the Open Source

Initiative had in 1998. The distinction between a robust market in UNIX operating systems and a standard UNIX-based *infrastructure* on which other markets and other activities can take place still remains unclear to even those closest to the money and machines. It does not yet exist, and may well never come to.

The growth of Free Software in the 1980s and 1990s depended on openness as a concept and component that was figured out during the UNIX wars. It was during these wars that the Free Software Foundation (and other groups, in different ways) began to recognize the centrality of the issue of intellectual property to the goal of creating an infrastructure for the successful creation of open systems.[27] The GNU (GNU's Not Unix) project in particular, but also the X Windows system at MIT, the Remote Procedure Call and Network File System (NFS) systems created by Sun, and tools like sendmail and BIND were each in their own way experiments with alternative licensing arrangements and were circulating widely on a variety of the UNIX versions in the late 1980s. Thus, the experience of open systems, while technically a failure as far as UNIX was concerned, was nonetheless a profound learning experience for an entire generation of engineers, hackers, geeks, and entrepreneurs. Just as the UNIX operating system had a pedagogic life of its own, inculcating itself into the minds of engineers as the paradigm of an operating system, open systems had much the same effect, realizing an inchoate philosophy of openness, interconnection, compatibility, interoperability—in short, *availability* and *modifiability*—that was in conflict with intellectual-property structures as they existed. To put it in Freudian terms: the neurosis of open systems wasn't cured, but the structure of its impossibility had become much clearer to everyone. UNIX, the operating system, did not disappear at all—but UNIX, the market, did.

Open Systems Two: Networks

The struggle to standardize UNIX as a platform for open systems was not the only open-systems struggle; alongside the UNIX wars, another "religious war" was raging. The attempt to standardize networks—in particular, protocols for the inter-networking of multiple, diverse, and autonomous networks of computers—was also a key aspect of the open-systems story of the 1980s.[28] The war between the TCP/IP and OSI was also a story of failure and surprising success: the story of a successful standard with international approval (the OSI protocols) eclipsed by the experimental, military-funded TCP/IP, which exemplified an alternative and unusual standards process. The moral-technical orders expressed by OSI and TCP/IP are, like that of UNIX, on the border between government, university, and industry; they represent conflicting social imaginaries in which power and legitimacy are organized differently and, as a result, expressed differently in the technology.

OSI and TCP/IP started with different goals: OSI was intended to satisfy everyone, to be the complete and comprehensive model against which all competing implementations would be validated; TCP/IP, by contrast, emphasized the easy and robust interconnection of diverse networks. TCP/IP is a protocol developed by bootstrapping between standard and implementation, a mode exemplified by the Requests for Comments system that developed alongside them as part of the Arpanet project. OSI was a "model" or reference standard developed by internationally respected standards organizations.

In the mid-1980s OSI was en route to being adopted internationally, but by 1993 it had been almost completely eclipsed by TCP/IP. The success of TCP/IP is significant for three reasons: (1) *availability* —TCP/IP was itself available via the network and development open to anyone, whereas OSI was a bureaucratically confined and expensive standard and participation was confined to state and corporate representatives, organized through ISO in Geneva; (2) *modifiability*—TCP/IP could be copied from an existing implementation (such as the BSD version of UNIX) and improved, whereas OSI was a complex standard that had few existing implementations available to copy; and (3) *serendipity*—new uses that took advantage of availability and modifiability sprouted, including the "killer app" that was the World Wide Web, which was built to function on existing TCP/IP-based networks, convincing many manufacturers to implement that protocol instead of, or in addition to, OSI.

The success of TCP/IP over OSI was also significant because of the difference in the standardization *processes* that it exemplified. The OSI standard (like all official international standards) is conceived and published as an aid to industrial growth: it was imagined according to the ground rules of intellectual property and as an attempt to facilitate the expansion of markets in networking.

OSI would be a "vendor-neutral" standard: vendors would create their own, secret implementations that could be validated by OSI and thereby be expected to interoperate with other OSI-validated systems. By stark contrast, the TCP/IP protocols were not published (in any conventional sense), nor were the implementations validated by a legitimate international-standards organization; instead, the protocols are themselves represented by implementations that allow connection to the network itself (where the TCP/IP protocols and implementations are themselves made available). The fact that one can only join the network if one possesses or makes an implementation of the protocol is generally seen as the ultimate in validation: it works.[29] In this sense, the struggle between TCP/IP and OSI is indicative of a very familiar twentieth-century struggle over the role and extent of government planning and regulation (versus entrepreneurial activity and individual freedom), perhaps best represented by the twin figures of Friedrich Hayek and Maynard Keynes. In this story, it is Hayek's aversion to planning and the subsequent privileging of spontaneous order that eventually triumphs, not Keynes's paternalistic view of the government as a neutral body that absorbs or encourages the swings of the market.

Bootstrapping Networks

The "religious war" between TCP/IP and OSI occurred in the context of intense competition among computer manufacturers and during a period of vibrant experimentation with computer networks worldwide. As with most developments in computing, IBM was one of the first manufacturers to introduce a networking system for its machines in the early 1970s: the System Network Architecture (SNA). DEC followed suit with Digital Network Architecture (DECnet or DNA), as did Univac with Distributed Communications Architecture (DCA), Burroughs with Burroughs Network Architecture (BNA), and others. These architectures were, like the proprietary operating systems of the same era, considered closed networks, networks that interconnected a centrally planned and specified number of machines of the same type or made by the same manufacturer. The goal of such networks was to make connections internal to a firm, even if that involved geographically widespread systems (e.g., from branch to headquarters). Networks were also to be products.

CONCEIVING OPEN SYSTEMS

The 1970s and 1980s saw extraordinarily vibrant experimentation with academic, military, and commercial networks. Robert Metcalfe had developed Ethernet at Xerox PARC in the mid-1970s, and IBM later created a similar technology called "token ring." In the 1980s the military discovered that the Arpanet was being used predominantly by computer scientists and not just for military applications, and decided to break it into MILNET and CSNET.[30] Bulletin Board Services, which connected PCs to each other via modems to download files, appeared in the late 1970s. Out of this grew Tom Jennings's very successful experiment called FidoNet.[31] In the 1980s an existing social network of university faculty on the East Coast of the United States started a relatively successful network called BITNET (Because It's There Network) in the mid-1980s.[32] The Unix to Unix Copy Protocol (uucp), which initially enabled the Usenet, was developed in the late 1970s and widely used until the mid-1980s to connect UNIX computers together. In 1984 the NSF began a program to fund research in networking and created the first large backbones for NSFNet, successor to the CSNET and Arpanet.[33]

In the 1970s telecommunications companies and spin-off start-ups experimented widely with what were called "videotex" systems, of which the most widely implemented and well-known is Minitel in France.[34] Such systems were designed for consumer users and often provided many of the now widespread services available on the Internet in a kind of embryonic form (from comparison shopping for cars, to directory services, to pornography).[35] By the late 1970s, videotex systems were in the process of being standardized by the Commité Consultative de Information, Technologie et Télé-communications (CCITT) at the International Telecommunications Union (ITU) in Geneva. These standards efforts would eventually be combined with work of the International Organization for Standardization (ISO) on OSI, which had originated from work done at Honeywell.[36]

One important feature united almost all of these experiments: the networks of the computer manufacturers were generally piggybacked, or bootstrapped, onto existing telecommunications infrastructures built by state-run or regulated monopoly telecommunications firms. This situation inevitably spelled grief, for telecommunications providers are highly regulated entities, while the computer industry has been almost totally unregulated from its inception. Since an increasingly core part of the computer industry's business involved transporting signals through telecommunications systems without being regulated to do so, the telecommunications industry naturally felt themselves at a disadvantage.[37] Telecommunications companies were not slow to respond to the need for data communications, but their ability to experiment with products and practices outside the scope of telephony and telegraphy was often hindered by concerns about antitrust and monopoly.[38] The unregulated computer industry, by contrast, saw the tentativeness of the telecommunications industry (or national PTTs) as either bureaucratic inertia or desperate attempts to maintain control and power over existing networks—though no computer manufacturer relished the idea of building their own physical network when so many already existed.

TCP/IP and OSI have become emblematic of the split between the worlds of telecommunications and computing; the metaphors of religious wars or of blood feuds and cold wars were common.[39] A particularly arch account from this period is Carl Malamud's *Exploring the Internet: A Technical Travelogue*, which documents Malamud's (physical) visits to Internet sites around the globe, discussions (and beer) with networking researchers on technical details of the networks they have created, and his own typically geeky, occasionally offensive takes on cultural difference.[40] A subtheme of the story is the religious war between Geneva (in particular the ITU) and the Internet: Malamud tells the story of asking the ITU to release its 19,000-page "blue book" of standards on the Internet, to facilitate its adoption and spread.

The resistance of the ITU and Malamud's heroic if quixotic attempts are a parable of the moral-technical imaginaries of openness— and indeed, his story draws specifically on the usable past of Giordano Bruno.[41] The "bruno" project demonstrates the gulf that exists between two models of legitimacy—those of ISO and the ITU—in which standards represent the legal and legitimate consensus of a regulated industry, approved by member nations, paid for and enforced by governments, and implemented and adhered to by corporations.

Opposite ISO is the ad hoc, experimental style of Arpanet and Internet researchers, in which standards are freely available and implementations represent the mode of achieving consensus, rather than the outcome of the consensus. In reality, such a rhetorical

CONCEIVING OPEN SYSTEMS

opposition is far from absolute: many ISO standards are used on the Internet, and ISO remains a powerful, legitimate standards organization. But the clash of established (telecommunications) and emergent (computer-networking) industries is an important context for understanding the struggle between OSI and TCP/IP.

The need for standard networking protocols is unquestioned: interoperability is the bread and butter of a network. Nonetheless, the goals of the OSI and the TCP/IP protocols differed in important ways, with profound implications for the shape of that interoperability. OSI's goals were completeness, control, and comprehensiveness. OSI grew out of the telecommunications industry, which had a long history of confronting the vicissitudes of linking up networks and facilitating communication around the world, a problem that required a strong process of consensus and negotiation among large, powerful, government-run entities, as well as among smaller manufacturers and providers. OSI's feet were firmly planted in the international standardization organizations like OSI and the ITU (an organization as old as telecommunications itself, dating to the 1860s).

Even if they were oft-mocked as slow, bureaucratic, or cumbersome, the processes of ISO and ITU—based in consensus, international agreement, and thorough technical specification—are processes of unquestioned legitimacy. The representatives of nations and corporations who attend ISO and ITU standards discussions, and who design, write, and vote on these standards, are usually not bureaucrats, but engineers and managers directly concerned with the needs of their constituency. The consensus-oriented process means that ISO and ITU standards attempt to satisfy all members' goals, and as such they tend to be very large, complex, and highly specific documents. They are generally sold to corporations and others who need to use them, rather than made freely available, a fact that until recently reflected their legitimacy, rather than lack thereof.

TCP/IP, on the other hand, emerged from very different conditions.[42] These protocols were part of a Department of Defense–funded experimental research project: Arpanet. The initial Arpanet protocols (the Network Control Protocol, or NCP) were insufficient, and TCP/IP was an experiment in interconnecting two different "packet-switched networks": the ground-line–based Arpanet network and a radio-wave network called Packet Radio.[43] The problem facing the designers was not how to accommodate everyone, but merely how to solve a specific problem: interconnecting two technically diverse networks, each with autonomous administrative boundaries, but forcing neither of them to give up the system or the autonomy.

Until the mid-1980s, the TCP/IP protocols were resolutely research-oriented, and not the object of mainstream commercial interest. Their development reflected a core set of goals shared by researchers and ultimately promoted by the central funding agency, the Department of Defense. The TCP/IP protocols are often referred to as enabling packet-switched networks, but this is only partially correct; the real innovation of this set of protocols was a design for an "inter-network," a system that would interconnect several diverse and autonomous networks (packet-switched or circuit-switched), without requiring them to be transformed, redesigned, or standardized—in short, by requiring only standardization of the intercommunication between networks, not standardization of the network itself. In the first paper describing the protocol Robert Kahn and Vint Cerf motivated the need for TCP/IP thus: "Even though many different and complex problems must be solved in the design of an individual packet-switching network, these problems are manifestly compounded when dissimilar networks are interconnected. Issues arise which may have no direct counterpart in an individual network and which strongly influence the way in which Internetwork communication can take place."[44]

The explicit goal of TCP/IP was thus to share computer resources, not necessarily to connect two individuals or firms together, or to create a competitive market in networks or networking software. Sharing between different kinds of networks implied allowing the different networks to develop autonomously (as their creators and maintainers saw best), but without sacrificing the ability to continue sharing. Years later, David Clark, chief Internet engineer for several years in the 1980s, gave a much more explicit explanation of the goals that led to the TCP/IP protocols. In particular, he suggested that the main overarching goal was not just to share resources but "to develop an effective technique for multiplexed utilization of existing interconnected networks," and he more explicitly stated the issue of control that faced the designers: "Networks represent administrative boundaries of control, and it was an ambition of this project to come to grips with the problem of integrating a number

CONCEIVING OPEN SYSTEMS

of separately administrated entities into a common utility."[45] By placing the goal of expandability first, the TCP/IP protocols were designed with a specific kind of simplicity in mind: the test of the protocols' success was simply the ability to connect.

By setting different goals, TCP/IP and OSI thus differed in terms of technical details; but they also differed in terms of their context and legitimacy, one being a product of international-standards bodies, the other of military-funded research experiments. The technical and organizational differences imply different *processes* for standardization, and it is the peculiar nature of the so-called Requests for Comments (RFC) process that gave TCP/IP one of its most distinctive features. The RFC system is widely recognized as a unique and serendipitous outcome of the research process of Arpanet.[46] In a thirty-year retrospective (published, naturally, as an RFC: RFC 2555), Vint Cerf says, "Hiding in the history of the RFCs is the history of human institutions for achieving cooperative work." He goes on to describe their evolution over the years: "When the RFCs were first produced, they had an almost 19th century character to them—letters exchanged in public debating the merits of various design choices for protocols in the ARPANET. As email and bulletin boards emerged from the fertile fabric of the network, the far-flung participants in this historic dialog began to make increasing use of the online medium to carry out the discussion—reducing the need for documenting the debate in the RFCs and, in some respects, leaving historians somewhat impoverished in the process. RFCs slowly became conclusions rather than debates."[47]

Increasingly, they also became part of a system of discussion *and implementation* in which participants created working software as part of an experiment in developing the standard, after which there was more discussion, then perhaps more implementation, and finally, a standard. The RFC process was a way to condense the process of standardization and validation into implementation; which is to say, the proof of open systems was in the successful connection of diverse networks, and the creation of a standard became a kind of ex post facto rubber-stamping of this demonstration. Any further improvement of the standard hinged on an improvement on the standard implementation because the standards that resulted were freely and widely available: "A user could request an RFC by email from his host computer and have it automatically delivered to his mailbox. . . . RFCs were also shared freely with official standards bodies, manufacturers and vendors, other working groups, and universities. None of the RFCs were ever restricted or classified. This was no mean feat when you consider that they were being funded by DoD during the height of the Cold War."[48]

The OSI protocols were not nearly so freely available. The ironic reversal—the transparency of a military-research program versus the opacity of a Geneva-based international-standards organization—goes a long way toward explaining the reasons why geeks might find the story of TCP/IP's success to be so appealing. It is not that geeks are secretly militaristic, but that they delight in such surprising reversals, especially when those reversals exemplify the kind of ad hoc, clever solution to problems of coordination that the RFC process does. The RFC process is not the only alternative to a consensus-oriented model of standardization pioneered in the international organizations of Geneva, but it is a specific response to a reorientation of power and knowledge that was perhaps more "intuitively obvious" to the creators of Arpanet and the Internet, with its unusual design goals and context, than it would have been to the purveyors of telecommunications systems with over a hundred years of experience in connecting people in very specific and established ways.

Success as Failure

By 1985, OSI was an official standard, one with widespread acceptance by engineers, by the government and military (the "GOSIP" standard), and by a number of manufacturers, the most significant of which was General Motors, with its Manufacturing Automation Protocol (MAP). In textbooks and handbooks of the late 1980s and early 1990s, OSI was routinely referred to as the inevitable standard—which is to say, it had widespread legitimacy as the standard that everyone *should* be implementing—but few implementations existed. Many of the textbooks on networking from the late 1980s, especially those slanted toward a theoretical introduction, give elaborate detail of the OSI reference model—a generation of students in networking was no doubt trained to understand the world in terms of OSI—but the ambivalence continued. Indeed, the most enduring legacy of the creation of the OSI protocols is not the protocols themselves (some of which, like ASN.1, are still widely used today), but the pedagogical model: the "7 layer stack" that is as ubiquitous in networking classes and textbooks as UNIX is in operating-systems classes.[49]

But in the late 1980s, the ambivalence turned to confusion. With OSI widely recognized as the standard, TCP/IP began to show up in more and more actually existing systems. For example, in *Computer Network Architectures and Protocols*, Carl Sunshine says, "Now in the late 1980s, much of the battling seems over. CCITT and ISO have aligned their efforts, and the research community seems largely to have resigned itself to OSI." But immediately afterward he adds: "It is ironic that while a consensus has developed that OSI is indeed inevitable, the TCP/IP protocol suite has achieved widespread deployment, and now serves as a *de facto* interoperability standard. . . . It appears that the vendors were unable to bring OSI products to market quickly enough to satisfy the demand for interoperable systems, and TCP/IP were there to fill the need."[50]

The more implementations that appeared, the less secure the legitimate standard seemed to be. By many accounts the OSI specifications were difficult to implement, and the yearly networking-industry "Interop" conferences became a regular locale for the religious war between TCP/IP and OSI. The success of TCP/IP over OSI reflects the reorientation of knowledge and power to which Free Software is also a response. The reasons for the success are no doubt complex, but the significance of the success of TCP/IP illustrates three issues: availability, modifiability, and serendipity.

Availability The TCP/IP standards themselves were free to anyone and available over TCP/IP networks, exemplifying one of the aspects of a recursive public: that the only test of participation in a TCP/IP-based internetwork is the fact that one possesses or has created a device that implements TCP/IP. Access to the network is contingent on the interoperability of the networks. The standards were not "published" in a conventional sense, but made available through the network itself, without any explicit intellectual property restrictions, and without any fees or restrictions on who could access them. By contrast, ISO standards are generally not circulated freely, but sold for relatively high prices, as a source of revenue, and under the general theory that only legitimate corporations or government agencies would need access to them.

Related to the availability of the standards is the fact that the standards *process* that governed TCP/IP was itself open to anyone, whether corporate, military or academic. The structure of governance of the Internet Engineering Task Force (the IETF) and the Internet Society (ISOC) allowed for anyone with the means available to attend the "working group" meetings that would decide on the standards that would be approved. Certainly this does not mean that the engineers and defense contractors responsible actively *sought out* corporate stakeholders or imagined the system to be "public" in any dramatic fashion; however, compared to the system in place at most standards bodies (in which members are usually required to be the representatives of corporations or governments), the IETF allowed individuals to participate qua individuals.[51]

Modifiability Implementations of TCP/IP were widely available, bootstrapped from machine to machine along with the UNIX operating system and other tools (e.g., the implementation of TCP/IP in BSD 4.2, the BSD version of UNIX), generally including the source code. An existing implementation is a much more expressive and usable object than a specification for an implementation, and though ISO generally prepares reference implementations for such standards, in the case of OSI there were many fewer implementations to work with or build on. Because multiple implementations of TCP/IP already existed, it was easy to validate: did your (modified) implementation work with the other existing implementations? By contrast, OSI would provide independent validation, but the in situ validation through connection to other OSI networks was much harder to achieve, there being too few of them, or access being restricted. It is far easier to build on an existing implementation and to improve on it piecemeal, or even to rewrite it completely, using its faults as a template (so to speak), than it is to create an implementation based solely on a standard. The existence of the TCP/IP protocols in BSD 4.2 not only meant that people who installed that operating system could connect to the Internet easily, at a time when it was by no means standard to be able to do so, but it also meant that manufacturers or tinkerers could examine the implementation in BSD 4.2 as the basis for a modified, or entirely new, implementation.

Serendipity Perhaps most significant, the appearance of widespread and popular applications that were dependent on TCP/IP gave those protocols an inertia that OSI, with relatively few such applications, did not have. The most important of these by far was the World Wide Web (the http protocol, the HTML mark-up language, and implementations of both servers, such as libwww, and clients, such as Mosaic and Netscape). The basic components of the Web were made to work on top of the TCP/IP networks, like other services that had already been designed (ftp, telnet, gopher, archie, etc.); thus, Tim Berners-Lee, who co-invented the World Wide Web, could also rely on the availability and openness of previous work for his own protocols. In addition, Berners-Lee and CERN (the European Organization for Nuclear Research) dedicated their work to the public domain more or less immediately, essentially allowing anyone to do anything they wished with the system they had cobbled together.[52] From the perspective of the tension between TCP/IP and OSI, the World Wide Web was thus what engineers call a "killer app," because its existence actually drove individuals and corporations to make decisions (in favor of TCP/IP) that it might not have made otherwise.

Conclusion

Openness and open systems are key to understanding the practices of Free Software: the open-systems battles of the 1980s set the context for Free Software, leaving in their wake a partially articulated infrastructure of operating systems, networks, and markets that resulted from figuring out open systems. The failure to create a standard UNIX operating system opened the door for Microsoft Windows NT, but it also set the stage for the emergence of the Linux-operating-system kernel to emerge and spread. The success of the TCP/IP protocols forced multiple competing networking schemes into a single standard—and a singular entity, the Internet—which carried with it a set of built-in goals that mirror the moral-technical order of Free Software.

This "infrastructure" is at once technical (protocols and standards and implementations) and moral (expressing ideas about the proper order and organization of commercial efforts to provide high-tech software, networks, and computing power). As with the invention of UNIX, the opposition commercial-noncommercial (or its doppelgangers public-private, profit-nonprofit, capitalist-socialist, etc.)

doesn't capture the context. Constraints on the ability to collaborate, compete, or withdraw are *in the making* here through the technical and moral imaginations of the actors involved: from the corporate behemoths like IBM to (onetime) startups like Sun to the independent academics and amateurs and geeks with stakes in the new high-tech world of networks and software.

The creation of a UNIX market failed. The creation of a legitimate international networking standard failed. But they were local failures only. They opened the doors to new forms of commercial practice (exemplified by Netscape and the dotcom boom) and new kinds of politicotechnical fractiousness (ICANN, IPv6, and "net neutrality"). But the blind spot of open systems—intellectual property—at the heart of these failures also provided the impetus for some geeks, entrepreneurs, and lawyers to start figuring out the legal and economic aspects of Free Software, and it initiated a vibrant experimentation with copyright licensing and with forms of innovative coordination and collaboration built on top of the rapidly spreading protocols of the Internet.

Writing Copyright Licenses 6.

To protect your rights, we need to make restrictions
that forbid anyone to deny you these rights or to ask you
to surrender the rights.—Preamble to the GNU
General Public License

The use of novel, unconventional copyright licenses is, without a
doubt, the most widely recognized and exquisitely refined component of Free Software. The GNU General Public License (GPL), written initially by Richard Stallman, is often referred to as a beautiful,
clever, powerful "hack" of intellectual-property law—when it isn't
being denounced as a viral, infectious object threatening the very
fabric of economy and society. The very fact that something so bor-
ing, so arcane, and so legalistic as a copyright license can become
an object of both devotional reverence and bilious scorn means
there is much more than fine print at stake.

By the beginning of the twenty-first century, there were hundreds of different Free Software licenses, each with subtle legal and technical differences, and an enormous legal literature to explain their details, motivation, and impact.[1] Free Software licenses differ from conventional copyright licenses on software because they usually restrict only the terms of distribution, while so-called End User License Agreements (EULAs) that accompany most proprietary software restrict what users can *do* with the software. Ethnographically speaking, licenses show up everywhere in the field, and contemporary hackers are some of the most legally sophisticated non-lawyers in the world. Indeed, apprenticeship in the world of hacking is now impossible, as Gabriella Coleman has shown, without a long, deep study of intellectual-property law.[2]

But how did it come to be this way? As with the example of sharing UNIX source code, Free Software licenses are often explained as a reaction to expanding intellectual-property laws and resistance to rapacious corporations. The text of the GPL itself begins deep in such assumptions: "The licenses for most software are designed to take away your freedom to share and change it."[3] But even if corporations are rapacious, sharing and modifying software are by no means natural human activities. The ideas of sharing and of common property and its relation to freedom must always be produced through specific practices of sharing, before being defended. The GPL is a precise example of how geeks fit together the practices of sharing and modifying software with the moral and technical orders—the social imaginaries—of freedom and autonomy. It is at once an exquisitely precise legal document and the expression of an idea of how software should be made available, shareable, and modifiable.

In this chapter I tell the story of the creation of the GPL, the first Free Software license, during a controversy over EMACS, a very widely used and respected piece of software; the controversy concerned the reuse of bits of copyrighted source code in a version of EMACS ported to UNIX. There are two reasons to retell this story carefully. The first is simply to articulate the details of the origin of the Free Software license itself, as a central component of Free Software, details that should be understood in the context of changing copyright law and the UNIX and open-systems struggles of the 1980s. Second, although the story of the GPL is also an oft-told story of the "hacker ethic," the GPL is not an "expression" of this

WRITING COPYRIGHT LICENSES

ethic, as if the ethic were genotype to a legal phenotype. Opposite the familiar story of ethics, I explain how the GPL was "figured out" in the controversy over EMACS, how it was formed in response to a complicated state of affairs, both legal and technical, and in a medium new to all the participants: the online mailing lists and discussion lists of Usenet and Arpanet.[4]

The story of the creation of the GNU General Public License ultimately affirms the hacker ethic, not as a story of the ethical hacker genius, but as a historically specific event with a duration and a context, as something that emerges in response to the reorientation of knowledge and power, and through the active modulation of existing practices among both human and nonhuman actors. While hackers themselves might understand the hacker ethic as an unchanging set of moral norms, their practices belie this belief and demonstrate how ethics and norms can emerge suddenly and sharply, undergo repeated transformations, and bifurcate into ideologically distinct camps (Free Software vs. Open Source), even as the practices remain stable relative to them. The hacker ethic does not descend from the heights of philosophy like the categorical imperative—hackers have no Kant, nor do they want one. Rather, as Manuel Delanda has suggested, the philosophy of Free Software is the fact of Free Software itself, its practices and its *things*. If there is a hacker ethic, it is Free Software itself, it is the recursive public itself, which is much more than a list of norms.[5] By understanding it in this way, it becomes possible to track the proliferation and differentiation of the hacker ethic into new and surprising realms, instead of assuming its static universal persistence as a mere procedure that hackers execute.

Free Software Licenses, Once More with Feeling

In lecturing on liberalism in 1935, John Dewey said the following of Jeremy Bentham: "He was, we might say, the first great muck-raker in the field of law . . . but he was more than that, whenever he saw a defect, he proposed a remedy. He was an inventor in law and administration, as much so as any contemporary in mechanical production."[6] Dewey's point was that the liberal reforms attributed to Bentham came not so much from his theories as from his direct involvement in administrative and legal reform—his experimentation.

Whether or not Bentham's influence is best understood this way, it nonetheless captures an important component of liberal reform in Europe and America that is also a key component in the story of Free Software: that the route to achieving change is through direct experiment with the system of law and administration.

A similar story might be told of Richard Stallman, hacker hero and founder of the Free Software Foundation, creator of (among many other things) the GNU C Compiler and GNU EMACS, two of the most widely used and tested Free Software tools in the world. Stallman is routinely abused for holding what many perceive to be "dogmatic" or "intractable" ideological positions about freedom and the right of individuals to do what they please with software. While it is no doubt quite true that his speeches and writings clearly betray a certain fervor and fanaticism, it would be a mistake to assume that his speeches, ideas, or belligerent demands concerning word choice constitute the real substance of his reform. In fact, it is the software he has created and the licenses he has written and rewritten which are the key to his Bentham-like inventiveness. Unlike Bentham, however, Stallman is not a creator of law and administrative structure, but a hacker.

Stallman's GNU General Public License "hacks" the federal copyright law, as is often pointed out. It does this by taking advantage of the very strong rights granted by federal law to actually *loosen* the restrictions normally associated with ownership. Because the statutes grant owners strong powers to create restrictions, Stallman's GPL contains the restriction that anybody can use the licensed material, for any purpose, so long as they subsequently offer the same restriction. Hacks (after which hackers are named) are clever solutions to problems or shortcomings in technology. Hacks are work-arounds, clever, shortest-path solutions that take advantage of characteristics of the system that may or may not have been obvious to the people who designed it. Hacks range from purely utilitarian to mischievously pointless, but they always depend on an existing system or tool through which they achieve their point. To call Free Software a hack is to point out that it would be nothing without the existence of intellectual-property law: it relies on the structure of U.S. copyright law (USC§17) in order to subvert it. Free Software licenses are, in a sense, immanent to copyright laws—there is nothing illegal or even legally arcane about what they accomplish—but there is nonetheless a kind of lingering sense that this particular use of copyright was not how the law was intended to function.

Like all software since the 1980 copyright amendments, Free Software is copyrightable—and what's more, automatically copyrighted as it is written (there is no longer any requirement to register). Copyright law grants the author (or the employer of the author) a number of strong rights over the dispensation of what has been written: rights to copy, distribute, and change the work.[7] Free Software's hack is to immediately make use of these rights in order to abrogate the rights the programmer has been given, thus granting all subsequent licensees rights to copy, distribute, modify, and use the copyrighted software. Some licenses, like the GPL, add the further restriction that every licensee must offer the same terms to any subsequent licensee, others make no such restriction on subsequent uses. Thus, while statutory law suggests that individuals need strong rights and grants them, Free Software licenses effectively annul them in favor of other activities, such as sharing, porting, and forking software. It is for this reason that they have earned the name "copyleft."[8]

This is a convenient ex post facto description, however. Neither Stallman nor anyone else started out with the intention of hacking copyright law. The hack of the Free Software licenses was a response to a complicated controversy over a very important invention, a tool that in turn enabled an invention called EMACS. The story of the controversy is well-known among hackers and geeks, but not often told, and not in any rich detail, outside of these small circles.[9]

EMACS, the Extensible, Customizable,
Self-documenting, Real-time Display Editor

EMACS is a text editor; it is also something like a religion. As one of the two most famous text editors, it is frequently lauded by its devoted users and attacked by detractors who prefer its competitor (Bill Joy's vi, also created in the late 1970s). EMACS is more than just a tool for writing text; for many programmers, it was (and still is) the principal interface to the operating system. For instance, it allows a programmer not only to write a program but also to debug it, to compile it, to run it, and to e-mail it to another user, all from within the same interface. What's more, it allows users to quickly and easily write extensions to EMACS itself, extensions that automate frequent tasks and in turn become core features of the software. It can do almost anything, but it can also frustrate almost anyone. The name itself is taken from its much admired extensibility: EMACS stands for "editing macros" because it allows programmers to quickly record a series of commands and bundle them into a macro that can be called with a simple key combination. In fact, it was one of the first editors (if not the first) to take advantage of keys like ctrl and meta, as in the now ubiquitous ctrl-S familiar to users of non-free word processors like Microsoft Word™.

Appreciate the innovation represented by EMACS: before the UNIX-dominated minicomputer era, there were very few programs for directly manipulating text on a display. To conceive of source code as independent of a program running on a machine meant first conceiving of it as typed, printed, or hand-scrawled code which programmers would scrutinize in its more tangible, paper-based form. Editors that allowed programmers to display the code in front of them on a screen, to manipulate it directly, and to save changes to those files were an innovation of the mid- to late 1960s and were not widespread until the mid-1970s (and this only for bleeding-edge academics and computer corporations). Along with a few early editors, such as QED (originally created by Butler Lampson and Peter Deutsch, and rewritten for UNIX by Ken Thompson), one of the most famous of these was TECO (text editor and corrector), written by Dan Murphy for DEC's PDP-1 computer in 1962–63. Over the years, TECO was transformed (ported and extended) to a wide variety of machines, including machines at Berkeley and MIT, and to other DEC hardware and operating systems. By the early 1970s, there was a version of TECO running on the Incompatible Time-sharing System (ITS), the system in use at MIT's Artificial Intelligence (AI) Lab, and it formed the basis for EMACS. (Thus, EMACS was itself conceived of as a series of macros for a separate editor: Editing MACroS for TECO.)

Like all projects on ITS at the AI Lab, many people contributed to the extension and maintenance of EMACS (including Guy Steele, Dave Moon, Richard Greenblatt, and Charles Frankston), but there is a clear recognition that Stallman made it what it was. The earliest AI Lab memo on EMACS, by Eugene Ciccarelli, says: "Finally, of all the people who have contributed to the development of EMACS, and the TECO behind it, special mention and appreciation go to Richard M. Stallman. He not only gave TECO the power and generality it has, but brought together the good ideas of many different Teco-function packages, added a tremendous amount of new ideas and environment, and created EMACS. Personally one of the joys of my avocational life has been writing Teco/EMACS functions; what makes this fun and not painful is the rich set of tools to work with, all but a few of which have an 'RMS' chiseled somewhere on them."[10]

At this point, in 1978, EMACS lived largely on ITS, but its reputation soon spread, and it was ported to DEC's TOPS-20 (Twenex) operating system and rewritten for Multics and the MIT's LISP machine, on which it was called EINE (Eine Is Not EMACS), and which was followed by ZWEI (Zwei Was Eine Initially).

The proliferation of EMACS was both pleasing and frustrating to Stallman, since it meant that the work fragmented into different projects, each of them EMACS-like, rather than building on one core project, and in a 1981 report he said, "The proliferation of such superficial facsimiles of EMACS has an unfortunate confusing effect: their users, not knowing that they are using an imitation of EMACS and never having seen EMACS itself, are led to believe they are enjoying all the advantages of EMACS. Since any real-time display editor is a tremendous improvement over what they probably had before, they believe this readily. To prevent such confusion, we urge everyone to refer to a nonextensible imitation of EMACS as an 'ersatz EMACS.'"[11]

Thus, while EMACS in its specific form on ITS was a creation of Stallman, the *idea* of EMACS or of any "real-time display editor" was proliferating in different forms and on different machines. The porting of EMACS, like the porting of UNIX, was facilitated by both its conceptual design integrity and its widespread availability.

The phrase "nonextensible imitation" captures the combination of design philosophy and moral philosophy that EMACS represented. Extensibility was not just a useful feature for the individual computer user; it was a way to make the improvements of each new user easily available equally to all by providing a standard way for users to add extensions and to learn how to use new extensions that were added (the "self-documenting" feature of the system). The program had a conceptual integrity that was compromised when it was copied imperfectly. EMACS has a modular, extensible design that by its very nature invites users to contribute to it, to extend it, and to make it perform all manner of tasks—to literally copy and modify it, instead of imitating it. For Stallman, this was not only a fantastic design for a text editor, but an expression of the way he had always done things in the small-scale setting of the AI Lab. The story of Stallman's moral commitments stresses his resistance to secrecy in software production, and EMACS is, both in its design and in Stallman's distribution of it an example of this resistance.

Not everyone shared Stallman's sense of communal order, however. In order to facilitate the extension of EMACS through sharing, Stallman started something he called the "EMACS commune." At the end of the 1981 report—"EMACS: The Extensible, Customizable Self-documenting Display Editor," dated 26 March—he explained the terms of distribution for EMACS: "It is distributed on a basis of communal sharing, which means that all improvements must be given back to me to be incorporated and distributed. Those who are interested should contact me. Further information about how EMACS works is available in the same way."[12]

In another report, intended as a user's manual for EMACS, Stallman gave more detailed and slightly more colorful instructions:

> EMACS does not cost anything; instead, you are joining the EMACS software-sharing commune. The conditions of membership are that you must send back any improvements you make to EMACS, including any libraries you write, and that you must not redistribute the system except exactly as you got it, complete. (You can also distribute your customizations, *separately*.) Please do not attempt to get a copy of EMACS, for yourself or anyone else, by dumping it off of your local system. It is almost certain to be incomplete or inconsistent. It is pathetic to hear from sites that received incomplete copies lacking the sources [source code], asking me years later whether sources are available. (All sources are distributed, and should be on line at every site so that users can read them and copy code from them). If you wish to give away a copy of EMACS, copy a distribution tape from MIT, or mail me a tape and get a new one.[13]

Because EMACS was so widely admired and respected, Stallman had a certain amount of power over this commune. If it had been an obscure, nonextensible tool, useful for a single purpose, no one would have heeded such demands, but because EMACS was by nature the kind of tool that was both useful for all kinds of tasks and customizable for specific ones, Stallman was not the only person who benefited from this communal arrangement. Two disparate sites may well have needed the same macro extension, and therefore many could easily see the social benefit in returning extensions for inclusion, as well as in becoming a kind of co-developer of such a powerful system. As a result, the demands of the EMACS commune, while unusual and autocratic, were of obvious value to the flock. In terms of the concept of recursive public, EMACS was itself the tool through which it was possible for users to extend EMACS, the medium of their affinity; users had a natural incentive to share their contributions so that all might receive the maximum benefit.

The terms of the EMACS distribution agreement were not quite legally binding; nothing compelled participation except Stallman's reputation, his hectoring, or a user's desire to reciprocate. On the one hand, Stallman had not yet chosen to, or been forced to, understand the details of the legal system, and so the EMACS commune was the next best thing. On the other hand, the state of intellectual-property law was in great flux at the time, and it was not clear to anyone, whether corporate or academic, exactly what kind of legal arrangements would be legitimate (the 1976 changes to copyright law were some of the most drastic in seventy years, and a 1980 amendment made software copyrightable, but no court cases had yet tested these changes). Stallman's "agreement" was a set of informal rules that expressed the general sense of mutual aid that was a feature of both the design of the system and Stallman's own experience at the AI Lab. It was an expression of the way Stallman expected others to behave, and his attempts to punish or shame people amounted to informal enforcement of these expectations. The small scale of the community worked in Stallman's favor.

At its small scale, Stallman's commune was confronting many of the same issues that haunted the open-systems debates emerging at the same time, issues of interoperability, source-code sharing, standardization, portability, and forking. In particular, Stallman was acutely aware of the blind spot of open systems: the conflict of moral-technical orders represented by intellectual property. While UNIX vendors left intellectual-property rules unchallenged and simply assumed that they were the essential ground rules of debate, Stallman made them the substance of his experiment and, like Bentham, became something of a legal muckraker as a result.

Stallman's communal model could not completely prevent the porting and forking of software. Despite Stallman's request that imitators refer to their versions of EMACS as ersatz EMACS, few did. In the absence of legal threats over a trademarked term there was not much to stop people from calling their ports and forks EMACS, a problem of success not unlike that of Kleenex or Xerox. Few people took the core ideas of EMACS, implemented them in an imitation, and then called it something else (EINE and ZWEI were exceptions). In the case of UNIX the proliferation of forked versions of the software did not render them any less UNIX, even when AT&T insisted on ownership of the trademarked name. But as time went on, EMACS was ported, forked, rewritten, copied, or imitated on different operating systems and different computer architectures in universities and corporations around the world; within five or six years, many versions of EMACS were in wide use. It was this situation of successful adoption that would provide the context for the controversy that occurred between 1983 and 1985.

The Controversy

In brief the controversy was this: in 1983 James Gosling decided to sell his version of EMACS—a version written in C for UNIX called GOSMACS—to a commercial software vendor called Unipress. GOSMACS, the second most famous implementation of EMACS (after Stallman's itself), was written when Gosling was a graduate student at Carnegie Mellon University. For years, Gosling had distributed GOSMACS by himself and had run a mailing list on Usenet, on which he answered queries and discussed extensions. Gosling had explicitly asked people not to redistribute the program, but to come back to him (or send interested parties to him directly) for new versions, making GOSMACS more of a benevolent dictatorship than a commune. Gosling maintained his authority, but graciously accepted revisions and bug-fixes and extensions from users, incorporating them into new releases. Stallman's system, by contrast, allowed users to distribute their extensions themselves, as well as have them included in the "official" EMACS. By 1983, Gosling had decided he was unable to effectively maintain and support GOSMACS—a task he considered the proper role of a corporation.

WRITING COPYRIGHT LICENSES

For Stallman, Gosling's decision to sell GOSMACS to Unipress was "software sabotage." Even though Gosling had been substantially responsible for writing GOSMACS, Stallman felt somewhat proprietorial toward this ersatz version—or, at the very least, was irked that no noncommercial UNIX version of EMACS existed. So Stallman wrote one himself (as part of a project he announced around the same time, called GNU [GNU's Not UNIX], to create a complete non-AT&T version of UNIX). He called his version GNU EMACS and released it under the same EMACS commune terms. The crux of the debate hinged on the fact that Stallman used, albeit ostensibly with permission, a small piece of Gosling's code in his new version of EMACS, a fact that led numerous people, including the new commercial suppliers of EMACS, to cry foul. Recriminations and legal threats ensued and the controversy was eventually resolved when Stallman rewrote the offending code, thus creating an entirely "Gosling-free" version that went on to become the standard UNIX version of EMACS.

The story raises several questions with respect to the changing legal context. In particular, it raises questions about the difference between "law on the books" and "law in action," that is, the difference between the actions of hackers and commercial entities, advised by lawyers and legally minded friends, and the text and interpretation of statutes as they are written by legislators and interpreted by courts and lawyers. The legal issues span trade secret, patent, and trademark, but copyright is especially significant. Three issues were undecided at the time: the copyrightability of software, the definition of what counts as software and what doesn't, and the meaning of copyright infringement. While the controversy did not resolve any of these issues (the first two would be resolved by Congress and the courts, the third remains somewhat murky), it did clarify the legal issues for Stallman sufficiently that he could leave behind the informal EMACS commune and create the first version of a Free Software license, the GNU General Public License, which first started appearing in 1985.

Gosling's decision to sell GOSMACS, announced in April of 1983, played into a growing EMACS debate being carried out on the GOSMACS mailing list, a Usenet group called net.emacs. Since net .emacs was forwarded to the Arpanet via a gateway maintained by John Gilmore at Sun Microsystems, a fairly large community of EMACS users were privy to Gosling's announcement. Prior to his declaration, there had been quite a bit of discussion regarding different versions of EMACS, including an already "commercial" version called CCA EMACS, written by Steve Zimmerman, of Computer Corporation of America (CCA).[14] Some readers wanted comparisons between CCA EMACS and GOSMACS; others objected that it was improper to discuss a commercial version on the list: was such activity legitimate, or should it be carried out as part of the commercial company's support activities? Gosling's announcement was therefore a surprise, since it was already perceived to be the "noncommercial" version.

Date: Tue Apr 12 04:51:12 1983
Subject: EMACS goes commercial
The version of EMACS that I wrote is now available commercially through a company called Unipress. . . . They will be doing development, maintenance and will be producing a real manual. EMACS will be available on many machines (it already runs on VAXen under Unix and VMS, SUNs, codatas, and Microsoft Xenix). Along with this, I regret to say that I will no longer be distributing it.

This is a hard step to take, but I feel that it is necessary. I can no longer look after it properly, there are too many demands on my time. EMACS has grown to be completely unmanageable. Its popularity has made it impossible to distribute free: just the task of writing tapes and stuffing them into envelopes is more than I can handle.

The alternative of abandoning it to the public domain is unacceptable. Too many other programs have been destroyed that way.

Please support these folks. The effort that they can afford to put into looking after EMACS is directly related to the support they get. Their prices are reasonable.
James.[15]

The message is worth paying careful attention to: Gosling's work of distributing the tapes had become "unmanageable," and the work of maintenance, upkeep, and porting (making it available on multiple architectures) is something he clearly believes should be done by a commercial enterprise. Gosling, it is clear, did not understand his effort in creating and maintaining EMACS to have emerged from a communal sharing of bits of code—even if he had done a Sisyphean amount of work to incorporate all the changes and suggestions his users had made—but he did long have a commit-

WRITING COPYRIGHT LICENSES

ment to distributing it for free, a commitment that resulted in many people contributing bits and pieces to GOSMACS.

"Free," however, did not mean "public domain," as is clear from his statement that "abandoning it" to the public domain would destroy the program. The distinction is an important one that was, and continues to be, lost on many sophisticated members of net .emacs. Here, *free* means without charge, but Gosling had no intention of letting that word suggest that he was not the author, owner, maintainer, distributor, and sole beneficiary of whatever value GOSMACS had. *Public domain*, by contrast, implied giving up all these rights.[16] His decision to sell GOSMACS to Unipress was a decision to transfer these rights to a company that would then charge for all the labor he had previously provided for no charge (for "free"). Such a distinction was not clear to everyone; many people considered the fact that GOSMACS was free to imply that it was in the public domain.[17] Not least of these was Richard Stallman, who referred to Gosling's act as "software sabotage" and urged people to avoid using the "semi-ersatz" Unipress version.[18]

To Stallman, the advancing commercialization of EMACS, both by CCA and by Unipress, was a frustrating state of affairs. The commercialization of CCA had been of little concern so long as GOSMACS remained free, but with Gosling's announcement, there was no longer a UNIX version of EMACS available. To Stallman, however, "free" meant something more than either "public domain" or "for no cost." The EMACS commune was designed to keep EMACS alive and growing as well as to provide it for free—it was an image of community stewardship, a community that had included Gosling until April 1983.

The disappearance of a UNIX version of EMACS, as well as the sudden commercial interest in making UNIX into a marketable operating system, fed into Stallman's nascent plan to create a completely new, noncommercial, non-AT&T UNIX operating system that he would give away free to anyone who could use it. He announced his intention on 27 September 1983:[19]

Free Unix!
Starting this Thanksgiving I am going to write a complete Unix-compatible software system called GNU (for Gnu's Not Unix), and give it away free to everyone who can use it. Contributions of time, money, programs and equipment are greatly needed.

His justifications were simple.

Why I Must Write GNU

I consider that the golden rule requires that if I like a program I must share it with other people who like it. I cannot in good conscience sign a nondisclosure agreement or a software license agreement.

So that I can continue to use computers without violating my principles, I have decided to put together a sufficient body of free software so that I will be able to get along without any software that is not free.[20]

At that point, it is clear, there was no "free software license." There was the word *free*, but not the term *public domain*. There was the "golden rule," and there was a resistance to nondisclosure and license arrangements in general, but certainly no articulated conception of copyleft of Free Software as a legally distinct entity. And yet Stallman hardly intended to "abandon it" to the public domain, as Gosling suggested. Instead, Stallman likely intended to require the same EMACS commune rules to apply to Free Software, rules that he would be able to control largely by overseeing (in a non-legal sense) who was sent or sold what and by demanding (in the form of messages attached to the software) that any modifications or improvements come in the form of donations. It was during the period 1983–85 that the EMACS commune morphed into the GPL, as Stallman began adding copyrights and appending messages that made explicit what people could do with the software.[21]

The GNU project initially received little attention, however; scattered messages to net.unix-wizards over the course of 1983–84 periodically ask about the status and how to contact them, often in the context of discussions of AT&T UNIX licensing practices that were unfolding as UNIX was divested and began to market its own version of UNIX.[22] Stallman's original plan for GNU was to start with the core operating system, the kernel, but his extensive work on EMACS and the sudden need for a free EMACS for UNIX led him to start with a UNIX version of EMACS. In 1984 and into 1985, he and others began work on a UNIX version of GNU EMACS. The two commercial versions of UNIX EMACS (CCA EMACS and Unipress EMACS) continued to circulate and improve in parallel. DEC users meanwhile used the original free version created by Stallman. And, as often happens, life went on: Zimmerman left CCA in August 1984, and Gosling moved to Sun, neither of them remaining closely involved in the software they had created, but leaving the new owners to do so.

By March 1985, Stallman had a complete version (version 15) of GNU EMACS running on the BSD 4.2 version of UNIX (the version Bill Joy had helped create and had taken with him to form the core of Sun's version of UNIX), running on DEC's VAX computers. Stallman announced this software in a characteristically flamboyant manner, publishing in the computer programmers' monthly magazine *Dr. Dobbs* an article entitled "The GNU Manifesto."[23]

Stallman's announcement that a free version of UNIX EMACS was available caused some concern among commercial distributors. The main such concern was that GNU EMACS 15.34 contained code marked "Copyright (c) James Gosling," code used to make EMACS display on screen.[24] The "discovery" (not so difficult, since Stallman always distributed the source code along with the binary) that this code had been reused by Stallman led to extensive discussion among EMACS users of issues such as the mechanics of copyright, the nature of infringement, the definition of software, the meaning of public domain, the difference between patent, copyright, and trade secret, and the mechanics of permission and its granting—in short, a discussion that would be repeatedly recapitulated in nearly every software and intellectual property controversy in the future.

The story of the controversy reveals the structure of rumor on the Usenet to be a bit like the child's game of Chinese Whispers, except that the translations are all archived. GNU EMACS 15.34 was released in March 1985. Between March and early June there was no mention of its legal status, but around June 3 messages on the subject began to proliferate. The earliest mention of the issue appeared not on net.emacs, but on fa.info-vax—a newsgroup devoted to discussions of VAX computer systems ("fa" stands for "from Arpanet")—and it included a dialogue, between Ron Natalie and Marty Sasaki, labeled "GNU EMACS: How Public Domain?": "FOO, don't expect that GNU EMACS is really in the public domain. UNIPRESS seems rather annoyed that there are large portions of it that are marked copyright James Gosling."[25] This message was reprinted on 4 June 1985 on net.emacs, with the addendum: "RMS's work is based on a version of Gosling code that existed before Unipress got it. Gosling had put that code into the public domain. Any work taking off from the early Gosling code is therefore also public domain."[26]

The addendum was then followed by an extensive reply from Zimmerman, whose CCA EMACS had been based on Warren Montgomery's Bell Labs EMACS but rewritten to avoid reusing the code, which may account for why his understanding of the issue seems to have been both deep and troubling for him.

> This is completely contrary to Gosling's public statements. Before he made his arrangements with Unipress, Gosling's policy was that he would send a free copy of his EMACS to anyone who asked, but he did not (publicly, at least) give anyone else permission to make copies. Once Unipress started selling Gosling's EMACS, Gosling stopped distributing free copies and still did not grant anyone else permission to make them; instead, he suggested that people buy EMACS from Unipress. All versions of Gosling's EMACS distributed by him carry his copyright notice, and therefore none of them are in the public domain. Removing copyright notices without the author's permission is, of course, illegal. Now, a quick check of my GNU EMACS sources shows that sure enough, a number of files have Gosling's copyright notice in them. What this all means is that unless RMS got written permission from Gosling to distribute his code, all copies of GNU EMACS constitute violations of the copyright law. All those people making such copies, including those people who allow them to be copied off their machines, could each be liable for large sums of money. I think that RMS had better tell us if he has Gosling's written permission to make these copies. If so, why has he not stated this earlier (preferably in the distribution itself) and thereby cleared up a potentially major point of confusion? If not, why has he gone ahead and made many, many people liable for criminal prosecution by recommending that they distribute this code without even warning them of their liability? (People who distribute this code would be liable even if they claim that they didn't see Gosling's notices; the fact that the notices are there is sufficient. "Ignorance of the law is no excuse.")
>
> Now, I have nothing against free software; it's a free country and people can do what they want. It's just that people who do distribute free software had better be sure that they have the legal right to do so, or be prepared to face the consequences. (Jun 9, 1985).[27]

Stallman replied the next day.

WRITING COPYRIGHT LICENSES

Nobody has any reason to be afraid to use or distribute GNU EMACS. It is well known that I do not believe any software is anyone's property. However, for the GNU project, I decided it was necessary to obey the law. I have refused to look at code I did not have permission to distribute. About 5% of GNU EMACS is close to (though quite a bit changed from) an old version of Gosling EMACS. I am distributing it for Fen Labalme, who received permission from Gosling to distribute it. It is therefore legal for me to do so. To be scrupulously legal, I put statements at the front of the files concerned, describing this situation.

I don't see anything I should warn people about—except that Zimmerman is going to try to browbeat them.[28]

Stallman's original defense for using Gosling's code was that he had permission to do so. According to him, Fen Labalme had received written permission—whether to make use of or to redistribute is not clear—the display code that was included in GNU EMACS 15.34. According to Stallman, versions of Labalme's version of Gosling's version of EMACS were in use in various places (including at Labalme's employer, Megatest), and Stallman and Labalme considered this a legally defensible position.[29]

Over the next two weeks, a slew of messages attempted to pick apart and understand the issues of copyright, ownership, distribution, and authorship. Gosling wrote to clarify that GOSMACS had never been in the public domain, but that "unfortunately, two moves have left my records in a shambles," and he is therefore silent on the question of whether he granted permission.[30] Gosling's claim could well be strategic: giving permission, had he done so, might have angered Unipress, which expected exclusive control over the version he had sold; by the same token, he may well have approved of Stallman's re-creation, but not have wanted to affirm this in any legally actionable way. Meanwhile, Zimmerman relayed an anonymous message suggesting that some lawyers somewhere found the "third hand redistribution" argument was legally "all wet."[31]

Stallman's biggest concern was not so much the legality of his own actions as the possibility that people would choose not to use the software because of legal threats (even if such threats were issued only as rumors by former employees of companies that distributed software they had written). Stallman wanted users not only to feel safe using his software but to adopt his view that software exists to be shared and improved and that anything that hinders this is a loss for everyone, which necessitates an EMACS commune.

Stallman's legal grounds for using Gosling's code may or may not have been sound. Zimmerman did his best throughout to explain in detail what kind of permission Stallman and Labalme would have needed, drawing on his own experience with the CCA lawyers and AT&T Bell Labs, all the while berating Stallman for not creating the display code himself. Meanwhile, Unipress posted an official message that said, "UniPress wants to inform the community that portions of the GNU EMACS program are most definitely not public domain, and that use and/or distribution of the GNU EMACS program is not necessarily proper."[32] The admittedly vague tone of the message left most people wondering what that meant—and whether Unipress intended to sue anyone. Strategically speaking, the company may have wished to maintain good will among hackers and readers of net.emacs, an audience likely composed of many potential customers. Furthermore, if Gosling had given permission to Stallman, then Unipress would themselves have been on uncertain legal ground, unable to firmly and definitively threaten users of GNU EMACS with legal action. In either case, the question of whether or not permission was needed was not in question—only the question of whether it had been granted.[33]

However, a more complicated legal issue also arose as a result, one concerning the status of code contributed to Gosling by others. Fen Labalme wrote a message to net.emacs, which, although it did not clarify the legal status of Gosling's code (Labalme was also unable to find his "permission" from Gosling), did raise a related issue: the fact that he and others had made significant contributions to GOSMACS, which Gosling had incorporated into his version, then sold to Unipress without their permission: "As one of the 'others' who helped to bring EMACS [GOSMACS] up to speed, I was distressed when Jim sold the editor to UniPress. This seemed to be a direct violation of the trust that I and others had placed in Jim as we sent him our improvements, modifications, and bug fixes. I am especially bothered by the general mercenary attitude surrounding EMACS which has taken over from the once proud 'hacker' ethic —EMACS is a tool that can make all of our lives better. Let's help it to grow!"[34]

WRITING COPYRIGHT LICENSES

Labalme's implication, though he may not even have realized this himself, is that Gosling may have infringed on the rights of others in selling the code to Unipress, as a separate message from Joaquim Martillo makes clear: "The differences between current version of Unipress EMACS and Gnu EMACS display.c (a 19 page module) is about 80%. For all the modules which Fen LeBalme [sic] gave RMS permission to use, the differences are similar. Unipress is not even using the disputed software anymore! Now, these modules contain code people like Chris Torek and others contributed when Gosling's emacs was in the public domain. I must wonder whether these people would have contributed had they known their freely-given code was going to become part of someone's product."[35]

Indeed, the general irony of this complicated situation was certainly not as evident as it might have been given the emotional tone of the debates: Stallman was using code from Gosling based on permission Gosling had given to Labalme, but Labalme had written code for Gosling which Gosling had commercialized without telling Labalme—conceivably, but not likely, the same code. Furthermore, all of them were creating software that had been originally conceived in large part by Stallman (but based on ideas and work on TECO, an editor written twenty years before EMACS), who was now busy rewriting the very software Gosling had rewritten for UNIX. The "once proud hacker ethic" that Labalme mentions would thus amount not so much to an explicit belief in sharing so much as a fast-and-loose practice of making contributions and fixes without documenting them, giving oral permission to use and reuse, and "losing" records that may or may not have existed—hardly a noble enterprise.

But by 27 June 1985, all of the legal discussion was rendered moot when Stallman announced that he would completely rewrite the display code in EMACS.

I have decided to replace the Gosling code in GNU EMACS, even though I still believe Fen and I have permission to distribute that code, in order to keep people's confidence in the GNU project.

I came to this decision when I found, this night, that I saw how to rewrite the parts that had seemed hard. I expect to have the job done by the weekend.[36]

On 5 July, Stallman sent out a message that said:

Celebrate our independence from Unipress!

EMACS version 16, 100% Gosling-free, is now being tested at several places. It appears to work solidly on Vaxes, but some other machines have not been tested yet.[37]

The fact that it only took one week to create the code is a testament to Stallman's widely recognized skills in creating great software—it doesn't appear to have indicated any (legal) threat or urgency. Indeed, even though Unipress seems also to have been concerned about their own reputation, and despite the implication made by Stallman that they had forced this issue to happen, they took a month to respond. At that point, the Unipress employee Mike Gallaher wrote to insist, somewhat after the fact, that Unipress had no intention of suing anyone—as long as they were using the Gosling-free EMACS version 16 and higher.

> UniPress has no quarrel with the Gnu project. It bothers me that people seem to think we are trying to hinder it. In fact, we hardly did or said much at all, except to point out that the Gnumacs code had James Gosling's copyright in it. We have not done anything to keep anyone from using Gnumacs, nor do we intend to now that it is "Gosling-free" (version 16.56).
>
> You can consider this to be an official statement from UniPress: There is nothing in Gnumacs version 16.56 that could possibly cause UniPress to get upset. If you were afraid to use Gnumacs because you thought we would hassle you, don't be, on the basis of version 16.56.[38]

Both Stallman and Unipress received various attacks and defenses from observers of the controversy. Many people pointed out that Stallman should get credit for "inventing" EMACS and that the issue of him infringing on his own invention was therefore ironic. Others proclaimed the innocence and moral character of Unipress, which, it was claimed, was providing more of a service (support for EMACS) than the program itself.

Some readers interpreted the fact that Stallman had rewritten the display code, whether under pressure from Unipress or not, as confirmation of the ideas expressed in "The GNU Manifesto," namely, that commercial software stifles innovation. According to this logic, precisely because Stallman was forced to rewrite the code, rather than build on something that he himself assumed he had permission to do, there was no innovation, only fear-induced caution.[39] On the other hand, latent within this discussion is a deep sense of propriety about what people had created; many people, not only Stallman and Gosling and Zimmerman, had contributed to making EMACS what it was, and most had done so under the assumption, legally correct or not, that it would not be taken away from them or, worse, that others might profit by it.

Gosling's sale of EMACS is thus of a different order from his participation in the common stewardship of EMACS. The distinction between creating software and maintaining it is a commercial fiction driven in large part by the structure of intellectual property. It mirrors the experience of open systems. Maintaining software can mean improving it, and improving it can mean incorporating the original work and ideas of others. To do so by the rules of a changing intellectual-property structure forces different choices than to do so according to an informal hacker ethic or an experimental "commune." One programmer's minor improvement is another programmer's original contribution.

The Context of Copyright

The EMACS controversy occurred in a period just after some of the largest changes to U.S. intellectual-property law in seventy years. Two aspects of this context are worth emphasizing: (1) practices and knowledge about the law change slowly and do not immediately reflect the change in either the law or the strategies of actors; (2) U.S. law creates a structural form of uncertainty in which the interplay between legislation and case law is never entirely certain. In the former aspect, programmers who grew up in the 1970s saw a commercial practice entirely dominated by trade secret and patent protection, and very rarely by copyright; thus, the shift to widespread use of copyright law (facilitated by the 1976 and 1980 changes to the law) to protect software was a shift in thinking that only slowly dawned on many participants, even the most legally astute, since it was a general shift in strategy as well as a statutory change. In the latter aspect, the 1976 and 1980 changes to the copyright law contained a number of uncertainties that would take over a decade to be worked out in case law, issues such as the copyrightability of software, the definition of software, and the meaning of infringement in software copyright, to say nothing of the impact of the codification of fair use and the removal of the requirement to register (issues that arguably went unnoticed until the turn of the millennium). Both aspects set the stage for the EMACS controversy and Stallman's creation of the GPL.

Legally speaking, the EMACS controversy was about copyright, permission, and the meanings of a public domain and the reuse of software (and, though never explicitly mentioned, fair use). Software patenting and trade-secret law are not directly concerned, but they nonetheless form a background to the controversy. Many of the participants expressed a legal and conventional orthodoxy that software was not patentable, that is, that algorithms, ideas, or fundamental equations fell outside the scope of patent, even though the 1981 case *Diamond v. Diehr* is generally seen as the first strong support by the courts for forcing the United States Patent and Trademark Office to grant patents on software.[40] Software, this orthodoxy went, was better protected by trade-secret law (a state-by-state law, not a federal statute), which provided protection for any intellectual property that an owner reasonably tried to maintain as a secret. The trade-secret status of UNIX, for example, meant that all the educational licensees who were given the source code of UNIX had agreed to keep it secret, even though it was manifestly circulating the world over; one could therefore run afoul of trade-secret rules if one looked at the source code (e.g., signed a nondisclosure license or was shown the code by an employee) and then implemented something similar.

By contrast, copyright law was rarely deployed in matters of software production. The first copyright registration of software occurred in 1964, but the desirability of relying on copyright over trade secret was uncertain well into the 1970s.[41] Some corporations, like IBM, routinely marked all source code with a copyright symbol. Others asserted it only on the binaries they distributed or in the license agreements. The case of software on the UNIX operating system and its derivatives is particularly haphazard, and the existence of copyright notices by the authors varies widely. An informal survey by Barry Gold singled out only James Gosling, Walter Tichy (author of rcs), and the RAND Corporation as having adequately labeled source code with copyright notices.[42] Gosling was also the first to register EMACS as copyrighted software in 1983,

WRITING COPYRIGHT LICENSES

while Stallman registered GNU EMACS just after version 15.34 was released in May 1985.[43]

The uncertainty of the change from reliance on trade secret to reliance on copyright is clear in some of the statements made by Stallman around the reuse of Gosling's code. Since neither Stallman nor Gosling sought to keep the program secret in any form—either by licensing it or by requiring users to keep it secret—there could be no claims of trade-secret status on either program. Nonetheless, there was frequent concern about whether one had seen any code (especially code from a UNIX operating system, which is covered by trade secret) and whether code that someone else had seen, rewritten, or distributed publicly was therefore "in the public domain."[44] But, at the same time, Stallman was concerned that rewriting Gosling's display code would be too difficult: "Any display code would have a considerable resemblance to that display code, just by virtue of doing the same job. Without any clear idea of exactly how much difference there would have to be to reassure you users, I cannot tell whether the rewrite would accomplish that. The law is not any guidance here. . . . Writing display code that is significantly different is not easy."[45]

Stallman's strategy for rewriting software, including his plan for the GNU operating system, also involved "not looking at" anyone else's code, so as to ensure that no trade-secret violations would occur. Although it was clear that Gosling's code was not a trade secret, it was also not obvious that it was "in the public domain," an assumption that might be made about other kinds of software protected by trade secret. Under trade-secret rules, Gosling's public distribution of GOSMACS appears to give the green light for its reuse, but under copyright law, a law of strict liability, any unauthorized use is a violation, regardless of how public the software may have been.[46]

The fact of trade-secret protection was nonetheless an important aspect of the EMACS controversy: the version of EMACS that Warren Montgomery had created at Bell Labs (and on which Zimmerman's CCA version would be based) *was* the subject of trade-secret protection by AT&T, by virtue of being distributed with UNIX and under a nondisclosure agreement. AT&T was at the time still a year away from divestiture and thus unable to engage in commercial exploitation of the software. When CCA sought to commercialize the version of UNIX Zimmerman had based on Montgomery's, it was necessary to remove any AT&T code in order to avoid violating their trade-secret status. CCA in turn distributed their EMACS as either binary or as source (the former costing about $1,000, the latter as much as $7,000) and relied on copyright rather than trade-secret protection to prevent unauthorized uses of their software.[47]

The uncertainty over copyright was thus in part a reflection of a changing strategy in the computer-software industry, a kind of uneven development in which copyright slowly and haphazardly came to replace trade secret as the main form of intellectual-property protection. This switch had consequences for how noncommercial programmers, researchers, and amateurs might interpret their own work, as well as for the companies whose lawyers were struggling with the same issues. Of course, copyright and trade-secret protection are not mutually exclusive, but they structure the need for secrecy in different ways, and they make different claims on issues like similarity, reuse, and modification.

The 1976 changes to copyright law were therefore extremely significant in setting out a new set of boundaries and possibilities for intellectual-property arguments, arguments that created a different kind of uncertainty from that of a changing commercial strategy: a structural uncertainty created by the need for a case law to develop around the statutory changes implemented by Congress.

The Copyright Act of 1976 introduced a number of changes that had been some ten years in the making, largely organized around new technologies like photocopier machines, home audiotaping, and the new videocassette recorders. It codified fair-use rights, it removed the requirement to register, and it expanded the scope of copyrightable materials considerably. It did not, however, explicitly address software, an oversight that frustrated many in the computer industry, in particular the young software industry. Pursuant to this oversight, the National Commission on New Technological Uses of Copyright (CONTU) was charged with making suggestions for changes to the law with respect to software. It was therefore only in 1980 that Congress implemented these changes, adding software to title 17 of the U.S. copyright statute as something that could be considered copyrightable by law.[48]

The 1980 amendment to the copyright law answered one of three lingering questions about the copyrightability of software: the simple question of whether it was copyrightable material at all. Congress answered yes. It did not, however, designate what constituted "software." During the 1980s, a series of court cases helped specify what counted as software, including source code, object code (binaries), screen display and output, look and feel, and microcode and firmware.[49] The final question, which the courts are still faced with adjudicating, concerns how much similarity constitutes an infringement in each of these cases. The implications of the codification of fair use and the requirement to register continue to unfold even into the present.

The EMACS controversy confronts all three of these questions. Stallman's initial creation of EMACS was accomplished under conditions in which it was unclear whether copyright would apply (i.e., before 1980). Stallman, of course, did not attempt to copyright the earliest versions of EMACS, but the 1976 amendments removed the requirement to register, thus rendering everything written after 1978 automatically copyrighted. Registration represented only an additional effort to assert ownership in cases of suspected infringement.

Throughout this period, the question of whether software was copyrightable—or copyrighted—was being answered differently in different cases: AT&T was relying on trade-secret status; Gosling, Unipress, and CCA negotiated over copyrighted material; and Stallman was experimenting with his "commune." Although the uncertainty was answered statutorily by the 1980 amendment, not everyone instantly grasped this new fact or changed practices based on it. There is ample evidence throughout the Usenet archive that the 1976 changes were poorly understood, especially by comparison with the legal sophistication of hackers in the 1990s and 2000s. Although the law changed in 1980, practices changed more slowly, and justifications crystallized in the context of experiments like that of GNU EMACS.

Further, a tension emerged between the meaning of source code and the meaning of software. On the one hand was the question of whether the source code or the binary code was copyrightable, and on the other was the question of defining the *boundaries* of software in a context wherein all software relies on other software in order to run at all. For instance, EMACS was originally built on top of TECO, which was referred to both as an editor and as a programming language; even seemingly obvious distinctions (e.g., application vs. programming language) were not necessarily always clear.

If EMACS was an application written in TECO qua programming language, then it would seem that EMACS should have its own copyright, distinct from any other program written in TECO. But if EMACS was an extension or modification of TECO qua editor, then it would seem that EMACS was a derivative work and would require the explicit permission of the copyright holder.

Further, each version of EMACS, in order to be EMACS, needed a LISP interpreter in order to make the extensible interface similar across all versions. But not all versions used the same LISP interpreter. Gosling's used an interpreter called MOCKLISP (mlisp in the trademarked Unipress version), for instance. The question of whether the LISP interpreter was a core component of the software or an "environment" needed in order to extend the application was thus also uncertain and unspecified in the law. While both might be treated as software suitable for copyright protection, both might also be understood as necessary components out of which copyrightable software would be built.[50]

What's more, both the 1976 and 1980 amendments are silent on the copyright status of source code vs. binary code. While all the versions of EMACS were distributed in binary, Stallman and Gosling both included the source to allow users to modify it and extend it, but they differed on the proper form of redistribution. The threshold between modifying software for oneself and copyright infringement was not yet clear, and it hung on the meaning of redistribution. Changing the software for use on a single computer might be necessary to get it to run, but by the early days of the Arpanet, innocently placing that code in a public directory on one computer could look like mass distribution.[51]

Finally, the question of what constitutes infringement was at the heart of this controversy and was not resolved by law or by legal adjudication, but simply by rewriting the code to avoid the question. Stallman's use of Gosling's code, his claim of third-hand permission, the presence or absence of written permission, the sale of GOSMACS to Unipress when it most likely contained code not written by Gosling but copyrighted in his name—all of these issues complicated the question of infringement to the point where Stallman's only feasible option for continuing to create software was to avoid using anyone else's code at all. Indeed, Stallman's decision to use Gosling's code (which he claims to have changed in significant portions) might have come to nothing if he had unethically and illegally chosen not to include the copyright notice at all (under the theory that the code was original to Stallman, or an imitation, rather than a portion of Gosling's work). Indeed, Chris Torek received Gosling's permission to remove Gosling's name and copyright from the version of display.c he had heavily modified, but he chose not to omit them: "The only reason I didn't do so is that I feel that he should certainly be credited as the inspiration (at the very least) for the code."[52] Likewise, Stallman was most likely concerned to obey the law and to give credit where credit was due, and therefore left the copyright notice attached—a clear case of blurred meanings of authorship and ownership.

In short, the interplay between new statutes and their settlement in court or in practice was a structural uncertainty that set novel constraints on the meaning of copyright, and especially on the norms and forms of permission and reuse. GNU EMACS 15.34 was the safest option—a completely new version that performed the same tasks, but in a different manner, using different algorithms and code.

Even as it resolved the controversy, however, GNU EMACS posed new problems for Stallman: how would the EMACS commune survive if it wasn't clear whether one could legally use another person's code, even if freely contributed? Was Gosling's action in selling work by others to Unipress legitimate? Would Stallman be able to enforce its opposite, namely, prevent people from commercializing EMACS code they contributed to him? How would Stallman avoid the future possibility of his own volunteers and contributors later asserting that he had infringed on their copyright?

By 1986, Stallman was sending out a letter that recorded the formal transfer of copyright to the Free Software Foundation (which he had founded in late 1985), with equal rights to nonexclusive use of the software.[53] While such a demand for the expropriation of copyright might seem contrary to the aims of the GNU project, in the context of the unfolding copyright law and the GOSMACS controversy it made perfect sense. Having been accused himself of not having proper permission to use someone else's copyrighted material in his free version of GNU EMACS, Stallman took steps to forestall such an event in the future.

The interplay between technical and legal issues and "ethical" concerns was reflected in the practical issues of fear, intimidation, and common-sense (mis)understandings of intellectual-property law. Zimmerman's veiled threats of legal liability were directed not only at Stallman but at anyone who was using the program Stallman had written; breaking the law was, for Zimmerman, an ethical lapse, not a problem of uncertainty and change. Whether or not such an interpretation of the law was correct, it did reveal the mechanisms whereby a low level of detailed knowledge about the law—and a law in flux, at that (not to mention the litigious reputation of the U.S. legal system worldwide)—often seemed to justify a sense that buying software was simply a less risky option than acquiring it for free. Businesses, not customers, it was assumed, would be liable for such infringements. By the same token, the sudden concern of software programmers (rather than lawyers) with the detailed mechanics of copyright law meant that a very large number of people found themselves asserting common-sense notions, only to be involved in a flame war over what the copyright law "actually says."

Such discussion has continued and grown exponentially over the last twenty years, to the point that Free Software hackers are now nearly as deeply educated about intellectual property law as they are about software code.[54] Far from representing the triumph of the hacker ethic, the GNU General Public License represents the concrete, tangible outcome of a relatively wide-ranging cultural conversation hemmed in by changing laws, court decisions, practices both commercial and academic, and experiments with the limits and forms of new media and new technology.

Conclusion

The rest of the story is quickly told: Stallman resigned from the AI Lab at MIT and started the Free Software Foundation in 1985; he created a raft of new tools, but ultimately no full UNIX operating system, and issued General Public License 1.0 in 1989. In 1990 he was awarded a MacArthur "genius grant." During the 1990s, he was involved in various high-profile battles among a new generation of hackers; those controversies included the debate around Linus Torvalds's creation of Linux (which Stallman insisted be referred to as GNU/Linux), the forking of EMACS into Xemacs, and Stallman's own participation in—and exclusion from—conferences and events devoted to Free Software.

Between 1986 and 1990, the Free Software Foundation and its software became extremely well known among geeks. Much of this had to do with the wealth of software that they produced and distributed via Usenet and Arpanet. And as the software circulated and was refined, so were the new legal constraints and the process of teaching users to understand what they could and could not do with the software—and why it was *not* in the public domain.

Each time a new piece of software was released, it was accompanied by one or more text files which explained what its legal status was. At first, there was a file called DISTRIB, which contained an explanation of the rights the new owner had to modify and redistribute the software.[55] DISTRIB referenced a file called COPYING, which contained the "GNU EMACS copying permission notice," also known as the GNU EMACS GPL. The first of these licenses listed the copyright holder as Richard Stallman (in 1985), but by 1986 all licenses referred to the Free Software Foundation as the copyright holder.

As the Free Software Foundation released other pieces of software, the license was renamed—GNU CC GPL, a GNU Bison GPL, a GNU GDB GPL, and so on, all of which were essentially the same terms—in a file called COPYING, which was meant to be distributed along with the software. In 1988, after the software and the licenses had become considerably more widely available, Stallman made a few changes to the license that relaxed some of the terms and specified others.[56] This new version would become the GNU GPL 1.0. By the time Free Software emerged into the public consciousness in the late 1990s, the GPL had reached version 2.0, and the Free Software Foundation had its own legal staff.

The creation of the GPL and the Free Software Foundation are often understood as expressions of the hacker ethic, or of Stallman's ideological commitment to freedom. But the story of EMACS and the complex technical and legal details that structure it illustrate how the GPL is more than just a hack: it was a novel, privately ordered legal "commune." It was a space thoroughly independent of, but insinuated into the existing bedrock of rules and practices of the world of corporate and university software, and carved out of the slippery, changing substance of intellectual-property statutes. At a time when the giants of the software industry were fighting to create a different kind of openness—one that preserved and would even strengthen existing relations of intellectual property—this hack was a radical alternative that emphasized the sovereignty not of a national or corporate status quo, but of self-fashioning individuals who sought to opt out of that national-corporate unity. The creation of the GNU GPL was not a return to a golden age of small-scale communities freed from the dominating structures of bureaucratic modernity, but the creation of something new out of those structures. It relied on and emphasized, not their destruction, but their stability—at least until they are no longer necessary.

The significance of the GPL is due to its embedding within and emergence from the legal and technical infrastructure. Such a practice of situated reworking is what gives Free Software—and perhaps all forms of engineering and creative practice—its warp and weft. Stallman's decision to resign from the AI Lab and start the Free Software Foundation is a good example; it allowed Stallman no only to devote energy to Free Software but also to formally differentiate the organizations, to forestall at least the potential threat that MIT (which still provided him with office space, equipment, and network connection) might decide to claim ownership over his work. One might think that the hacker ethic and the image of self-determining free individuals would demand the total absence of organizations, but it requires instead their proliferation and modulation. Stallman himself was never so purely free: he relied on the largesse of MIT's AI Lab, without which he would have had no office, no computer, no connection to the network, and indeed, for a while, no home.

The Free Software Foundation represents a recognition on his part that individual and communal independence would come at the price of a legally and bureaucratically recognizable entity, set apart from MIT and responsible only to itself. The Free Software Foundation took a classic form: a nonprofit organization with a hierarchy. But by the early 1990s, a new set of experiments would begin that questioned the look of such an entity. The stories of Linux and Apache reveal how these ventures both depended on the work of the Free Software Foundation and departed from the hierarchical tradition it represented, in order to innovate new similarly embedded sociotechnical forms of coordination.

The EMACS text editor is still widely used, in version 22.1 as of 2007, and ported to just about every conceivable operating system. The controversy with Unipress has faded into the distance, as newer and more intense controversies have faced Stallman and Free Software, but the GPL has become the most widely used and most finely scrutinized of the legal licenses. More important, the EMACS controversy was by no means the only one to have erupted in the lives of software programmers; indeed, it has become virtually a rite of passage for young geeks to be involved in such debates, because it is the only way in which the technical details and the legal details that confront geeks can be explored in the requisite detail. Not all such arguments end in the complete rewriting of source code, and today many of them concern the attempt to convince or evangelize for the release of source code under a Free Software license. The EMACS controversy was in some ways a primal scene—a traumatic one, for sure—that determined the outcome of many subsequent fights by giving form to the Free Software license and its uses.

7. Coordinating Collaborations

The final component of Free Software is coordination. For many participants and observers, this is the central innovation and essential significance of Open Source: the possibility of enticing potentially huge numbers of volunteers to work freely on a software project, leveraging the law of large numbers, "peer production," "gift economies," and "self-organizing social economies."[1] Coordination in Free Software is of a distinct kind that emerged in the 1990s, directly out of the issues of sharing source code, conceiving open systems, and writing copyright licenses—all necessary precursors to the practices of coordination. The stories surrounding these issues find continuation in those of the Linux operating-system kernel, of the Apache Web server, and of Source Code Management tools (SCMs); together these stories reveal how coordination worked and what it looked like in the 1990s.

Coordination is important because it collapses and resolves the distinction between technical and social forms into a meaningful whole for participants. On the one hand, there is the coordination and management of people; on the other, there is the coordination of source code, patches, fixes, bug reports, versions, and distributions—but together there is a meaningful technosocial practice of managing, decision-making, and accounting that leads to the collaborative production of complex software and networks. Such coordination would be unexceptional, essentially mimicking long-familiar corporate practices of engineering, except for one key fact: it has no goals. Coordination in Free Software privileges *adaptability* over *planning*. This involves more than simply allowing any kind of modification; the structure of Free Software coordination actually gives precedence to a generalized openness to change, rather than to the following of shared plans, goals, or ideals dictated or controlled by a hierarchy of individuals.[2]

Adaptability does not mean randomness or anarchy, however; it is a very specific way of resolving the tension between the individual curiosity and virtuosity of hackers, and the collective coordination necessary to create and use complex software and networks. No man is an island, but no archipelago is a nation, so to speak. Adaptability preserves the "joy" and "fun" of programming without sacrificing the careful engineering of a stable product. Linux and Apache should be understood as the *results* of this kind of coordination: experiments with adaptability that have worked, to the surprise of many who have insisted that complexity requires planning and hierarchy. Goals and planning are the province of governance—the practice of goal-setting, orientation, and definition of control—but adaptability is the province of critique, and this is why Free Software is a recursive public: it stands outside power and offers powerful criticism in the form of working alternatives. It is not the domain of the new—after all Linux is just a rewrite of UNIX—but the domain of critical and responsive public direction of a collective undertaking.

Linux and Apache are more than pieces of software; they are organizations of an unfamiliar kind. My claim that they are "recursive publics" is useful insofar as it gives a name to a practice that is neither corporate nor academic, neither profit nor nonprofit, neither governmental nor nongovernmental. The concept of recursive public includes, within the spectrum of political activity, the creation, modification, and maintenance of software, networks, and legal documents. While a "public" in most theories is a body of people and a discourse that give expressive form to some concern, "recursive public" is meant to suggest that geeks not only give expressive form to some set of concerns (e.g., that software should be free or that intellectual property rights are too expansive) but also give concrete *infrastructural* form to the means of expression itself. Linux and Apache are tools for creating networks by which expression of new kinds can be guaranteed and by which further infrastructural experimentation can be pursued. For geeks, hacking and programming are variants of free speech and freedom of assembly.

From UNIX to Minix to Linux

Linux and Apache are the two paradigmatic cases of Free Software in the 1990s, both for hackers and for scholars of Free Software. Linux is a UNIX-like operating-system kernel, bootstrapped out of the Minix operating system created by Andrew Tanenbaum.[3] Apache is the continuation of the original National Center for Supercomputing Applications (NCSA) project to create a Web server (Rob McCool's original program, called httpd), bootstrapped out of a distributed collection of people who were using and improving that software.

Linux and Apache are both experiments in coordination. Both projects evolved decision-making systems through experiment: a voting system in Apache's case and a structured hierarchy of decision-makers, with Linus Torvalds as benevolent dictator, in Linux's case. Both projects also explored novel technical tools for coordination, especially Source Code Management (SCM) tools such as Concurrent Versioning System (cvs). Both are also cited as exemplars of how "fun," "joy," or interest determine individual participation and of how it is possible to maintain and encourage that participation and mutual aid instead of narrowing the focus or eliminating possible routes for participation.

Beyond these specific experiments, the stories of Linux and Apache are detailed here because both projects were actively central to the construction and expansion of the Internet of the 1990s by allowing a massive number of both corporate and noncorporate sites to cheaply install and run servers on the Internet. Were Linux and Apache nothing more than hobbyist projects with a few thousand interested tinkerers, rather than the core technical components of an emerging planetary network, they would probably not represent the same kind of revolutionary transformation ultimately branded a "movement" in 1998–99.

Linus Torvalds's creation of the Linux kernel is often cited as the first instance of the real "Open Source" development model, and it has quickly become the most studied of the Free Software projects.[4] Following its appearance in late 1991, Linux grew quickly from a small, barely working kernel to a fully functional replacement for the various commercial UNIX systems that had resulted from the UNIX wars of the 1980s. It has become versatile enough to be used on desktop PCs with very little memory and small CPUs, as well as in "clusters" that allow for massively parallel computing power.

When Torvalds started, he was blessed with an eager audience of hackers keen on seeing a UNIX system run on desktop computers and a personal style of encouragement that produced enormous positive feedback. Torvalds is often given credit for creating, through his "management style," a "new generation" of Free Software— a younger generation than that of Stallman and Raymond. Linus and Linux are not in fact the causes of this change, but the results of being at the right place at the right time and joining together a number of existing components. Indeed, the title of Torvalds's semi-autobiographical reflection on Linux—*Just for Fun: The Story of an Accidental Revolutionary*—captures some of the character of its genesis.

The "fun" referred to in the title reflects the privileging of adaptability over planning. Projects, tools, people, and code that were fun were those that were not dictated by existing rules and ideas. Fun, for geeks, was associated with the sudden availability, especially for university students and amateur hackers, of a rapidly expanding underground world of networks and software—Usenet and the Internet especially, but also university-specific networks, online environments and games, and tools for navigating information of all kinds. Much of this activity occurred without the benefit of any explicit theorization, with the possible exception of the discourse of "community" (given print expression by Howard Rheingold in 1993 and present in nascent form in the pages of *Wired* and *Mondo 2000*) that took place through much of the 1990s.[5] The late 1980s and early 1990s gave rise to vast experimentation with the collaborative possibilities of the Internet as a medium. Particularly attractive was that this medium was built using freely available tools, and the tools themselves were open to modification and creative reuse. It was a style that reflected the quasi-academic and quasi-commercial environment, of which the UNIX operating system was an exemplar— not pure research divorced from commercial context, nor entirely the domain of commercial rapacity and intellectual property.

Fun included the creation of mailing lists by the spread of software such as list-serv and majordomo; the collaborative maintenance and policing of Usenet; and the creation of Multi-User Dungeons (MUDs) and MUD Object Orienteds (MOOs), both of which gave game players and Internet geeks a way to co-create software environments and discover many of the problems of management and policing that thereby emerged.[6] It also included the increasing array of "information services" that were built on top of the Internet, like archie, gopher, Veronica, WAIS, ftp, IRC—all of which were necessary to access the growing information wealth of the underground community lurking on the Internet. The meaning and practice of coordination in all of these projects was up for grabs: some were organized strictly as university research projects (gopher), while others were more fluid and open to participation and even control by contributing members (MOOs and MUDs). Licensing issues were explicit in some, unclear in some, and completely ignored in others. Some projects had autocratic leaders, while others experimented with everything from representative democracy to anarchism.

During this period (roughly 1987 to 1993), the Free Software Foundation attained a mythic cult status—primarily among UNIX and EMACS users. Part of this status was due to the superiority of the tools Stallman and his collaborators had already created: the GNU C Compiler (gcc), GNU EMACS, the GNU Debugger (gdb), GNU Bison, and loads of smaller utilities that replaced the original AT&T UNIX versions. The GNU GPL had also acquired a life of its own by this time, having reached maturity as a license and become the de facto choice for those committed to Free Software and the Free Software Foundation. By 1991, however, the rumors of the imminent appearance of Stallman's replacement UNIX operating system had started to sound empty—it had been six years since his public announcement of his intention. Most hackers were skeptical of Stallman's operating-system project, even if they acknowledged the success of all the other tools necessary to create a full-fledged operating system, and Stallman himself was stymied by the development of one particular component: the kernel itself, called GNU Hurd.

Linus Torvalds's project was not initially imagined as a contribution to the Free Software Foundation: it was a Helsinki university student's late-night project in learning the ins and outs of the relatively new Intel 386/486 microprocessor. Torvalds, along with tens of thousands of other computer-science students, was being schooled in UNIX through the pedagogy of Andrew Tanenbaum's Minix, Douglas Comer's Xinu-PC, and a handful of other such teaching versions designed to run on IBM PCs. Along with the classroom pedagogy in the 1980s came the inevitable connection to, lurking on, and posting to the Usenet and Arpanet mailing lists devoted to technical (and nontechnical) topics of all sorts.[7] Torvalds was subscribed, naturally, to comp.os.minix, the newsgroup for users of Minix.

The fact of Linus Torvalds's pedagogical embedding in the world of UNIX, Minix, the Free Software Foundation, and the Usenet should not be underestimated, as it often is in hagiographical accounts of the Linux operating system. Without this relatively robust moral-technical order or infrastructure within which it was possible to be at the right place at the right time, Torvalds's late-night dorm-room project would have amounted to little more than that—but the pieces were all in place for his modest goals to be transformed into something much more significant.

Consider his announcement on 25 August 1991:

Hello everybody out there using minix—I'm doing a (free) operating system (just a hobby, won't be big and professional like gnu) for 386(486) AT clones. This has been brewing since april, and is starting to get ready. I'd like any feedback on things people like/dislike in minix, as my OS resembles it somewhat (same physical layout of the file-system (due to practical reasons) among other things). I've currently ported bash(1.08) and gcc(1.40), and things seem to work. This implies that I'll get something practical within a few months, and I'd like to know what features most people would want. Any suggestions are welcome, but I won't promise I'll implement them :-)

Linus . . .

PS. Yes—it's free of any minix code, and it has a multi-threaded fs. It is NOT portable (uses 386 task switching etc), and it probably never will support anything other than AT-harddisks, as that's all I have :-(.[8]

Torvalds's announcement is telling as to where his project fit into the existing context: "just a hobby," not "big and professional like gnu" (a comment that suggests the stature that Stallman and the Free Software Foundation had achieved, especially since they were in reality anything but "big and professional"). The announcement was posted to the Minix list and thus was essentially directed at Minix users; but Torvalds also makes a point of insisting that the system would be free of cost, and his postscript furthermore indicates that it would be free of Minix code, just as Minix had been free of AT&T code.

Torvalds also mentions that he has ported "bash" and "gcc," software created and distributed by the Free Software Foundation and tools essential for interacting with the computer and compiling new versions of the kernel. Torvalds's decision to use these utilities, rather than write his own, reflects both the boundaries of his project (an operating-system kernel) and his satisfaction with the availability and reusability of software licensed under the GPL.

So the system is based on Minix, just as Minix had been based on UNIX—piggy-backed or bootstrapped, rather than rewritten in an entirely different fashion, that is, rather than becoming a different kind of operating system. And yet there are clearly concerns about the need to create something that is not Minix, rather than simply extending or "debugging" Minix. This concern is key to understanding what happened to Linux in 1991.

Tanenbaum's Minix, since its inception in 1984, was always intended to allow students to see and change the source code of Minix in order to learn how an operating system worked, but it was not Free Software. It was copyrighted and owned by Prentice Hall, which distributed the textbooks. Tanenbaum made the case—similar to Gosling's case for Unipress—that Prentice Hall was distributing the system far wider than if it were available only on the Internet: "A point which I don't think everyone appreciates is that making something available by FTP is not necessarily the way to provide the widest distribution. The Internet is still a highly elite group. Most computer users are NOT on it. . . . MINIX is also widely used in Eastern Europe, Japan, Israel, South America, etc. Most of these people would never have gotten it if there hadn't been a company selling it."[9]

By all accounts, Prentice Hall was not restrictive in its sublicensing of the operating system, if people wanted to create an "enhanced" version of Minix. Similarly, Tanenbaum's frequent presence on comp.os.minix testified to his commitment to sharing his knowledge about the system with anyone who wanted it—not just paying customers. Nonetheless, Torvalds's pointed use of the word *free* and his decision not to reuse any of the code is a clear indication of his desire to build a system completely unencumbered by restrictions, based perhaps on a kind of intuitive folkloric sense of the dangers associated with cases like that of EMACS.[10]

The most significant aspect of Torvalds's initial message, however, is his request: "I'd like to know what features most people would want. Any suggestions are welcome, but I won't promise I'll implement them." Torvalds's announcement and the subsequent interest it generated clearly reveal the issues of coordination and organization that would come to be a feature of Linux. The reason Torvalds had so many eager contributors to Linux, from the very start, was because he enthusiastically took them off of Tanenbaum's hands.

Design and Adaptability

Tanenbaum's role in the story of Linux is usually that of the straw man—a crotchety old computer-science professor who opposes the revolutionary young Torvalds. Tanenbaum did have a certain revolutionary reputation himself, since Minix was used in classrooms around the world and could be installed on IBM PCs (something no other commercial UNIX vendors had achieved), but he was also a natural target for people like Torvalds: the tenured professor espousing the textbook version of an operating system. So, despite the fact that a very large number of people were using or knew of Minix as a UNIX operating system (estimates of comp.os.minix subscribers were at 40,000), Tanenbaum was emphatically not interested in collaboration or collaborative debugging, especially if debugging also meant creating extensions and adding features that would make the system bigger and harder to use as a stripped-down tool for teaching. For Tanenbaum, this point was central: "I've been repeatedly offered virtual memory, paging, symbolic links, window systems, and all manner of features. I have usually declined because I am still trying to keep the system simple enough for students to understand. You can put all this stuff in your version, but I won't put it in mine. I think it is this point which irks the people who say 'MINIX is not free,' not the $60."[11]

So while Tanenbaum was in sympathy with the Free Software Foundation's goals (insofar as he clearly wanted people to be able to use, update, enhance, and learn from software), he was not in sympathy with the idea of having 40,000 strangers make his software "better." Or, to put it differently, the goals of Minix remained those of a researcher and a textbook author: to be useful in classrooms and cheap enough to be widely available and usable on the largest number of cheap computers.

By contrast, Torvalds's "fun" project had no goals. Being a cocky nineteen-year-old student with little better to do (no textbooks to write, no students, grants, research projects, or committee meetings), Torvalds was keen to accept all the ready-made help he could find to make his project better. And with 40,000 Minix users, he had a more or less instant set of contributors. Stallman's audience for EMACS in the early 1980s, by contrast, was limited to about a hundred distinct computers, which may have translated into thousands, but certainly not tens of thousands of users. Tanenbaum's work in creating a generation of students who not only understood the internals of an operating system but, more specifically, understood the internals of the UNIX operating system created a huge pool of competent and eager UNIX hackers. It was the work of porting UNIX not only to various machines but to a generation of minds as well that set the stage for this event—and this is an essential, though often overlooked component of the success of Linux.

Many accounts of the Linux story focus on the fight between Torvalds and Tanenbaum, a fight carried out on comp.os.minix with the subject line "Linux is obsolete."[12] Tanenbaum argued that Torvalds was reinventing the wheel, writing an operating system that, as far as the state of the art was concerned, was now obsolete. Torvalds, by contrast, asserted that it was better to make something quick and dirty that worked, invite contributions, and worry about making it state of the art later. Far from illustrating some kind of outmoded conservatism on Tanenbaum's part, the debate highlights the distinction between forms of coordination and the meanings of collaboration. For Tanenbaum, the goals of Minix were either pedagogical or academic: to teach operating-system essentials or to explore new possibilities in operating-system design. By this model, Linux could do neither; it couldn't be used in the classroom because it would quickly become too complex and feature-laden to teach, and it wasn't pushing the boundaries of research because it was an out-of-date operating system. Torvalds, by contrast, had no goals. What drove his progress was a commitment to fun and to a largely inarticulate notion of what interested him and others, defined at the outset almost entirely against Minix and other free operating systems, like FreeBSD. In this sense, it could only emerge out of the context—which set the constraints on its design—of UNIX, open systems, Minix, GNU, and BSD.

Both Tanenbaum and Torvalds operated under a model of coordination in which one person was ultimately responsible for the entire project: Tanenbaum oversaw Minix and ensured that it remained true to its goals of serving a pedagogical audience; Torvalds would oversee Linux, but he would incorporate as many different features as users wanted or could contribute. Very quickly—with a pool of 40,000 potential contributors—Torvalds would be in the same position Tanenbaum was in, that is, forced to make decisions about the goals of Linux and about which enhancements would go into it and which would not. What makes the story of Linux so interesting to observers is that it appears that Torvalds made no decision: he accepted almost everything.

Tanenbaum's goals and plans for Minix were clear and autocratically formed. Control, hierarchy, and restriction are after all appropriate in the classroom. But Torvalds wanted to do more. He wanted to go on learning and to try out alternatives, and with Minix as the only widely available way to do so, his decision to part ways starts to make sense; clearly he was not alone in his desire to explore and extend what he had learned. Nonetheless, Torvalds faced the problem of coordinating a new project and making similar decisions about its direction. On this point, Linux has been the subject of much reflection by both insiders and outsiders. Despite images of Linux as either an anarchic bazaar or an autocratic dictatorship, the reality is more subtle: it includes a hierarchy of contributors, maintainers, and "trusted lieutenants" and a sophisticated, informal, and intuitive sense of "good taste" gained through reading and incorporating the work of co-developers.

While it was possible for Torvalds to remain in charge as an individual for the first few years of Linux (1991–95, roughly), he eventually began to delegate some of that control to people who would make decisions about different subcomponents of the kernel.

It was thus possible to incorporate more of the "patches" (pieces of code) contributed by volunteers, by distributing some of the work of evaluating them to people other than Torvalds. This informal hierarchy slowly developed into a formal one, as Steven Weber points out: "The final *de facto* 'grant' of authority came when Torvalds began publicly to reroute relevant submissions to the lieutenants. In 1996 the decision structure became more formal with an explicit differentiation between 'credited developers' and 'maintainers.' . . . If this sounds very much like a hierarchical decision structure, that is because it is one—albeit one in which participation is strictly voluntary."[13]

Almost all of the decisions made by Torvalds and lieutenants were of a single kind: whether or not to incorporate a piece of code submitted by a volunteer. Each such decision was technically complex: insert the code, recompile the kernel, test to see if it works or if it produces any bugs, decide whether it is worth keeping, issue a new version with a log of the changes that were made. Although the various official leaders were given the authority to make such changes, coordination was still technically informal. Since they were all working on the same complex technical object, one person (Torvalds) ultimately needed to verify a final version, containing all the subparts, in order to make sure that it worked without breaking.

Such decisions had very little to do with any kind of design goals or plans, only with whether the submitted patch "worked," a term that reflects at once technical, aesthetic, legal, and design criteria that are not explicitly recorded anywhere in the project—hence, the privileging of adaptability over planning. At no point were the patches assigned or solicited, although Torvalds is justly famous for encouraging people to work on particular problems, but only if they wanted to. As a result, the system morphed in subtle, unexpected ways, diverging from its original, supposedly backwards "monolithic" design and into a novel configuration that reflected the interests of the volunteers and the implicit criteria of the leaders.

By 1995–96, Torvalds and lieutenants faced considerable challenges with regard to hierarchy and decision-making, as the project had grown in size and complexity. The first widely remembered response to the ongoing crisis of benevolent dictatorship in Linux was the creation of "loadable kernel modules," conceived as a way to release some of the constant pressure to decide which patches would be incorporated into the kernel. The decision to modularize

COORDINATING COLLABORATIONS

Linux was simultaneously technical and social: the software-code base would be rewritten to allow for external loadable modules to be inserted "on the fly," rather than all being compiled into one large binary chunk; at the same time, it meant that the responsibility to ensure that the modules worked devolved from Torvalds to the creator of the module. The decision repudiated Torvalds's early opposition to Tanenbaum in the "monolithic vs. microkernel" debate by inviting contributors to separate core from peripheral functions of an operating system (though the Linux kernel remains monolithic compared to classic microkernels). It also allowed for a significant proliferation of new ideas and related projects. It both contracted and distributed the hierarchy; now Linus was in charge of a tighter project, but more people could work with him according to structured technical and social rules of responsibility.

Creating loadable modules changed the look of Linux, but not because of any planning or design decisions set out in advance. The choice is an example of the privileged adaptability of the Linux, resolving the tension between the curiosity and virtuosity of individual contributors to the project and the need for hierarchical control in order to manage complexity. The commitment to adaptability dissolves the distinction between the technical means of coordination and the social means of management. It is about producing a meaningful whole by which both people and code can be coordinated—an achievement vigorously defended by kernel hackers.

The adaptable organization and structure of Linux is often described in evolutionary terms, as something without teleological purpose, but responding to an environment. Indeed, Torvalds himself has a weakness for this kind of explanation.

Let's just be honest, and admit that it [Linux] wasn't designed.

Sure, there's design too—the design of UNIX made a scaffolding for the system, and more importantly it made it easier for people to communicate because people had a mental *model* for what the system was like, which means that it's much easier to discuss changes.

But that's like saying that you know that you're going to build a car with four wheels and headlights—it's true, but the real bitch is in the details.

And I know better than most that what I envisioned 10 years ago has *nothing* in common with what Linux is today. There was certainly no premeditated design there.[14]

Adaptability does not answer the questions of intelligent design. Why, for example, does a car have four wheels and two headlights? Often these discussions are polarized: either technical objects are designed, or they are the result of random mutations. What this opposition overlooks is the fact that design and the coordination of collaboration go hand in hand; one reveals the limits and possibilities of the other. Linux represents a particular example of such a problematic—one that has become the paradigmatic case of Free Software—but there have been many others, including UNIX, for which the engineers created a system that reflected the distributed collaboration of users around the world even as the lawyers tried to make it conform to legal rules about licensing and practical concerns about bookkeeping and support.

Because it privileges adaptability over planning, Linux is a recursive public: operating systems and social systems. It privileges openness to new directions, at every level. It privileges the right to propose changes by actually creating them and trying to convince others to use and incorporate them. It privileges the right to fork the software into new and different kinds of systems. Given what it privileges, Linux ends up evolving differently than do systems whose life and design are constrained by corporate organization, or by strict engineering design principles, or by legal or marketing definitions of products—in short, by clear goals. What makes this distinction between the goal-oriented design principle and the principle of adaptability important is its relationship to politics. Goals and planning are the subject of negotiation and consensus, or of autocratic decision-making; adaptability is the province of critique. It should be remembered that Linux is by no means an attempt to create something radically new; it is a rewrite of a UNIX operating system, as Torvalds points out, but one that through adaptation can end up becoming something new.

Patch and Vote

The Apache Web server and the Apache Group (now called the Apache Software Foundation) provide a second illuminating example of the how and why of coordination in Free Software of the 1990s. As with the case of Linux, the development of the Apache project illustrates how adaptability is privileged over planning and, in particular, how this privileging is intended to resolve the tensions between individual curiosity and virtuosity and collective control and decision-making. It is also the story of the progressive evolution of coordination, the simultaneously technical and social mechanisms of coordinating people and code, patches and votes.

The Apache project emerged out of a group of users of the original httpd (HyperText Transmission Protocol Daemon) Web server created by Rob McCool at NCSA, based on the work of Tim Berners-Lee's World Wide Web project at CERN. Berners-Lee had written a specification for the World Wide Web that included the mark-up language HTML, the transmission protocol http, and a set of libraries that implemented the code known as libwww, which he had dedicated to the public domain.[15]

The NCSA, at the University of Illinois, Urbana-Champaign, picked up both www projects, subsequently creating both the first widely used browser, Mosaic, directed by Marc Andreessen, and httpd. Httpd was public domain up until version 1.3. Development slowed when McCool was lured to Netscape, along with the team that created Mosaic. By early 1994, when the World Wide Web had started to spread, many individuals and groups ran Web servers that used httpd; some of them had created extensions and fixed bugs. They ranged from university researchers to corporations like Wired Ventures, which launched the online version of its magazine (HotWired.com) in 1994. Most users communicated primarily through Usenet, on the comp.infosystems.www.* newsgroups, sharing experiences, instructions, and updates in the same manner as other software projects stretching back to the beginning of the Usenet and Arpanet newsgroups.

When NCSA failed to respond to most of the fixes and extensions being proposed, a group of several of the most active users of httpd began to communicate via a mailing list called new-httpd in 1995. The list was maintained by Brian Behlendorf, the webmaster for HotWired, on a server he maintained called hyperreal; its participants were those who had debugged httpd, created extensions, or added functionality. The list was the primary means of association and communication for a diverse group of people from various locations around the world. During the next year, participants hashed out issues related to coordination, to the identity of and the processes involved in patching the "new" httpd, version 1.3.[16]

Patching a piece of software is a peculiar activity, akin to debugging, but more like a form of ex post facto design. Patching covers the spectrum of changes that can be made: from fixing security holes and bugs that prevent the software from compiling to feature and performance enhancements. A great number of the patches that initially drew this group together grew out of needs that each individual member had in making a Web server function. These patches were not due to any design or planning decisions by NCSA, McCool, or the assembled group, but most were useful enough that everyone gained from using them, because they fixed problems that everyone would or could encounter. As a result, the need for a coordinated new-httpd release was key to the group's work. This new version of NCSA httpd had no name initially, but *apache* was a persistent candidate; the somewhat apocryphal origin of the name is that it was "a patchy webserver."[17]

At the outset, in February and March 1995, the pace of work of the various members of new-httpd differed a great deal, but was in general extremely rapid. Even before there was an official release of a new httpd, process issues started to confront the group, as Roy Fielding later explained: "Apache began with a conscious attempt to solve the process issues first, before development even started, because it was clear from the very beginning that a geographically distributed set of volunteers, without any traditional organizational ties, would require a unique development process in order to make decisions."[18]

The need for process arose more or less organically, as the group developed mechanisms for managing the various patches: assigning them IDs, testing them, and incorporating them "by hand" into the main source-code base. As this happened, members of the list would occasionally find themselves lost, confused by the process or the efficiency of other members, as in this message from Andrew Wilson concerning Cliff Skolnick's management of the list of bugs:

> Cliff, can you concentrate on getting an uptodate copy of the bug/improvement list please. I've already lost track of just what the heck is meant to be going on. Also what's the status of this pre-pre-pre release Apache stuff. It's either a pre or it isn't surely? AND is the pre-pre-etc thing the same as the thing Cliff is meant to be working on?
>
> Just what the fsck is going on anyway? Ay, ay ay! Andrew Wilson.[19]

To which Rob Harthill replied, "It is getting messy. I still think we should all implement one patch at a time together. At the rate (and hours) some are working we can probably manage a couple of patches a day. . . . If this is acceptable to the rest of the group, I think we should order the patches, and start a systematic processes of discussion, implementations and testing."[20]

Some members found the pace of work exciting, while others appealed for slowing or stopping in order to take stock. Cliff Skolnick created a system for managing the patches and proposed that list-members vote in order to determine which patches be included.[21] Rob Harthill voted first.

> Here are my votes for the current patch list shown at
> http://www.hyperreal.com/httpd/patchgen/list.cgi
> I'll use a vote of
> -1 have a problem with it
> 0 haven't tested it yet (failed to understand it or whatever)
> +1 tried it, liked it, have no problem with it.
> [Here Harthill provides a list of votes on each patch.]
> If this voting scheme makes sense, lets use it to filter out the stuff we're happy with. A "-1" vote should veto any patch. There seems to be about 6 or 7 of us actively commenting on patches, so I'd suggest that once a patch gets a vote of +4 (with no vetos), we can add it to an alpha.[22]

Harthill's votes immediately instigated discussion about various patches, further voting, and discussion about the process (i.e., how many votes or vetoes were needed), all mixed together in a flurry of e-mail messages. The voting process was far from perfect, but it did allow some consensus on what "apache" would be, that is, which patches would be incorporated into an "official" (though not very public) release: Apache 0.2 on 18 March.[23] Without a voting system, the group of contributors could have gone on applying patches individually, each in his own context, fixing the problems that ailed each user, but ignoring those that were irrelevant or unnecessary in that context. With a voting process, however, a convergence on a tested and approved new-httpd could emerge. As the process was refined, members sought a volunteer to take votes, to open and close the voting once a week, and to build a new version of Apache when the voting was done. (Andrew Wilson was the first volunteer, to which Cliff Skolnick replied, "I guess the first vote is voting Andrew as the vote taker :-).")[24] The patch-and-vote process that emerged in the early stages of Apache was not entirely novel; many contributors noted that the FreeBSD project used a similar process, and some suggested the need for a "patch coordinator" and others worried that "using patches gets very ugly, very quickly."[25]

The significance of the patch-and-vote system was that it clearly represented the tension between the virtuosity of individual developers and a group process aimed at creating and maintaining a common piece of software. It was a way of balancing the ability of each separate individual's expertise against a common desire to ship and promote a stable, bug-free, public-domain Web server. As Roy Fielding and others would describe it in hindsight, this tension was part of Apache's advantage.

> Although the Apache Group makes decisions as a whole, all of the actual work of the project is done by individuals. The group does not write code, design solutions, document products, or provide support to our customers; individual people do that. The group provides an environment for collaboration and an excellent trial-by-fire for ideas and code, but the creative energy needed to solve a particular problem, redesign a piece of the system, or fix a given bug is almost always contributed by individual volunteers working on their own, for their own purposes, and not at the behest of the group. Competitors mistakenly assume Apache will be unable to take on new or unusual tasks because of the perception that we act as a group rather than follow a single leader. What they fail to see is that, by remaining open to new contributors, the group has an unlimited supply of innovative ideas, and it is the individuals who chose to pursue their own ideas who are the real driving force for innovation.[26]

Although openness is widely touted as the key to the innovations of Apache, the claim is somewhat disingenuous: patches are just that, patches. Any large-scale changes to the code could not be accomplished by applying patches, especially if each patch must be subjected to a relatively harsh vote to be included. The only way to make sweeping changes—especially changes that require iteration and testing to get right—is to engage in separate "branches" of a project or to differentiate between internal and external releases— in short, to fork the project temporarily in hopes that it would soon rejoin its stable parent. Apache encountered this problem very early on with the "Shambhala" rewrite of httpd by Robert Thau.

Shambhala was never quite official: Thau called it his "noodling" server, or a "garage" project. It started as his attempt to rewrite httpd as a server which could handle and process multiple requests at the same time. As an experiment, it was entirely his own project, which he occasionally referred to on the new-httpd list: "Still hacking Shambhala, and laying low until it works well enough to talk about."[27] By mid-June of 1995, he had a working version that he announced, quite modestly, to the list as "a garage project to explore some possible new directions I thought *might* be useful for the group to pursue."[28] Another list member, Randy Terbush, tried it out and gave it rave reviews, and by the end of June there were two users exclaiming its virtues. But since it hadn't ever really been officially identified as a fork, or an alternate development pathway, this led Rob Harthill to ask: "So what's the situation regarding Shambhala and Apache, are those of you who have switched to it giving up on Apache and this project? If so, do you need a separate list to discuss Shambhala?"[29]

Harthill had assumed that the NCSA code-base was "tried and tested" and that Shambhala represented a split, a fork: "The question is, should we all go in one direction, continue as things stand or Shambahla [sic] goes off on its own?"[30] His query drew out the miscommunication in detail: that Thau had planned it as a "drop-in" replacement for the NCSA httpd, and that his intentions were to make it the core of the Apache server, if he could get it to work. Harthill, who had spent no small amount of time working hard at patching the existing server code, was not pleased, and made the core issues explicit.

Maybe it was rst's [Robert Thau's] choice of phrases, such as "garage project" and it having a different name, maybe I didn't read his mailings thoroughly enough, maybe they weren't explicit enough, whatever. . . . It's a shame that nobody using Shambhala (who must have realized what was going on) didn't raise these issues weeks ago. I can only presume that rst was too modest to push Shambhala, or at least discussion of it, onto us more vigourously. I remember saying words to the effect of "this is what I plan to do, stop me if you think this isn't a good idea." Why the hell didn't anyone say something? . . . [D]id others get the same impression about rst's work as I did? Come on people, if you want to be part of this group, collaborate![31]

Harthill's injunction to collaborate seems surprising in the context of a mailing list and project created to facilitate collaboration, but the injunction is specific: collaborate by making plans and sharing goals. Implicit in his words is the tension between a project with clear plans and goals, an overarching design to which everyone contributes, as opposed to a group platform without clear goals that provides individuals with a setting to try out alternatives. Implicit in his words is the spectrum between debugging an existing piece of software with a stable identity and rewriting the fundamental aspects of it to make it something new. The meaning of collaboration bifurcates here: on the one hand, the privileging of the autonomous work of individuals which is submitted to a group peer review and then incorporated; on the other, the privileging of a set of shared goals to which the actions and labor of individuals is subordinated.[32]

Indeed, the very design of Shambhala reflects the former approach of privileging individual work: like UNIX and EMACS before it, Shambhala was designed as a modular system, one that could "make some of that process [the patch-and-vote process] obsolete, by allowing stuff which is not universally applicable (e.g., database back-ends), controversial, or just half-baked, to be shipped anyway as optional modules."[33] Such a design separates the core platform from the individual experiments that are conducted on it, rather than creating a design that is modular in the hierarchical sense of each contributor working on an assigned section of a project. Undoubtedly, the core platform requires coordination, but extensions and modifications can happen without needing to transform the whole project.[34] Shambhala represents a certain triumph of the "shut up and show me the code" aesthetic: Thau's "modesty" is instead a recognition that he should be quiet until it "works well enough to talk about," whereas Harthill's response is frustration that no one has talked about what Thau was planning to do before it was even attempted. The consequence was that Harthill's work seemed to be in vain, replaced by the work of a more virtuosic hacker's demonstration of a superior direction.

In the case of Apache one can see how coordination in Free Software is not just an afterthought or a necessary feature of distributed work, but is in fact at the core of software production itself, governing the norms and forms of life that determine what will count as good software, how it will progress with respect to a context and background, and how people will be expected to interact around the topic of design decisions. The privileging of adaptability brings with it a choice in the mode of collaboration: it resolves the tension between the agonistic competitive creation of software, such as Robert Thau's creation of Shambhala, and the need for collective coordination of complexity, such as Harthill's plea for collaboration to reduce duplicated or unnecessary work.

Check Out and Commit

The technical and social forms that Linux and Apache take are enabled by the tools they build and use, from bug-tracking tools and mailing lists to the Web servers and kernels themselves. One such tool plays a very special role in the emergence of these organizations: Source Code Management systems (SCMs). SCMs are tools for coordinating people and code; they allow multiple people in dispersed locales to work simultaneously on the same object, the same source code, without the need for a central coordinating overseer and without the risk of stepping on each other's toes. The history of SCMs—especially in the case of Linux—also illustrates the recursive-depth problem: namely, is Free Software still free if it is created with non-free tools?

SCM tools, like the Concurrent Versioning System (cvs) and Subversion, have become extremely common tools for Free Software programmers; indeed, it is rare to find a project, even a project conducted by only one individual, which does not make use of these tools. Their basic function is to allow two or more programmers to work on the same files at the same time and to provide feedback on where their edits conflict. When the number of programmers grows large, an SCM can become a tool for managing complexity. It keeps track of who has "checked out" files; it enables users to lock files if they want to ensure that no one else makes changes at the same time; it can keep track of and display the conflicting changes made by two users to the same file; it can be used to create "internal" forks or "branches" that may be incompatible with each other, but still allows programmers to try out new things and, if all goes well, merge the branches into the trunk later on. In sophisticated forms it can be used to "animate" successive changes to a piece of code, in order to visualize its evolution.

Beyond mere coordination functions, SCMs are also used as a form of distribution; generally SCMs allow anyone to check out the code, but restrict those who can check in or "commit" the code. The result is that users can get instant access to the most up-to-date version of a piece of software, and programmers can differentiate between stable releases, which have few bugs, and "unstable" or experimental versions that are under construction and will need the help of users willing to test and debug the latest versions. SCM tools automate certain aspects of coordination, not only reducing the labor involved but opening up new possibilities for coordination.

The genealogy of SCMs can be seen in the example of Ken Thompson's creation of a diff tape, which he used to distribute changes that had been contributed to UNIX. Where Thompson saw UNIX as a spectrum of changes and the legal department at Bell Labs saw a series of versions, SCM tools combine these two approaches by minutely managing the revisions, assigning each change (each diff) a new version number, and storing the history of all of those changes so that software changes might be precisely undone in order to discover which changes cause problems. Written by Douglas McIlroy, "diff" is itself a piece of software, one of the famed small UNIX tools that do one thing well. The program diff compares two files, line by line, and prints out the differences between them in a structured format (showing a series of lines with codes that indicate changes, additions, or removals). Given two versions of a text, one could run diff to find the differences and make the appropriate changes to synchronize them, a task that is otherwise tedious and, given the exactitude of source code, prone to human error. A useful side-effect of diff (when combined with an editor like ed or EMACS) is that when someone makes a set of changes to a file and runs diff on both the original and the changed file, the output (i.e., the changes only) can be used to reconstruct the original file from the changed file. Diff thus allows for a clever, space-saving way to save all the changes ever made to a file, rather than retaining full copies of every new version, one saves only the changes. Ergo, version control. diff—and programs like it—became the basis for managing the complexity of large numbers of programmers working on the same text at the same time.

One of the first attempts to formalize version control was Walter Tichy's Revision Control System (RCS), from 1985.[35] RCS kept track of the changes to different files using diff and allowed programmers to see all of the changes that had been made to that file. RCS, however, could not really tell the difference between the work of one programmer and another. All changes were equal, in that sense, and any questions that might arise about why a change was made could remain unanswered.

In order to add sophistication to RCS, Dick Grune, at the Vrije Universiteit, Amsterdam, began writing scripts that used RCS as a multi-user, Internet-accessible version-control system, a system that eventually became the Concurrent Versioning System. cvs allowed multiple users to check out a copy, make changes, and then commit those changes, and it would check for and either prevent or flag conflicting changes. Ultimately, cvs became most useful when programmers could use it remotely to check out source code from anywhere on the Internet. It allowed people to work at different speeds, different times, and in different places, without needing a central person in charge of checking and comparing the changes. cvs created a form of decentralized version control for very-large-scale collaboration; developers could work offline on software, and always on the most updated version, yet still be working on the same object.

Both the Apache project and the Linux kernel project use SCMs. In the case of Apache the original patch-and-vote system quickly began to strain the patience, time, and energy of participants as the number of contributors and patches began to grow. From the very beginning of the project, the contributor Paul Richards had urged the group to make use of cvs. He had extensive experience with the system in the Free-BSD project and was convinced that it provided a superior alternative to the patch-and-vote system. Few other contributors had much experience with it, however, so it wasn't until over a year after Richards began his admonitions that cvs was eventually adopted. However, cvs is not a simple replacement for a patch-and-vote system; it necessitates a different kind of organization. Richards recognized the trade-off. The patch-and-vote system created a very high level of quality assurance and peer review of the patches that people submitted, while the cvs system allowed individuals to make more changes that might not meet the same level of quality assurance. The cvs system allowed branches—stable, testing, experimental—with different levels of quality assurance, while the patch-and-vote system was inherently directed at one final and stable version. As the case of Shambhala exhibited, under the patch-and-vote system experimental versions would remain unofficial garage projects, rather than serve as official branches with people responsible for committing changes.

While SCMs are in general good for managing conflicting changes, they can do so only up to a point. To allow *anyone* to commit a change, however, could result in a chaotic mess, just as difficult to disentangle as it would be without an SCM. In practice, therefore, most projects designate a handful of people as having the right to "commit" changes. The Apache project retained its voting scheme, for instance, but it became a way of voting for "committers" instead for patches themselves. Trusted committers—those with the mysterious "good taste," or technical intuition—became the core members of the group.

The Linux kernel has also struggled with various issues surrounding SCMs and the management of responsibility they imply. The story of the so-called VGER tree and the creation of a new SCM called Bitkeeper is exemplary in this respect.[36] By 1997, Linux developers had begun to use cvs to manage changes to the source code, though not without resistance. Torvalds was still in charge of the changes to the official stable tree, but as other "lieutenants" came on board, the complexity of the changes to the kernel grew. One such lieutenant was Dave Miller, who maintained a "mirror" of the stable Linux kernel tree, the VGER tree, on a server at Rutgers. In September 1998 a fight broke out among Linux kernel developers over two related issues: one, the fact that Torvalds was failing to incorporate (patch) contributions that had been forwarded to him by various people, including his lieutenants; and two, as a result, the VGER cvs repository was no longer in synch with the stable tree maintained by Torvalds. Two different versions of Linux threatened to emerge.

A great deal of yelling ensued, as nicely captured in Moody's *Rebel Code*, culminating in the famous phrase, uttered by Larry McVoy: "Linus does not scale." The meaning of this phrase is that the ability of Linux to grow into an ever larger project with increasing complexity, one which can handle myriad uses and functions (to "scale" up), is constrained by the fact that there is only one Linus Torvalds. By all accounts, Linus was and is excellent at what he does—but there is only one Linus. The danger of this situation is the danger of a fork. A fork would mean one or more new versions would proliferate under new leadership, a situation much like the spread of UNIX. Both the licenses and the SCMs are designed to facilitate this, but only as a last resort. Forking also implies dilution and confusion—competing versions of the same thing and potentially unmanageable incompatibilities.

The fork never happened, however, but only because Linus went on vacation, returning renewed and ready to continue and to be more responsive. But the crisis had been real, and it drove developers into considering new modes of coordination. Larry McVoy offered to create a new form of SCM, one that would allow a much more flexible response to the problem that the VGER tree represented. However, his proposed solution, called Bitkeeper, would create far more controversy than the one that precipitated it.

McVoy was well-known in geek circles before Linux. In the late stages of the open-systems era, as an employee of Sun, he had penned an important document called "The Sourceware Operating System Proposal." It was an internal Sun Microsystems document that argued for the company to make its version of UNIX freely available. It was a last-ditch effort to save the dream of open systems. It was also the first such proposition within a company to "go open source," much like the documents that would urge Netscape to Open Source its software in 1998. Despite this early commitment, McVoy chose *not* to create Bitkeeper as a Free Software project, but to make it quasi-proprietary, a decision that raised a very central question in ideological terms: can one, or should one, create Free Software using non-free tools?

On one side of this controversy, naturally, was Richard Stallman and those sharing his vision of Free Software. On the other were pragmatists like Torvalds claiming no goals and no commitment to "ideology"—only a commitment to "fun." The tension laid bare the way in which recursive publics negotiate and modulate the core components of Free Software from within. Torvalds made a very strong and vocal statement concerning this issue, responding to Stallman's criticisms about the use of non-free software to create Free Software: "Quite frankly, I don't _want_ people using Linux for ideological reasons. I think ideology sucks. This world would be a much better place if people had less ideology, and a whole lot more 'I do this because it's FUN and because others might find it useful, not because I got religion.'"[37]

Torvalds emphasizes pragmatism in terms of *coordination*: the right tool for the job is the right tool for the job. In terms of *licenses,* however, such pragmatism does not play, and Torvalds has always been strongly committed to the GPL, refusing to let non-GPL software into the kernel. This strategic pragmatism is in fact a recognition of where experimental changes might be proposed, and where practices are settled. The GPL was a stable document, sharing source code widely was a stable practice, but coordinating a project using SCMs was, during this period, still in flux, and thus Bitkeeper was a tool well worth using so long as it remained suitable to Linux development. Torvalds was experimenting with the meaning of coordination: could a non-free tool be used to create Free Software?

McVoy, on the other hand, was on thin ice. He was experimenting with the meaning of Free Software licenses. He created three separate licenses for Bitkeeper in an attempt to play both sides: a commercial license for paying customers, a license for people who sell Bitkeeper, and a license for "free users." The free-user license allowed Linux developers to use the software for free—though it required them to use the latest version—and prohibited them from working on a competing project at the same time. McVoy's attempt to have his cake and eat it, too, created enormous tension in the developer community, a tension that built from 2002, when Torvalds began using Bitkeeper in earnest, to 2005, when he announced he would stop.

The tension came from two sources: the first was debates among developers addressing the moral question of using non-free software to create Free Software. The moral question, as ever, was also a technical one, as the second source of tension, the license restrictions, would reveal.

The developer Andrew Trigdell, well known for his work on a project called Samba and his reverse engineering of a Microsoft networking protocol, began a project to reverse engineer Bitkeeper by looking at the metadata it produced in the course of being used for the Linux project. By doing so, he crossed a line set up by McVoy's experimental licensing arrangement: the "free as long as you don't copy me" license. Lawyers advised Trigdell to stay silent on the topic while Torvalds publicly berated him for "willful destruction" and a moral lapse of character in trying to reverse engineer Bitkeeper. Bruce Perens defended Trigdell and censured Torvalds for his seemingly contradictory ethics.[38] McVoy never sued Trigdell, and Bitkeeper has limped along as a commercial project, because, much like the EMACS controversy of 1985, the Bitkeeper controversy of 2005 ended with Torvalds simply deciding to create his own SCM, called git.

The story of the VGER tree and Bitkeeper illustrate common tensions within recursive publics, specifically, the depth of the meaning of *free*. On the one hand, there is Linux itself, an exemplary Free Software project made freely available; on the other hand, however, there is the ability to contribute to this process, a process that is potentially constrained by the use of Bitkeeper. So long as the function of Bitkeeper is completely circumscribed—that is, completely planned—there can be no problem. However, the moment one user sees a way to change or improve the process, and not just the kernel itself, then the restrictions and constraints of Bitkeeper can come into play. While it is not clear that Bitkeeper actually prevented anything, it is also clear that developers clearly recognized it as a potential drag on a generalized commitment to adaptability. Or to put it in terms of recursive publics, only one *layer* is properly open, that of the kernel itself; the layer beneath it, the process of its construction, is not free in the same sense. It is ironic that Torvalds—otherwise the spokesperson for antiplanning and adaptability—willingly adopted this form of constraint, but not at all surprising that it was collectively rejected.

The Bitkeeper controversy can be understood as a kind of experiment, a modulation on the one hand of the kinds of acceptable licenses (by McVoy) and on the other of acceptable forms of coordination (Torvalds's decision to use Bitkeeper). The experiment was a failure, but a productive one, as it identified one kind of nonfree software that is not safe to use in Free Software development: the SCM that coordinates the people and the code they contribute. In terms of recursive publics the experiment identified the proper depth of recursion. Although it might be possible to create Free Software using some kinds of non-free tools, SCMs are not among them; both the software created and the software used to create it need to be free.[39]

The Bitkeeper controversy illustrates again that adaptability is not about radical invention, but about critique and response. Whereas controlled design and hierarchical planning represent the domain of governance—control through goal-setting and orientation of a collective or a project—adaptability privileges politics, properly speaking, the ability to critique existing design and to propose alternatives without restriction. The tension between goal-setting and adaptability is also part of the dominant ideology of intellectual property. According to this ideology, IP laws promote invention of new products and ideas, but restrict the re-use or transformation of existing ones; defining where novelty begins is a core test of the law. McVoy made this tension explicit in his justifications for Bitkeeper: "Richard [Stallman] might want to consider the fact that developing *new* software is extremely expensive. He's very proud of the collection of free software, but that's a collection of re-implementations, but no profoundly new ideas or products. . . . What if the free software model simply can't support the costs of developing new ideas?"[40]

Novelty, both in the case of Linux and in intellectual property law more generally, is directly related to the interplay of social and technical coordination: goal direction vs. adaptability. The ideal of adaptability promoted by Torvalds suggests a radical alternative to the dominant ideology of creation embedded in contemporary intellectual-property systems. If Linux is "new," it is new through adaptation and the coordination of large numbers of creative contributors who challenge the "design" of an operating system from the bottom up, not from the top down. By contrast, McVoy represents a moral imagination of design in which it is impossible to achieve novelty without extremely expensive investment in top-down, goal-directed, *unpolitical* design—and it is this activity that the intellectual-property system is designed to reward. Both are engaged, however, in an experiment; both are engaged in "figuring out" what the limits of Free Software are.

Coordination Is Design

Many popular accounts of Free Software skip quickly over the details of its mechanism to suggest that it is somehow inevitable or obvious that Free Software should work—a self-organizing, emergent system that manages complexity through distributed contributions by hundreds of thousands of people. In *The Success of Open Source* Steven Weber points out that when people refer to Open Source as a self-organizing system, they usually mean something more like "I don't understand how it works."[41]

Eric Raymond, for instance, suggests that Free Software is essentially the emergent, self-organizing result of "collaborative debugging": "Given enough eyeballs, all bugs are shallow."[42] The phrase implies that the core success of Free Software is the distributed, isolated, labor of debugging, and that design and planning happen elsewhere (when a developer "scratches an itch" or responds to a personal need). On the surface, such a distinction seems quite obvious: designing is designing, and debugging is removing bugs from software, and presto!—Free Software. At the extreme end, it is an understanding by which only individual geniuses are capable of planning and design, and if the initial conditions are properly set, then collective wisdom will fill in the details.

However, the actual practice and meaning of collective or collaborative debugging is incredibly elastic. Sometimes debugging means fixing an error; sometimes it means making the software do something different or new. (A common joke, often made at Microsoft's expense, captures some of this elasticity: whenever something doesn't seem to work right, one says, "That's a feature, not a bug.") Some programmers see a design decision as a stupid mistake and take action to correct it, whereas others simply learn to use the software as designed. Debugging can mean something as simple as reading someone else's code and helping them understand why it does not work; it can mean finding bugs in someone else's software; it can mean reliably reproducing bugs; it can mean pinpointing the cause of the bug in the source code; it can mean changing the source to eliminate the bug; or it can, at the limit, mean changing or even re-creating the software to make it do something different or better.[43] For academics, debugging can be a way to build a career: "Find bug. Write paper. Fix bug. Write paper. Repeat."[44] For commercial software vendors, by contrast, debugging is part of a battery of tests intended to streamline a product.

Coordination in Free Software is about adaptability over planning. It is a way of resolving the tension between individual virtuosity in creation and the social benefit in shared labor. If all software were created, maintained, and distributed only by individuals, coordination would be superfluous, and software would indeed be part of the domain of poetry. But even the paradigmatic cases of virtuosic creation—EMACS by Richard Stallman, UNIX by Ken Thompson and Dennis Ritchie—clearly represent the need for creative forms

of coordination and the fundamental practice of reusing, reworking, rewriting, and imitation. UNIX was not created de novo, but
was an attempt to streamline and rewrite Multics, itself a system
that evolved out of Project MAC and the early mists of time-sharing
and computer hacking.[45] EMACS was a reworking of the TECO
editor. Both examples are useful for understanding the evolution of
modes of coordination and the spectrum of design and debugging.

UNIX was initially ported and shared through mixed academic
and commercial means, through the active participation of computer scientists who both received updates and contributed fixes
back to Thompson and Ritchie. No formal system existed to manage this process. When Thompson speaks of his understanding of
UNIX as a "spectrum" and not as a series of releases (V1, V2, etc.),
the implication is that work on UNIX was continuous, both within
Bell Labs and among its widespread users. Thompson's use of the
diff tape encapsulates the core problem of coordination: how to collect and redistribute the changes made to the system by its users.

Similarly, Bill Joy's distribution of BSD and James Gosling's
distribution of GOSMACS were both ad hoc, noncorporate experiments in "releasing early and often." These distribution schemes
had a purpose (beyond satisfying demand for the software). The
frequent distribution of patches, fixes, and extensions eased the
pain of debugging software and satisfied users' demands for new
features and extensions (by allowing them to do both themselves).
Had Thompson and Ritchie followed the conventional corporate
model of software production, they would have been held responsible for thoroughly debugging and testing the software they distributed, and AT&T or Bell Labs would have been responsible for
coming up with all innovations and extensions as well, based on
marketing and product research. Such an approach would have
sacrificed adaptability in favor of planning. But Thompson's and
Ritchie's model was different: both the extension and the debugging of software became shared responsibilities of the users and
the developers. Stallman's creation of EMACS followed a similar
pattern; since EMACS was by design extensible and intended to
satisfy myriad unforeseen needs, the responsibility rested on the
users to address those needs, and sharing their extensions and fixes
had obvious social benefit.

The ability to see development of software as a spectrum implies
more than just continuous work on a product; it means seeing the

product itself as something fluid, built out of previous ideas and products and transforming, differentiating into new ones. Debugging, from this perspective, is not separate from design. Both are part of a spectrum of changes and improvements whose goals and direction are governed by the users and developers themselves, and the patterns of coordination they adopt. It is in the space between debugging and design that Free Software finds its niche.

Conclusion: Experiments and Modulations

Coordination is a key component of Free Software, and is frequently identified as the central component. Free Software is the result of a complicated story of experimentation and construction, and the forms that coordination takes in Free Software are specific outcomes of this longer story. Apache and Linux are both experiments— not scientific experiments per se but collective social experiments in which there are complex technologies and legal tools, systems of coordination and governance, and moral and technical orders already present.

Free Software is an experimental system, a practice that changes with the results of new experiments. The privileging of adaptability makes it a peculiar kind of experiment, however, one not directed by goals, plans, or hierarchical control, but more like what John Dewey suggested throughout his work: the experimental praxis of science extended to the social organization of governance in the service of improving the conditions of freedom. What gives this experimentation significance is the centrality of Free Software—and specifically of Linux and Apache—to the experimental expansion of the Internet. As an infrastructure or a milieu, the Internet is changing the conditions of social organization, changing the relationship of knowledge to power, and changing the orientation of collective life toward governance. Free Software is, arguably, the best example of an attempt to make this transformation public, to ensure that it uses the advantages of adaptability as critique to counter the power of planning as control. Free Software, as a recursive public, proceeds by proposing *and providing* alternatives. It is a bit like Kant's version of enlightenment: insofar as geeks speak (or hack) as *scholars*, in a public realm, they have a right to propose criticisms and changes of any sort; as soon as they relinquish that commitment, they become private employees or servants of the sovereign, bound by conscience and power to carry out the duties of their given office. The constitution of a public realm is not a universal activity, however, but a historically specific one: Free Software confronts the specific contemporary technical and legal infrastructure by which it is possible to propose criticisms and offer alternatives. What results is a recursive public filled not only with individuals who govern their own actions but also with code and concepts and licenses and forms of coordination that turn these actions into viable, concrete technical forms of life useful to inhabitants of the present.

PART III MODULATIONS

The question cannot be answered by argument. Experimental method means experiment, and the question can be answered only by trying, by organized effort. The reasons for making the trial are not abstract or recondite. They are found in the confusion, uncertainty and conflict that mark the modern world. . . . The task is to go on, and not backward, until the method of intelligence and experimental control is the rule in social relations and social direction.—JOHN DEWEY, *Liberalism and Social Action*

"If We Succeed, We Will Disappear" 8.

In early 2002, after years of reading and learning about Open Source and Free Software, I finally had a chance to have dinner with famed libertarian, gun-toting, Open Source–founding impresario Eric Raymond, author of *The Cathedral and the Bazaar* and other amateur anthropological musings on the subject of Free Software. He had come to Houston, to Rice University, to give a talk at the behest of the Computer and Information Technology Institute (CITI). Visions of a mortal confrontation between two anthropologists-manqué filled my head. I imagined explaining point by point why his references to self-organization and evolutionary psychology were misguided, and how the long tradition of economic anthropology contradicted basically everything he had to say about gift-exchange. Alas, two things conspired against this epic, if bathetic, showdown.

First, there was the fact that (as so often happens in meetings among geeks) there was only one woman present at dinner; she was young, perhaps unmarried, but not a student—an interested female hacker. Raymond seated himself beside this woman, turned toward her, and with a few one-minute-long exceptions proceeded to lavish her with all of his available attention. The second reason was that I was seated next to Richard Baraniuk and Brent Hendricks. All at once, Raymond looked like the past of Free Software, arguing the same arguments, using the same rhetoric of his online publications, while Baraniuk and Hendricks looked like its future, posing questions about the transformation—the *modulation*—of Free Software into something surprising and new.

Baraniuk, a professor of electrical engineering and a specialist in digital signal processing, and Hendricks, an accomplished programmer, had started a project called Connexions, an "open content repository of educational materials." Far more interesting to me than Raymond's amateur philosophizing was this extant project to extend the ideas of Free Software to the creation of educational materials—textbooks, in particular.

Rich and Brent were, by the looks of it, equally excited to be seated next to me, perhaps because I was answering their questions, whereas Raymond was not, or perhaps because I was a new hire at Rice University, which meant we could talk seriously about collaboration. Rich and Brent (and Jan Odegard, who, as director of CITI, had organized the dinner) were keen to know what I could add to help them understand the "social" aspects of what they wanted to do with Connexions, and I, in return, was equally eager to learn how they conceptualized their Free Software–like project: what had they kept the same and what had they changed in their own experiment? Whatever they meant by "social" (and sometimes it meant ethical, sometimes legal, sometimes cultural, and so on), they were clear that there were domains of expertise in which they felt comfortable (programming, project management, teaching, and a particular kind of research in computer science and electrical engineering) and domains in which they did not (the "norms" of academic life outside their disciplines, intellectual-property law, "culture"). Although I tried to explain the nature of my own expertise in social theory, philosophy, history, and ethnographic research, the academic distinctions were far less important than the fact that I could ask detailed and pointed questions about the project, questions that indicated to them that I must have some kind of stake in the domains that they needed filled—in particular, around the question of whether Connexions was the same thing as Free Software, and what the implications of that might be.

Raymond courted and chattered on, then left, the event of his talk and dinner of fading significance, but over the following weeks, as I caught up with Brent and Rich, I became (surprisingly quickly) part of their novel experiment.

After Free Software

My nonmeeting with Raymond is an allegory of sorts: an allegory of what comes after Free Software. The excitement around that table was not so much about Free Software or Open Source, but about a certain possibility, a kind of genotypic urge of which Free Software seemed a fossil phenotype and Connexions a live one. Rich and Brent were people in the midst of figuring something out. They were engaged in modulating the practices of Free Software. By *modulation* I mean exploring in detail the concrete practices— the how—of Free Software in order to ask what can be changed, and what cannot, in order to maintain something (openness?) that no one can quite put his finger on. What drew me immediately to Connexions was that it was related to Free Software, not meta-phorically or ideologically, but concretely, practically, and experi-mentally, a relationship that was more about *emergence out of* than it was about the *reproduction of* forms. But the opposition between emergence and reproduction immediately poses a question, not un-like that of the identity of species in evolution: if Free Software is no longer software, what exactly is it?

In part III I confront this question directly. Indeed, it was this question that necessitated part II, the analytic decomposition of the practices and histories of Free Software. In order to answer the question "Is Connexions Free Software?" (or vice versa) it was necessary to rethink Free Software as itself a collective, technical experiment, rather than as an expression of any ideology or culture. To answer yes, or no, however, merely begs the question "What is Free Software?" What is the *cultural significance* of these practices? The concept of a recursive public is meant to reveal in part the significance of both Free Software and emergent projects like Con-nexions; it is meant to help chart when these emergent projects branch off absolutely (cease to be public) and when they do not, by focusing on how they modulate the five components that give Free Software its contemporary identity.

Connexions modulates all of the components except that of the movement (there is, as of yet, no real "Free Textbook" movement, but the "Open Access" movement is a close second cousin).[1] Perhaps the most complex modulation concerns coordination—changes to the practice of coordination and collaboration in academic-textbook creation in particular, and more generally to the nature of collaboration and coordination of knowledge in science and scholarship generally.

Connexions emerged out of Free Software, and not, as one might expect, out of education, textbook writing, distance education, or any of those areas that are topically connected to pedagogy. That is to say, the people involved did not come to their project by attempting to deal with a problem salient to education and teaching as much as they did so through the problems raised by Free Software and the question of how those problems apply to university textbooks. Similarly, a second project, Creative Commons, also emerged out of a direct engagement with and exploration of Free Software, and not out of any legal movement or scholarly commitment to the critique of intellectual-property law or, more important, out of any desire to transform the entertainment industry. Both projects are resolutely committed to experimenting with the given practices of Free Software—to testing their limits and changing them where they can—and this is what makes them vibrant, risky, and potentially illuminating as cases of a recursive public.

While both initiatives are concerned with conventional subject areas (educational materials and cultural productions), they enter the fray orthogonally, armed with anxiety about the social and moral order in which they live, and an urge to transform it by modulating Free Software. This binds such projects across substantive domains, in that they are forced to be oppositional, not because they want to be (the movement comes last), but because they enter the domains of education and the culture industry as outsiders. They are in many ways intuitively troubled by the existing state of affairs, and their organizations, tools, legal licenses, and movements are seen as alternative imaginations of social order, especially concerning creative freedom and the continued existence of a commons of scholarly knowledge. To the extent that these projects remain in an orthogonal relationship, they are making a recursive public appear—something the textbook industry and the entertainment industry are, by contrast, not at all interested in doing, for obvious financial and political reasons.

Stories of Connexion

I'm at dinner again. This time, a windowless hotel conference room in the basement maybe, or perhaps high up in the air. Lawyers, academics, activists, policy experts, and foundation people are semi-excitedly working their way through the hotel's steam-table fare. I'm trying to tell a story to the assembled group—a story that I have heard Rich Baraniuk tell a hundred times—but I'm screwing it up. Rich always gets enthusiastic stares of wonder, light-bulbs going off everywhere, a subvocalized "Aha!" or a vigorous nod. I, on the other hand, am clearly making it too complicated. Faces and foreheads are squirmed up into lines of failed comprehension, people stare at the gravy-sodden food they're soldiering through, weighing the option of taking another bite against listening to me complicate an already complicated world. I wouldn't be doing this, except that Rich is on a plane, or in a taxi, delayed by snow or engineers or perhaps at an eponymous hotel in another city. Meanwhile, our co-organizer Laurie Racine, has somehow convinced herself that I have the childlike enthusiasm necessary to channel Rich. I'm flattered, but unconvinced. After about twenty minutes, so is she, and as I try to answer a question, she stops me and interjects, "Rich really needs to be here. He should really be telling this story."

Miraculously, he shows up and, before he can even say hello, is conscripted into telling his story properly. I sigh in relief and pray that I've not done any irreparable damage and that I can go back to my role as straight man. I can let the superaltern speak for himself. The downside of participant observation is being asked to participate in what one had hoped first of all to observe. I do know the story—I have heard it a hundred times. But somehow what I hear, ears tuned to academic questions and marveling at some of the stranger claims he makes, somehow this is not the ear for enlightenment that his practiced and boyish charm delivers to those hearing it for the first time; it is instead an ear tuned to questions of why: why this project? Why now? And even, somewhat convo-lutedly, why are people so fascinated when he tells the story? How could I tell it like Rich?

Rich is an engineer, in particular, a specialist in Digital Signal Processing (DSP). DSP is the science of signals. It is in everything, says Rich: your cell phones, your cars, your CD players, all those devices. It is a mathematical discipline, but it is also an intensely practical one, and it's connected to all kinds of neighboring fields of knowledge. It is the kind of discipline that can connect calculus, bioinformatics, physics, and music. The statistical and analytical techniques come from all sorts of research and end up in all kinds of interesting devices. So Rich often finds himself trying to teach students to make these kinds of connections—to understand that a Fourier transform is not just another chapter in calculus but a tool for manipulating signals, whether in bioinformatics or in music.

Around 1998 or 1999, Rich decided that it was time for him to write a textbook on DSP, and he went to the dean of engineering, Sidney Burris, to tell him about the idea. Burris, who is also a DSP man and longtime member of the Rice University community, said something like, "Rich, why don't you do something useful?" By which he meant: there are a hundred DSP textbooks out there, so why do you want to write the hundred and first? Burris encouraged Rich to do something bigger, something ambitious enough to put Rice on the map. I mention this because it is important to note that even a university like Rice, with a faculty and graduate students on par with the major engineering universities of the country, per-ceives that it gets no respect. Burris was, and remains, an inveterate supporter of Connexions, precisely because it might put Rice "in the history books" for having invented something truly novel.

At about the same time as his idea for a textbook, Rich's research group was switching over to Linux, and Rich was first learning about Open Source and the emergence of a fully free operating system created entirely by volunteers. It isn't clear what Rich's aha! moment was, other than simply when he came to an understand-ing that such a thing as Linux was actually possible. Nonetheless, at some point, Rich had the idea that his textbook could be an Open Source textbook, that is, a textbook created not just by him, but by DSP researchers all over the world, and made available to everyone to make use of and modify and improve as they saw fit, just like Linux. Together with Brent Hendricks, Yan David Erlich, and Ross Reedstrom, all of whom, as geeks, had a deep familiarity with the history and practices of Free and Open Source Software, Rich started to conceptualize a system; they started to think about modulations of different components of Free and Open Source Software. The idea of a Free Software textbook repository slowly took shape.

Thus, Connexions: an "open content repository of high-quality educational materials." These "textbooks" very quickly evolved into something else: "modules" of content, something that has never been sharply defined, but which corresponds more or less to a small chunk of teachable information, like two or three pages in a textbook. Such modules are much easier to conceive of in sciences like mathematics or biology, in which textbooks are often multiauthored collections, finely divided into short chapters with diagrams, exercises, theorems, or programs. Modules lend themselves much less well to a model of humanities or social-science scholarship based in reading texts, discussion, critique, and comparison—and this bias is a clear reflection of what Brent, Ross, and Rich knew best in terms of teaching and writing. Indeed, the project's frequent recourse to the image of an assembly-line model of knowledge production often confirms the worst fears of humanists and educators when they first encounter Connexions. The image suggests that knowledge comes in prepackaged and colorfully branded tidbits for the delectation of undergrads, rather than characterizing knowledge as a state of being or as a process.

The factory image (figure 7) is a bit misleading. Rich's and Brent's imaginations are in fact much broader, which shows whenever they demo the project, or give a talk, or chat at a party about it. Part of my failure to communicate excitement when I tell the story of Connexions is that I skip the examples, which is where Rich starts: what if, he says, you are a student taking Calculus 101 and, at the same time, Intro to Signals and Systems—no one is going to explain to you how Fourier transforms form a bridge, or connection, between them. "Our brains aren't organized in linear, chapter-by-chapter ways," Rich always says, "so why are our textbooks?" How can we give students a way to see the connection between statistics and genetics, between architecture and biology, between intellectual-property law and chemical engineering? Rich is always looking for new examples: a music class for kids that uses information from physics, or vice versa, for instance. Rich's great hope is that the

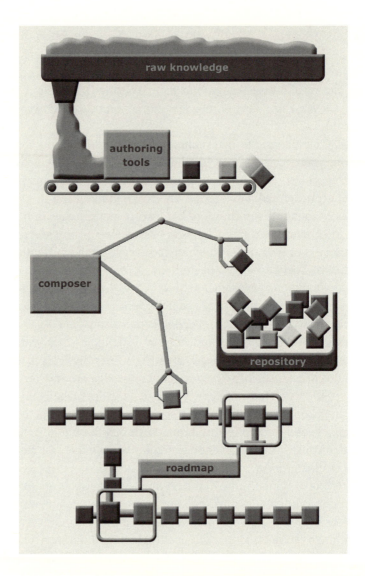

7. The Connexions textbook as a factory. Illustration by Jenn Drummond, Ross Reedstrom, Max Starkenberg, and others, 1999–2004. Used with permission.

more modules there are in the Connexions commons, the more fantastic and fascinating might be the possibilities for such novel—and natural—connections.

Free Software—and, in particular, Open Source in the guise of "self-organizing" distributed systems of coordination—provide a particular promise of meeting the challenges of teaching and learning that Rich thinks we face. Rich's commitment is not to a certain kind of pedagogical practice, but to the "social" or "community" benefits of thousands of people working "together" on a textbook. Indeed, Connexions did not emerge out of education or educational technology; it was not aligned with any particular theory of learning (though Rich eventually developed a rhetoric of linked, networked, connected knowledge—hence the name Connexions—that he uses often to sell the project). There is no school of education at Rice, nor a particular constituency for such a project (teacher-training programs, say, or administrative requirements for online education). What makes Rich's sell even harder is that the project emerged at about the same time as the high-profile failure of dot-com bubble–fueled schemes to expand university education into online education, distance education, and other systems of expanding the paying student body without actually inviting them onto campus. The largest of these failed experiments by far was the project at Columbia, which had reached the stage of implementation at the time the bubble burst in 2000.[2]

Thus, Rich styled Connexions as more than just a factory of knowledge—it would be a community or culture developing richly associative and novel kinds of textbooks—and as much more than just distance education. Indeed, Connexions was not the only such project busy differentiating itself from the perceived dangers of distance education. In April 2001 MIT had announced that it would make the content of all of its courses available for free online in a project strategically called OpenCourseWare (OCW). Such news could only bring attention to MIT, which explicitly positioned the announcement as a kind of final death blow to the idea of distance education, by saying that what students pay $35,000 and up for per year is not "knowledge"—which is free—but the experience of being at MIT. The announcement created pure profit from the perspective of MIT's reputation as a generator and disseminator of scientific knowledge, but the project did not emerge directly out of an interest in mimicking the success of Open Source. That angle was provided ultimately by the computer-science professor Hal Abelson, whose deep understanding of the history and growth of Free Software came from his direct involvement in it as a long-standing member of the computer-science community at MIT. OCW emerged most proximately from the strange result of a committee report, commissioned by the provost, on how MIT should position itself in the "distance/e-learning" field. The surprising response: don't do it, give the content away and add value to the campus teaching and research experience instead.[3]

OCW, Connexions, and distance learning, therefore, while all ostensibly interested in combining education with the networks and software, emerged out of different demands and different places. While the profit-driven demand of distance learning fueled many attempts around the country, it stalled in the case of OCW, largely because the final MIT Council on Educational Technology report that recommended OCW was issued at the same time as the first plunge in the stock market (April 2000). Such issues were not a core factor in the development of Connexions, which is not to say that the problems of funding and sustainability have not always been important concerns, only that genesis of the project was not at the administrative level or due to concerns about distance education. For Rich, Brent, and Ross the core commitment was to openness and to the success of Open Source as an experiment with massive, distributed, Internet-based, collaborative production of software—their commitment to this has been, from the beginning, completely and adamantly unwavering. Neverthless, the project has involved modulations of the core features of Free Software. Such modulations depend, to a certain extent, on being a project that emerges out of the ideas and practices of Free Software, rather than, as in the case of OCW, one founded as a result of conflicting goals (profit and academic freedom) and resulting in a strategic use of public relations to increase the symbolic power of the university over its fiscal growth.

When Rich recounts the story of Connexions, though, he doesn't mention any of this background. Instead, like a good storyteller, he waits for the questions to pose themselves and lets his demonstration answer them. Usually someone asks, "How is Connexions different from OCW?" And, every time, Rich says something similar: Connexions is about "communities," about changing the way scholars collaborate and create knowledge, whereas OCW is simply an attempt to transfer existing courses to a Web format in order to make the content of those courses widely available. Connexions is a radical experiment in the collaborative creation of educational materials, one that builds on the insights of Open Source and that actually encompasses the OCW project. In retrospective terms, it is clear that OCW was interested only in modulating the meaning of source code and the legal license, whereas Connexions seeks also to modulate the practice of coordination, with respect to academic textbooks.

Rich's story of the origin of Connexions usually segues into a demonstration of the system, in which he outlines the various technical, legal, and educational concepts that distinguish it. Connexions uses a standardized document format, the eXtensible Mark-up Language (XML), and a Creative Commons copyright license on every module; the Creative Commons license allows people not only to copy and distribute the information but to modify it and even to use it for commercial gain (an issue that causes repeated discussion among the team members). The material ranges from detailed explanations of DSP concepts (naturally) to K-12 music education (the most popular set of modules). Some contributors have added entire courses; others have created a few modules here and there. Contributors can set up workgroups to manage the creation of modules, and they can invite other users to join. Connexions uses a version-control system so that all of the changes are recorded; thus, if a module used in one class is changed, the person using it for another class can continue to use the older version if they wish. The number of detailed and clever solutions embodied in the system never ceases to thoroughly impress anyone who takes the time to look at it.

But what always animates people is the idea of random and flexible connection, the idea that a textbook author might be able to build on the work of hundreds of others who have already contributed, to create new classes, new modules, and creative connections between them, or surprising juxtapositions—from the biologist teaching a class on bioinformatics who needs to remind students of certain parts of calculus without requiring a whole course; to the architect who wants a studio to study biological form, not necessarily in order to do experiments in biology, but to understand buildings differently; to the music teacher who wants students to understand just enough physics to get the concepts of pitch and timbre; to or the physicist who needs a concrete example for the explanation of waves and oscillation.

The idea of such radical recombinations is shocking for some (more often for humanities and social-science scholars, rather than scientists or engineers, for reasons that clearly have to do with an ideology of authentic and individualized creative ability). The questions that result—regarding copyright, plagiarism, control, unauthorized use, misuse, misconstrual, misreading, defamation, and so on—generally emerge with surprising force and speed. If Rich were trying to sell a version of "distance learning," skepticism and suspicion would quickly overwhelm the project; but as it is, Connexions inverts almost all of the expectations people have developed about textbooks, classroom practice, collaboration, and copyright. More often than not people leave the discussion converted—no doubt helped along by Rich's storytelling gift.

Modulations: From Free Software to Connexions

Connexions surprises people for some of the same reasons as Free Software surprises people, emerging, as it does, directly out of the same practices and the same components. Free Software provides a template made up of the five components: shared source code, a concept of openness, copyleft licenses, forms of coordination, and a movement or ideology. Connexions starts with the idea of modulating a shared "source code," one that is not software, but educational textbook modules that academics will share, port, and fork. The experiment that results has implications for the other four components as well. The implications lead to new questions, new constraints, and new ideas.

The modulation of source code introduces a specific and potentially confusing difference from Free Software projects: Connexions is *both* a conventional Free Software project *and* an unconventional experiment based on Free Software. There is, of course, plenty of normal source code, that is, a number of software components that need to be combined in order to allow the creation of digital documents (the modules) and to display, store, transmit, archive, and measure the creation of modules. The creation and management of this software is expected to function more or less like all Free Software projects: it is licensed using Free Software licenses, it is built on open standards of various kinds, and it is set up to take contributions from other users and developers. The software system for managing modules is itself built on a variety of other Free Software components (and a commitment to using only Free Software). Connexions has created various components, which are either released like conventional Free Software or contributed to another Free Software project. The economy of contribution and release is a complex one; issues of support and maintenance, as well as of reputation and recognition, figure into each decision. Others are invited to contribute, just as they are invited to contribute to any Free Software project.[4]

At the same time, there is "content," the ubiquitous term for digital creations that are not software. The creation of content modules, on the other hand (which the software system makes technically possible), is intended to function *like* a Free Software project, in which, for instance, a group of engineering professors might get together to collaborate on pieces of a textbook on DSP. The Connexions project does not encompass or initiate such collaborations, but, rather, proceeds from the assumption that such activity is already happening and that Connexions can provide a kind of alternative platform—an alternative *infrastructure* even—which textbook-writing academics can make use of instead of the current infrastructure of publishing. The current infrastructure and technical model of textbook writing, this implies, is one that both prevents people from taking advantage of the Open Source model of collaborative development and makes academic work "non-free." The shared objects of content are not source code that can be compiled, like source code in C, but documents marked up with XML and filled with "educational" content, then "displayed" either on paper or onscreen.

The modulated meaning of *source code* creates all kinds of new questions—specifically with respect to the other four components. In terms of openness, for instance, Connexions modulates this component very little; most of the actors involved are devoted to the ideals of open systems and open standards, insofar as it is a Free Software project of a conventional type. It builds on UNIX (Linux) and the Internet, and the project leaders maintain a nearly fanatical devotion to openness at every level: applications, programming languages, standards, protocols, mark-up languages, interface tools. Every place where there is an open (as opposed to a proprietary) solution—that choice trumps all others (with one note-worthy exception).[5] James Boyle recently stated it well: "Wherever possible, design the system to run with open content, on open protocols, to be potentially available to the largest possible number of users, and to accept the widest possible range of experimental modifications from users who can themselves determine the development of the technology."[6]

With respect to content, the devotion to openness is nearly identical, because conventional textbook publishers "lock in" customers (students) through the creation of new editions and useless "enhanced" content, which jacks up prices and makes it difficult for educators to customize their own courses. "Openness" in this sense trades on the same reasoning as it did in the 1980s: the most important aspect of the project is the information people create, and any proprietary system locks up content and prevents people from taking it elsewhere or using it in a different context.

Indeed, so firm is the commitment to openness that Rich and Brent often say something like, "If we are successful, we will disappear." They do not want to become a famous online textbook publisher; they want to become a famous publishing *infrastructure*. Being radically open means that any other competitor can use your system—but it means they are *using your system*, and this is the goal. Being open means not only sharing the "source code" (content and modules), but devising ways to ensure the perpetual openness of that content, that is, to create a recursive public devoted to the maintenance and modifiability of the medium or infrastructure by which it communicates. Openness trumps "sustainability" (i.e., the self-perpetuation of the financial feasibility of a particular organization), and where it fails to, the commitment to openness has been compromised.

The commitment to openness and the modulation of the meaning of source code thus create implications for the meaning of Free Software licenses: do such licenses cover this kind of content? Are new licenses necessary? What should they look like? Connexions was by no means the first project to stimulate questions about the applicability of Free Software licenses to texts and documents. In the case of EMACS and the GPL, for example, Richard Stallman had faced the problem of licensing the manual at the same time as the source code for the editor. Indeed, such issues would ultimately result in a GNU Free Documentation License intended narrowly to

"IF WE SUCCEED, WE WILL DISAPPEAR"

cover software manuals. Stallman, due to his concern, had clashed during the 1990s with Tim O'Reilly, publisher and head of O'Reilly Press, which had long produced books and manuals for Free Software programs. O'Reilly argued that the principles reflected in Free Software licenses should not be applied to instructional books, because such books provided a service, a way for more people to learn how to use Free Software, and in turn created a larger audience. Stallman argued the opposite: manuals, just like the software they served, needed to be freely modifiable to remain useful.

By the late 1990s, after Free Software and Open Source had been splashed across the headlines of the mainstream media, a number of attempts to create licenses modeled on Free Software, but applicable to other things, were under way. One of the earliest and most general was the Open Content License, written by the educational-technology researcher David Wiley. Wiley's license was intended for use on any kind of content. Content could include text, digital photos, movies, music, and so on. Such a license raises new issues. For example, can one designate some parts of a text as "invariant" in order to prevent them from being changed, while allowing other parts of the text to be changed (the model eventually adopted by the GNU Free Documentation License)? What might the relationship between the "original" and the modified version be? Can one expect the original author to simply incorporate suggested changes? What kinds of forking are possible? Where do the "moral rights" of an author come into play (regarding the "integrity" of a work)?

At the same time, the modulation of source code to include academic textbooks has extremely complex implications for the meaning and context of coordination: scholars do not write textbooks like programmers write code, so should they coordinate in the same ways? Coordination of a textbook or a course in Connexions requires novel experiments in textbook writing. Does it lend itself to academic styles of work, and in which disciplines, for what kinds of projects? In order to cash in on the promise of distributed, collaborative creation, it would be necessary to find ways to coordinate scholars.

So, when Rich and Brent recognized in me, at dinner, someone who might know how to think about these issues, they were acknowledging that the experiment they had started had created a certain turbulence in their understanding of Free Software and, in turn, a need to examine the kinds of legal, cultural, and social practices that would be at stake.[7]

Modulations: From Connexions to Creative Commons

I'm standing in a parking lot in 100 degree heat and 90 percent humidity. It is spring in Houston. I am looking for my car, and I cannot find it. James Boyle, author of *Shamans, Software, and Spleens* and distinguished professor of law at Duke University, is standing near me, staring at me, wearing a wool suit, sweating and watching me search for my car under the blazing sun. His look says simply, "If I don't disembowel you with my Palm Pilot stylus, I am going to relish telling this humiliating story to your friends at every opportunity I can." Boyle is a patient man, with the kind of arch Scottish humor that can make you feel like his best friend, even as his stories of the folly of man unfold with perfect comic pitch and turn out to be about you. Having laughed my way through many an uproarious tale of the foibles of my fellow creatures, I am aware that I have just taken a seat among them in Boyle's theater of human weakness. I repeatedly press the panic button on my key chain, in the hopes that I am near enough to my car that it will erupt in a frenzy of honking and flashing that will end the humiliation.

The day had started well. Boyle had folded himself into my Volkswagen (he is tall), and we had driven to campus, parked the car in what no doubt felt like a memorable space at 9 A.M., and happily gone to the scheduled meeting—only to find that it had been mistakenly scheduled for the following day. Not my fault, though now, certainly, my problem. The ostensible purpose of Boyle's visit was to meet the Connexions team and learn about what they were doing. Boyle had proposed the visit himself, as he was planning to pass through Houston anyway. I had intended to pester him with questions about the politics and possibilities of licensing the content in Connexions and with comparisons to MIT's OCW and other such commons projects that Boyle knew of.

Instead of attending the meeting, I took him back to my office, where I learned more about why he was interested in Connexions. Boyle's interest was not entirely altruistic (nor was it designed to spend valuable quarter hours standing in a scorched parking lot as I looked for my subcompact car). What interested Boyle was find-

"IF WE SUCCEED, WE WILL DISAPPEAR"

ing a constituency of potential users for Creative Commons, the nonprofit organization he was establishing with Larry Lessig, Hal Abelson, Michael Carroll, Eric Eldred, and others—largely because he recognized the need for a ready constituency in order to make Creative Commons work. The constituency was needed both to give the project legitimacy and to allow its founders to understand what exactly was needed, legally speaking, for the creation of a whole new set of Free Software-like licenses.

Creative Commons, as an organization and as a movement, had been building for several years. In some ways, Creative Commons represented a simple modulation of the Free Software license: a broadening of the license's concept to cover other types of content. But the impetus behind it was not simply a desire to copy and extend Free Software. Rather, all of the people involved in Creative Commons were those who had been troubling issues of intellectual property, information technology, and notions of commons, public domains, and freedom of information for many years. Boyle had made his name with a book on the construction of the information society by its legal (especially intellectual property) structures. Eldred was a publisher of public-domain works and the lead plaintiff in a court case that went to the Supreme Court in 2002 to determine whether the recent extension of copyright term limits was constitutional. Abelson was a computer scientist with an active interest in issues of privacy, freedom, and law "on the electronic frontier." And Larry Lessig was originally interested in constitutional law, a clerk for Judge Richard Posner, and a self-styled cyberlaw scholar, who was, during the 1990s, a driving force for the explosion of interest in cyberlaw, much of it carried out at the Berkman Center for Internet and Society at Harvard University.

With the exception of Abelson—who, in addition to being a famous computer scientist, worked for years in the same building that Richard Stallman camped out in and chaired the committee that wrote the report recommending OCW—none of the members of Creative Commons cut their teeth on Free Software projects (they were lawyers and activists, primarily) and yet the emergence of Open Source into the public limelight in 1998 was an event that made more or less instant and intuitive sense to all of them. During this time, Lessig and members of the Berkman Center began an "open law" project designed to mimic the Internet-based collaboration of the Open Source project among lawyers who might want to contribute to the Eldred case. Creative Commons was thus built as much on a commitment to a notion of collaborative creation—the use of the Internet especially—but more generally on the ability of individuals to work together to create new things, and especially to coordinate the creation of these things by the use of novel licensing agreements.

Creative Commons provided more than licenses, though. It was part of a social imaginary of a moral and technical order that extended beyond software to include creation of all kinds; notions of technical and moral freedom to make use of one's own "culture" became more and more prominent as Larry Lessig became more and more involved in struggles with the entertainment industry over the "control of culture." But for Lessig, Creative Commons was a fall-back option; the direct route to a transformation of the legal structure of intellectual property was through the Eldred case, a case that built huge momentum throughout 2001 and 2002, was granted cert by the Supreme Court, and was heard in October of 2002. One of the things that made the case remarkable was the series of strange bedfellows it produced; among the economists and lawyers supporting the repeal of the 1998 "Sonny Bono" Copyright Term Extension Act were the arch free-marketeers and Nobel Prize winners Milton Friedman, James Buchanan, Kenneth Arrow, Ronald Coase, and George Akerlof. As Boyle pointed out in print, conservatives and liberals and libertarians all have reasons to be in favor of scaling back copyright expansion.[8] Lessig and his team lost the case, and the Supreme Court essentially affirmed Congress's interpretation of the Constitution that "for limited times" meant only that the time period be limited, not that it be short.

Creative Commons was thus a back-door approach: if the laws could not be changed, then people should be given the tools they needed to work around those laws. Understanding how Creative Commons was conceived requires seeing it as a modulation of both the notion of "source code" and the modulation of "copyright licenses." But the modulations take place in that context of a changing legal system that was so unfamiliar to Stallman and his EMACS users, a legal system responding to new forms of software, networks, and devices. For instance, the changes to the Copyright Act of 1976 created an unintended effect that Creative Commons would ultimately seize on. By eliminating the requirement to register copyrighted works (essentially granting copyright as soon as the work is "fixed in a tangible medium"), the copyright law created a situation wherein there was no explicit way in which a work could be intentionally placed in the public domain. Practically speaking an author could declare that a work was in the public domain, but legally speaking the risk would be borne entirely by the person who sought to make use of that work: to copy it, transform it, sell it, and so on. With the explosion of interest in the Internet, the problem ramified exponentially; it became impossible to know whether someone who had placed a text, an image, a song, or a video online intended for others to make use of it—*even if the author explicitly declared it "in the public domain."* Creative Commons licenses were thus conceived and rhetorically positioned as tools for making explicit exactly what uses could be made of a specific work. They protected the rights of people who sought to make use of "culture" (i.e., materials and ideas and works they had not authored), an approach that Lessig often summed up by saying, "Culture always builds on the past."

The background to and context of the emergence of Creative Commons was of course much more complicated and fraught. Concerns ranged from the plights of university libraries with regard to high-priced journals, to the problem of documentary filmmakers unable to afford, or even find the owners of, rights to use images or snippets in films, to the high-profile fights over online music trading, Napster, and the RIAA. Over the course of four years, Lessig and the other founders of Creative Commons would address all of these issues in books, in countless talks and presentations and conferences around the world, online and off, among audiences ranging from software developers to entrepreneurs to musicians to bloggers to scientists.

Often, the argument for Creative Commons draws heavily on the concept of culture besieged by the content industries. A story which Lessig enjoys telling—one that I heard on several occasions when I saw him speak at conferences—was that of Mickey Mouse. An interesting, quasi-conspiratorial feature of the twentieth-century expansion of intellectual-property law is that term limits seem to have been extended right around the time Mickey Mouse was about to become public property. True or not, the point Lessig likes to make is that the Mouse is not the de novo creation of the mind of Walt Disney that intellectual-property law likes to pretend it is, but built on the past of culture, in particular, on Steamboat Willie,

Charlie Chaplin, Krazy Kat, and other such characters, some as inspiration, some as explicit material. The greatness in Disney's creation comes not from the mind of Disney, but from the culture from which it emerged. Lessig will often illustrate this in videos and images interspersed with black-typewriter-font–bestrewn slides and a machine-gun style that makes you think he's either a beat-poet manqué or running for office, or maybe both.

Other examples of intellectual-property issues fill the books and talks of Creative Commons advocates, stories of blocked innovation, stifled creativity, and—the scariest point of all (at least for economist-lawyers)—inefficiency due to over-expansive intellectual-property laws and overzealous corporate lawyer-hordes.[9] Lessig often preaches to the converted (at venues like South by Southwest Interactive and the O'Reilly Open Source conferences), and the audiences are always outraged at the state of affairs and eager to learn what they can do. Often, getting involved in the Creative Commons is the answer. Indeed, within a couple of years, Creative Commons quickly became more of a movement (a modulation of the Free/Open Source movement) than an experiment in writing licenses.

On that hot May day in 2002, however, Creative Commons was still under development. Later in the day, Boyle did get a chance to meet with the Connexions project team members. The Connexions team had already realized that in pursuing an experimental project in which Free Software was used as a template they created a need for new kinds of licenses. They had already approached the Rice University legal counsel, who, though well-meaning, were not grounded at all in a deep understanding of Free Software and were thus naturally suspicious of it. Boyle's presence and his detailed questions about the project were like a revelation—a revelation that there were already people out there thinking about the very problem the Connexions team faced and that the team would not need to solve the problem themselves or make the Rice University legal counsel write new open-content licenses. What Boyle offered was the possibility for Connexions, as well as for myself as intermediary, to be involved in the detailed planning and license writing that was under way at Creative Commons. At the same time, it gave Creative Commons an extremely willing "early-adopter" for the license, and one from an important corner of the world: scholarly research and teaching.[10] My task, after recovering from the shame of being unable to find my car, was to organize a workshop in August at which members of Creative Commons, Connexions, MIT's OCW, and any other such projects would be invited to talk about license issues.

Participant Figuring Out

The workshop I organized in August 2002 was intended to allow Creative Commons, Connexions, and MIT's OCW project to try to articulate what each might want from the other. It was clear what Creative Commons wanted: to convince as many people as possible to use their licenses. But what Connexions and OCW might have wanted, from each other as well as from Creative Commons, was less clear. Given the different goals and trajectories of the two projects, their needs for the licenses differed in substantial ways—enough so that the very idea of using the same license was, at least temporarily, rendered impossible by MIT. While OCW was primarily concerned about obtaining permissions to place existing copyrighted work on the Web, Connexions was more concerned about ensuring that new work remain available and modifiable.

In retrospect, this workshop clarified the novel questions and problems that emerged from the process of modulating the components of Free Software for different domains, different kinds of content, and different practices of collaboration and sharing. Since then, my own involvement in this activity has been aimed at resolving some of these issues in accordance with an imagination of openness, an imagination of social order, that I had learned from my long experience with geeks, and not from my putative expertise as an anthropologist or a science-studies scholar. The fiction that I had at first adopted—that I was bringing scholarly knowledge to the table—became harder and harder to maintain the more I realized that it was my understanding of Free Software, gained through ongoing years of ethnographic apprenticeship, that was driving my involvement.

Indeed, the research I describe here was just barely undertaken as a research project. I could not have conceived of it as a fundable activity in advance of discovering it; I could not have imagined the course of events in any of the necessary detail to write a proper proposal for research. Instead, it was an outgrowth of thinking and participating that was already under way, participation that was driven largely by intuition and a feeling for the problem represented by Free Software. I wanted to help figure something out. I wanted to see how "figuring out" happens. While I could have organized a fundable research project in which I picked a mature Free Software project, articulated a number of questions, and spent time answering them among this group, such a project would not have answered the questions I was trying to form at the time: what is happening to Free Software as it spreads beyond the world of hackers and software? How is it being modulated? What kinds of limits are breached when software is no longer the central component? What other domains of thought and practice were or are "readied" to receive and understand Free Software and its implications?[11]

My experience—my participant-observation—with Creative Commons was therefore primarily done as an intermediary between the Connexions project (and, by implication, similar projects under way elsewhere) and Creative Commons with respect to the writing of licenses. In many ways this detailed, specific practice was the most challenging and illuminating aspect of my participation, but in retrospect it was something of a red herring. It was not only the modulation of the meaning of source code and of legal licenses that differentiated these projects, but, more important, the meaning of collaboration, reuse, coordination, and the cultural practice of sharing and building on knowledge that posed the trickiest of the problems.

My contact at Creative Commons was not James Boyle or Larry Lessig, but Glenn Otis Brown, the executive director of that organization (as of summer 2002). I first met Glenn over the phone, as I tried to explain to him what Connexions was about and why he should join us in Houston in August to discuss licensing issues related to scholarly material. Convincing him to come to Texas was an easier sell than explaining Connexions (given my penchant for complicating it unnecessarily), as Glenn was an Austin native who had been educated at the University of Texas before heading off to Harvard Law School and its corrupting influence at the hands of Lessig, Charlie Nesson, and John Perry Barlow.

Glenn galvanized the project. With his background as a lawyer, and especially his keen interest in intellectual-property law, and his long-standing love of music of all kinds Glenn lent incredible enthusiasm to his work. Prior to joining Creative Commons, he had

"IF WE SUCCEED, WE WILL DISAPPEAR"

clerked for the Hon. Stanley Marcus on the Eleventh Circuit Court of Appeals, in Miami, where he worked on the so-called Wind Done Gone case.[12] His participation in the workshop was an experiment of his own; he was working on a story that he would tell countless times and which would become one of the core examples of the kind of practice Creative Commons wanted to encourage.

A *New York Times* story describes how the band the White Stripes had allowed Steven McDonald, the bassist from Redd Kross, to lay a bass track onto the songs that made up the album *White Blood Cells.* In a line that would eventually become a kind of mantra for Creative Commons, the article stated: "Mr. McDonald began putting these copyrighted songs online without permission from the White Stripes or their record label; during the project, he bumped into Jack White, who gave him spoken assent to continue. It can be that easy when you skip the intermediaries."[13] The ease with which these two rockers could collaborate to create a modified work (called, of course, *Redd Blood Cells*) without entering a studio, or, more salient, a law firm, was emblematic of the notion that "culture builds on the past" and that it need not be difficult to do so.

Glenn told the story with obvious and animated enthusiasm, ending with the assertion that the White Stripes didn't have to give up all their rights to do this, but they didn't have to keep them all either; instead of "All Rights Reserved," he suggested, they could say "Some Rights Reserved." The story not only manages to capture the message and aims of Creative Commons, but is also a nice indication of the kind of dual role that Glenn played, first as a lawyer, and second as a kind of marketing genius and message man. The possibility of there being more than a handful of people like Glenn around was not lost on anyone, and his ability to switch between the language of law and that of nonprofit populist marketing was phenomenal.[14]

At the workshop, participants had a chance to hash out a number of different issues related to the creation of licenses that would be appropriate to scholarly content: questions of attribution and commercial use, modification and warranty; differences between federal copyright law concerning licenses and state law concerning commercial contracts. The starting point for most people was Free Software, but this was not the only starting point. There were at least two other broad threads that fed into the discussion and into the general understanding of the state of affairs facing projects like

Connexions or OCW. The first thread was that of digital libraries, hypertext, human-computer interaction research, and educational technology. These disciplines and projects often make common reference to two pioneers, Douglas Englebart and Theodore Nelson, and more proximately to things like Apple's HyperCard program and a variety of experiments in personal academic computing. The debates and history that lead up to the possibility of Connexions are complex and detailed, but they generally lack attention to legal detail. With the exception of a handful of people in library and information science who have made "digital" copyright into a subspecialty, few such projects, over the last twenty-five years, have made the effort to understand, much less incorporate, issues of intellectual property into their purview.

The other thread combines a number of more scholarly interests that come out of the disciplines of economics and legal theory: institutional economics, critical legal realism, law and economics—these are the scholastic designations. Boyle and Lessig, for example, are both academics; Boyle does not practice law, and Lessig has tried few cases. Nonetheless, they are both inheritors of a legal and philosophical pragmatism in which value is measured by the transformation of policy and politics, not by the mere extension or specification of conceptual issues. Although both have penned a large number of complicated theoretical articles (and Boyle is well known in several academic fields for his book *Shamans, Software, and Spleens* and his work on authorship and the law), neither, I suspect, would ever sacrifice the chance to make a set of concrete changes in legal or political practice given the choice. This point was driven home for me in a conversation I had with Boyle and others at dinner on the night of the launch of Creative Commons, in December 2002. During that conversation, Boyle said something to the effect of, "We actually *made* something; we didn't just sit around writing articles and talking about the dangers that face us—we made something." He was referring as much to the organization as to the legal licenses they had created, and in this sense Boyle qualifies very much as a polymathic geek whose understanding of technology is that it is an intervention into an already constituted state of affairs, one that demonstrates its value by being created and installed, not by being assessed in the court of scholarly opinions.

Similarly, Lessig's approach to writing and speaking is unabashedly aimed at transforming the way people approach intellectual-property law and, even more generally, the way they understand the relationship between their rights and their culture.[15] Lessig's approach, at a scholarly level, is steeped in the teachings of law and economics (although, as he has playfully pointed out, a "second" Chicago school) but is focused more on the understanding and manipulation of norms and customs ("culture") than on law narrowly conceived.[16]

Informing both thinkers is a somewhat heterodox economic consensus drawn primarily from institutional economics, which is routinely used to make policy arguments about the efficacy or efficiency of the intellectual-property system. Both are also informed by an emerging consensus on treating the public domain in the same manner in which environmentalists treated the environment in the 1960s.[17] These approaches begin with long-standing academic and policy concerns about the status and nature of "public goods," not directly with the problem of Free Software or the Internet. In some ways, the concern with public goods, commons, the public domain, and collective action are part of the same "reorientation of power and knowledge" I identify throughout *Two Bits*: namely, the legitimation of the media of knowledge creation, communication, and circulation. Most scholars of institutional economics and public policy are, however, just as surprised and bewildered by the fact of Free Software as the rest of the world has been, and they have sought to square the existing understanding of public goods and collective action with this new phenomenon.[18]

All of these threads form the weft of the experiment to modulate the components of Free Software to create different licenses that cover a broader range of objects and that deal with people and organizations that are not software developers. Rather than attempt to carry on arguments at the level of theory, however, my aim in participating was to see how and what was argued in practice by the people constructing these experiments, to observe what constraints, arguments, surprises, or bafflements emerged in the course of thinking through the creation of both new licenses and a new form of authorship of scholarly material. Like those who study "science in action" or the distinction between "law on the books" and "law in action," I sought to observe the realities of a practice heavily determined by textual and epistemological frameworks of various sorts.[19]

In my years with Connexions I eventually came to see it as something in between a natural experiment and a thought experiment: it was conducted in the open, and it invited participation from working scholars and teachers (a natural experiment, in that it was not a closed, scholarly endeavor aimed at establishing specific results, but an essentially unbounded, functioning system that people could and would come to depend on), and yet it proceeded by making a series of strategic guesses (a thought experiment) about three related things: (1) what it is (and will be) possible to do technically; (2) what it is (and will be) possible to do legally; and (3) what scholars and educators have done and now do in the normal course of their activities.

At the same time, this experiment gave shape to certain legal questions that I channeled in the direction of Creative Commons, issues that ranged from technical questions about the structure of digital documents, requirements of attribution, and URLs to questions about moral rights, rights of disavowal, and the meaning of "modification." The story of the interplay between Connexions and Creative Commons was, for me, a lesson in a particular mode of legal thinking which has been described in more scholarly terms as the difference between the Roman or, more proximately, the Napoleonic tradition of legal rationalism and the Anglo-American common-law tradition.[20] It was a practical experience of what exactly the difference is between legal code and software code, with respect to how those two things can be made flexible or responsive.

Reuse, Modification, and the Nonexistence of Norms 9.

The Connexions project was an experiment in modulating the practices of Free Software. It was not inspired by so much as it was based on a kind of template drawn from the experience of people who had some experience with Free Software, including myself. But how exactly do such templates get used? What is traced and what is changed? In terms of the cultural significance of Free Software, what are the implications of these changes? Do they maintain the orientation of a recursive public, or are they attempts to apply Free Software for other private concerns? And if they are successful, what are the implications for the domains they affect: education, scholarship, scientific knowledge, and cultural production? What effects do these changes have on the norms of work and the meaning and shape of knowledge in these domains?

In this chapter I explore in ethnographic detail how the modulations of Free Software undertaken by Connexions and Creative Commons are related to the problems of reuse, modification, and the norms of scholarly production. I present these two projects as responses to the contemporary reorientation of knowledge and power; they are recursive publics just as Free Software is, but they expand the domain of practice in new directions, that is, into the scholarly world of textbooks and research and into the legal domains of cultural production more generally.

In the course of "figuring out" what they are doing, these two projects encounter a surprising phenomenon: the changing meaning of the *finality* of a scholarly or creative work. Finality is not certainty. While certainty is a problematic that is well and often studied in the philosophy of science and in science studies, finality is less so. What makes a work stay a work? What makes a fact stay a fact? How does something, certain or not, achieve stability and identity? Such finality, the very paradigm of which is the published book, implies stability. But Connexions and Creative Commons, through their experiments with Free Software, confront the problem of how to stabilize a work in an unstable context: that of shareable source code, an open Internet, copyleft licenses, and new forms of coordination and collaboration.[1] The meaning of finality will have important effects on the ability to constitute a politics around any given work, whether a work of art or a work of scholarship and science. The actors in Creative Commons and Connexions realize this, and they therefore form yet another instance of a recursive public, precisely because they seek ways to define the meaning of finality publicly and openly—and to make modifiability an irreversible aspect of the process of stabilizing knowledge.

The modulations of Free Software performed by Connexions and Creative Commons reveal two significant issues. The first is the troublesome matter of the meaning of *reuse*, as in the reuse of concepts, ideas, writings, articles, papers, books, and so on for the creation of new objects of knowledge. Just as software source code can be shared, ported, and forked to create new versions with new functions, and just as software and people can be coordinated in new ways using the Internet, so too can scholarly and scientific content. I explore the implications of this comparison in this chapter. The central gambit of both Connexions and Creative Commons (and much of scientific practice generally) is that new work builds on previous work. In the sciences the notion that science is cumulative is not at issue, but exactly how scientific knowledge accumulates is far from clear. Even if "standing on the shoulders of giants" can be revealed to hide machinations, secret dealings, and Machiavellian maneuvering of the most craven sort, the very concept of cumulative knowledge is sound. Building a fact, a result, a machine, or a theory out of other, previous works—this kind of reuse as progress is not in question. But the actual material practice of writing, publication, and the reuse of other results and works is something that, until very recently, has been hidden from view, or has been so naturalized that the norms of practice are nearly invisible to practitioners themselves.

This raises the other central concern of this chapter: that of the existence or nonexistence of norms. For an anthropologist to query whether or not norms exist might seem to theorize oneself out of a job; one definition of anthropology is, after all, the making explicit of cultural norms. But the turn to "practices" in anthropology and science studies has in part been a turn away from "norms" in their classic sociological and specifically Mertonian fashion. Robert Merton's suggestion that science has been governed by norms—disinterestedness, communalism, organized skepticism, objectivity—has been repeatedly and roundly criticized by a generation of scholars in the sociology of scientific knowledge who note that even if such norms are asserted by actors, they are often subverted in the doing.[2] But a striking thing has happened recently; those Mertonian norms of science have in fact become the more or less explicit *goals in practice* of scientists, engineers, and geeks in the wake of Free Software. If Mertonian norms do not exist, then they are being invented. This, of course, raises novel questions: can one *create* norms? What exactly would this mean? How are norms different from culture or from legal and technical constraints? Both Connexions and Creative Commons explicitly pose this question and search for ways to identify, change, or work with norms as they understand them, in the context of reuse.

Whiteboards: What Was Publication?

More than once, I have found myself in a room with Rich Baraniuk and Brent Hendricks and any number of other employees of the

Connexions project, staring at a whiteboard on which a number of issues and notes have been scrawled. Usually, the notes have a kind of palimpsestic quality, on account of the array of previous conversations that are already there, rewritten in tiny precise script in a corner, or just barely erased beneath our discussion. These conversations are often precipitated by a series of questions that Brent, Ross Reedstrom, and the development team have encountered as they build and refine the system. They are never simple questions. A visitor staring at the whiteboard might catch a glimpse of the peculiar madness that afflicts the project: a mixture of legal terms, technical terms, and terms like *scholarly culture* or *DSP communities*. I'm consulted whenever this mixture of terms starts to worry the developers in terms of legality, culture, or the relationship between the two. I'm generally put in the position of speaking either as a lawyer (which, legally speaking, I am not supposed to do) or as an anthropologist (which I do mainly by virtue of holding a position in an anthropology department). Rarely are the things I say met with assent: Brent and Ross, like most hackers, are insanely well versed in the details of intellectual-property law, and they routinely correct me when I make bold but not-quite-true assertions about it. Nonetheless, they rarely feel well versed enough to make decisions about legal issues on their own, and often I have been called—on again as a thoughtful sounding board, and off again as intermediary with Creative Commons.

This process, I have come to realize, is about figuring something out. It is not just a question of solving technical problems to which I might have some specific domain knowledge. Figuring out is modulation; it is template-work. When Free Software functions as a template for projects like Connexions, it does so literally, by allowing us to trace a known form of practice (Free Software) onto a less well known, seemingly chaotic background and to see where the forms match up and where they do not. One very good way to understand what this means in a particular case—that is, to see more clearly the modulations that Connexions has performed—is to consider the practice and institution of scholarly publication through the template of Free Software.

Consider the ways scholars have understood the meaning and significance of print and publication in the past, prior to the Internet and the contemporary reorientation of knowledge and power. The list of ambitious historians and theorists of the relationship of media to knowledge is long: Lucien Febvre, Walter Ong, Marshall McLuhan, Jack Goody, Roger Chartier, Friedrich Kittler, Elizabeth Eisenstein, Adrian Johns, to name a few.[3] With the exception of Johns, however, the history of publication does not start with the conventional, legal, and formal practices of publication so much as it does with the material practices and structure of the media themselves, which is to say the mechanics and technology of the printed book.[4] Ong's theories of literacy and orality, Kittler's re-theorization of the structure of media evolution, Goody's anthropology of the media of accounting and writing—all are focused on the tangible media as the dependent variable of change. By contrast, Johns's *The Nature of the Book* uncovers the contours of the massive endeavor involved in making the book a reliable and robust form for the circulation of knowledge in the seventeenth century and after.

Prior to Johns's work, arguments about the relationship of print and power fell primarily into two camps: one could overestimate the role of print and the printing press by suggesting that the "fixity" of a text and the creation of multiple copies led automatically to the spread of ideas and the rise of enlightenment. Alternately, one could underestimate the role of the book by suggesting that it was merely a transparent media form with no more or less effect on the circulation or evaluation of ideas than manuscripts or television. Johns notes in particular the influence of Elizabeth Eisenstein's scholarship on the printing press (and Bruno Latour's dependence on this in turn), which very strongly identified the characteristics of the printed work with the cultural changes seen to follow, including the success of the scientific revolution and the experimental method.[5] For example, Eisenstein argued that fixity—the fact that a set of printed books can be exact copies of each other—implied various transformations in knowledge. Johns, however, is at pains to show just how unreliable texts are often perceived to be. From which sources do they come? Are they legitimate? Do they have the backing or support of scholars or the crown? In short, fixity can imply sound knowledge only if there is a system of evaluation already in place. Johns suggests a reversal of this now commonsense notion: "We may consider fixity not as an *inherent* quality, but as a *transitive* one. . . . We may adopt the principle that fixity exists only inasmuch as it is recognized and acted upon by people—and not otherwise. The consequence of this change in perspective is that print culture itself is immediately laid open to analysis. It becomes a *result* of manifold representations, practices and conflicts, rather than just the manifold *cause* with which we are often presented. In contrast to talk of a 'print logic' imposed on humanity, this approach allows us to recover the construction of different print cultures in particular historical circumstances."[6]

Johns's work focuses on the elaborate and difficult cultural, social, and economic work involved, in the sixteenth and seventeenth centuries, in transforming the European book into the kind of authority it is taken to be across the globe today. The creation and standardization not just of books but of a *publishing infrastructure* involved the kind of careful social engineering, reputation management, and skills of distinction, exclusion, and consensus that science studies has effectively explored in science and engineering. Hence, Johns focuses on "print-in-the-making" and the relationship of the print culture of that period to the reliability of knowledge. Instead of making broad claims for the transformation of knowledge by print (eerily similar in many respects to the broad claims made for the Internet), Johns explores the clash of representations and practices necessary to create the sense, in the twentieth century, that there really is or was only *one* print culture.

The problem of publication that Connexions confronts is thus not simply caused by the invention or spread of the Internet, much less that of Free Software. Rather, it is a confrontation with the problems of producing stability and finality under very different technical, legal, and social conditions—a problem more complex even than the "different print cultures in particular historical circumstances" that Johns speaks of in regard to the book. Connexions faces two challenges: that of figuring out the difference that today introduces with respect to yesterday, and that of creating or modifying an infrastructure in order to satisfy the demands of a properly authoritative knowledge. Connexions textbooks of necessity look different from conventional textbooks; they consist of digital documents, or "modules," that are strung together and made available through the Web, under a Creative Commons license that allows for free use, reuse, and modification. This version of "publication" clearly has implications for the meaning of authorship, ownership, stewardship, editing, validation, collaboration, and verification.

The conventional appearance of a book—in bookstores, through mail-order, in book clubs, libraries, or universities—was an event that signified, as the name suggests, its official *public* appearance in the world. Prior to this event, the text circulated only *privately*, which is to say only among the relatively small network of people who could make copies of it or who were involved in its writing, editing, proofreading, reviewing, typesetting, and so on. With the Internet, the same text can be made instantly available *at each of these stages* to just as many or more potential readers. It effectively turns the event of publication into a *notional* event—the click of a button—rather than a highly organized, material event. Although it is clear that the practice of publication has become denaturalized or destabilized by the appearance of new information technologies, this hardly implies that the work of stabilizing the meaning of publication—and producing authoritative knowledge as a result—has ceased. The tricky part comes in understanding how Free Software is used as a template by which the authority of publication in the Gutenberg Galaxy is being transformed into the authority of publication in the Turing Universe.

Publication in Connexions

In the case of Connexions there are roughly three stages to the creation of content. The first, temporally speaking, is whatever happens before Connexions is involved, that is, the familiar practices of what I would call *composition*, rather than simply writing. Some project must be already under way, perhaps started under the constraints of and in the era of the book, perhaps conceived as a digital textbook or an online textbook, but still, as of yet, written on paper or saved in a Word document or in LaTeX, on a scholar's desktop. It could be an individual project, as in the case of Rich's initial plan to write a DSP textbook, or it could be a large collaborative project to write a textbook.

The second stage is the one in which the document or set of documents is translated ("Connexified") into the mark-up system used by Connexions. Connexions uses the eXtensible Mark-up Language (XML), in particular a subset of tags that are appropriate to textbooks. These "semantic" tags (e.g., <term>) refer only to the meaning of the text they enclose, not to the "presentation" or syntactic look of what they enclose; they give the document the necessary structure it needs to be transformed in a number of creative ways. Because XML is related only to content, and not to presentation (it is sometimes referred to as "agnostic"), the same document in Connexions can be automatically made to look a number of different ways, as an onscreen presentation in a browser, as a pdf document, or as an on-demand published work that can be printed out as a book, complete with continuous page numbering, footnotes (instead of links), front and back matter, and an index. Therein lies much of Connexions's technical wizardry.

During the second stage, that of being marked up in XML, the document is not quite public, although it is on the Internet; it is in what is called a workgroup, where only those people with access to the particular workgroup (and those have been invited to collaborate) can see the document. It is only when the document is finished, ready to be distributed, that it will enter the third, "published" stage—the stage at which anyone on the Internet can ask for the XML document and the software will display it, using style sheets or software converters, as an HTML page, a pdf document for printing, or as a section of a larger course. However, publication does not here signify finality; indeed, one of the core advantages of Connexions is that the document is rendered less stable than the book-object it mimics: it can be updated, changed, corrected, deleted, copied, and so on, all without any of the rigmarole associated with changing a published book or article. Indeed, the very powerful notion of fixity theorized by McLuhan and Eisenstein is rendered moot here. The fact that a document has been printed (and printed as a *book*) no longer means that all copies will be the same; indeed, it may well change from hour to hour, depending on how many people contribute (as in the case of Free Software, which can go through revisions and updates as fast, or faster, than one can download and install new versions). With Wikipedia entries that are extremely politicized or active, for example, a "final" text is impossible, although the dynamics of revision and counter-revision do suggest outlines for the emergence of some kinds of stability. But Connexions differs from Wikipedia with respect to this finality as well, because of the insertion of the second stage, during which a self-defined group of people can work on a nonpublic text before committing changes that a public can see.

It should be clear, given the example of Connexions, or any similar project such as Wikipedia, that the changing meaning of "publication" in the era of the Internet has significant implications, both practical (they affect the way people can both write and publish their works) and legal (they fit uneasily into the categories established for previous media). The tangibility of a textbook is quite obviously transformed by these changes, but so too is the cultural significance of the practice of writing a textbook. And if textbooks are written differently, using new forms of collaboration and allowing novel kinds of transformation, then the validation, certification, and structure of authority of textbooks also change, inviting new forms of open and democratic participation in writing, teaching, and learning. No longer are all of the settled practices of authorship, collaboration, and publication configured around the same institutional and temporal scheme (e.g., the book and its publishing infrastructure). In a colloquial sense, this is obvious, for instance, to any musician today: recording and releasing a song to potentially millions of listeners is now technically possible for anyone, but how that fact changes the cultural significance of music creation is not yet clear. For most musicians, creating music hasn't changed much with the introduction of digital tools, since new recording and composition technologies largely mimic the recording practices that preceded them (for example, a program like Garage Band literally looks like a four-track recorder on the screen). Similarly, much of the practice of digital publication has been concerned with recreating something that looks like traditional publication.[7]

Perhaps unsurprisingly, the Connexions team spent a great deal of time at the outset of the project creating a pdf-document-creation system that would essentially mimic the creation of a conventional textbook, with the push of a button.[8] But even this process causes a subtle transformation: the concept of "edition" becomes much harder to track. While a conventional textbook is a stable entity that goes through a series of printings and editions, each of which is marked on its publication page, a Connexions document can go through as many versions as an author wants to make changes, all the while without necessarily changing editions. In this respect, the modulation of the concept of source code translates the practices of updating and "versioning" into the realm of textbook writing. Recall the cases ranging from the "continuum" of UNIX versions discussed by Ken Thompson to the complex struggles over version control in the Linux and Apache projects. In the case of writing source code, exactitude demands that the change of even a single character be tracked and labeled as a version change, whereas a conventional-textbook spelling correction or errata issuance would hardly create the need for a new edition.

In the Connexions repository all changes to a text are tracked and noted, but the identity of the module does not change. "Editions" have thus become "versions," whereas a substantially revised or changed module might require not reissuance but a forking of that module to create one with a new identity. Editions in publishing are not a feature of the medium per se; they are necessitated by the temporal and spatial practices of publication as an event, though this process is obviously made visible only in the book itself. In the same way, versioning is now used to manage a process, but it results in a very different configuration of the medium and the material available in that medium. Connexions traces the template of software production (sharing, porting, and forking and the norms and forms of coordination in Free Software) directly onto older forms of publication. Where the practices match, no change occurs, and where they don't, it is the reorientation of knowledge and power and the emergence of recursive publics that serves as a guide to the development of the system.

Legally speaking, the change from editions to versions and forks raises troubling questions about the boundaries and status of a copyrighted work. It is a peculiar feature of copyright law that it needs to be updated regularly each time the *media* change, in order to bring certain old practices into line with new possibilities. Scattered throughout the copyright statutes is evidence of old new media: gramophones, jukeboxes, cable TV, photocopiers, peer-to-peer file-sharing programs, and so on. Each new form of communication shifts the assumptions of past media enough that they require a reevaluation of the putative underlying balance of the constitutional mandate that gives (U.S.) intellectual-property law its inertia. Each new device needs to be understood in terms of creation, storage, distribution, production, consumption, and tangibility, in order to assess the dangers it poses to the rights of inventors and artists.

Because copyright law "hard codes" the particular media into the statutes, copyright law is comfortable with, for example, book editions or musical recordings. But in Connexions, new questions arise: how much change constitutes a new work, and thus demands a new copyright license? If a licensee receives one copy of a work, to which versions will he or she retain rights after changes? Because of the complexity of the software involved, there are also questions that the law simply cannot deal with (just as it had not been able to do in the late 1970s with respect to the definition of software): is the XML document equivalent to the viewable document, or must the style sheet also be included? Where does the "content" begin and the "software" end? Until the statutes either incorporate these new technologies or are changed to govern a more general process, rather than a particular medium, these questions will continue to emerge as part of the practice of writing.

This denaturalization of the notion of "publication" is responsible for much of the surprise and concern that greets Connexions and projects like it. Often, when I have shown the system to scholars, they have displayed boredom mixed with fear and frustration: "It can never replace the book." On the one hand, Connexions has made an enormous effort to make its output look as much like conventional books as possible; on the other hand, the anxiety evinced is justified, because what Connexions seeks to replace is not the book, which is merely ink and paper, but *the entire publishing process*. The fact that it is not replacing the book per se, but the entire process whereby manuscripts are made into stable and tangible objects called books is too overwhelming for most scholars to contemplate— especially scholars who have already mastered the existing process of book writing and creation. The fact that the legal system is built to safeguard something prior to and not fully continuous with the practice of Connexions only adds to the concern that such a transformation is immodest and risky, that it endangers a practice with centuries of stability behind it. Connexions, however, is not the cause of destabilization; rather, it is a response to or recognition of a problem. It is not a new problem, but one that periodically reemerges: a reorientation of knowledge and power that includes questions of enlightenment and rationality, democracy and self-governance, liberal values and problems of the authority and validation of knowledge. The salient moments of correlation are not the invention of the printing press and the Internet, but the struggle to make published books into a source of authoritative knowledge in the seventeenth and eighteenth centuries and the struggle to find ways to do the same with the Internet today.[9]

Connexions is, in many ways, understood by its practitioners to be both a response to the changing relations of knowledge and power, one that reaffirms the fundamental values of academic freedom and the circulation of knowledge, and also an experiment with, even a radicalization of, the ideals of both Free Software and Mertonian science. The transformation of the meaning of publication implies a fundamental shift in the status, in the finality of knowledge. It seeks to make of knowledge (knowledge in print, not in minds) something living and constantly changing, as opposed to something static and final. The fact that publication no longer signifies finality—that is, no longer signifies a state of fixity that is assumed in theory (and frequently in practice) to account for a text's reliability—has implications for how the text is used, reused, interpreted, valued, and trusted.[10] Whereas the traditional form of the book is the same across all printed versions or else follows an explicit practice of appearing in editions (complete with new prefaces and forewords), a Connexions document might very well look different from week to week or year to year.[11] While a textbook might also change significantly to reflect the changing state of knowledge in a given field, it is an explicit goal of Connexions to allow this to happen "in real time," which is to say, to allow educators to update textbooks as fast as they do scientific knowledge.[12]

These implications are not lost on the Connexions team, but neither are they understood as goals or as having simple solutions. There is a certain immodest, perhaps even reckless, enthusiasm surrounding these implications, an enthusiasm that can take both polymath and transhumanist forms. For instance, the destabilization of the contemporary textbook-publishing system that Connexions represents is (according to Rich) a more accurate way to represent the connections between concepts than a linear textbook format. Connexions thus represents a use of technology as an intervention into an existing context of practice. The fact that Connexions could also render the reliability or trustworthiness of scholarly knowledge unstable is sometimes discussed as an inevitable outcome of technical change—something that the world at large, not Connexions, must learn to deal with.

To put it differently, the "goal" of Connexions was never to destroy publishing, but it has been structured by the same kind of imaginations of moral and technical order that pervade Free Software and the construction of the Internet. In this sense Rich, Brent, and others are geeks in the same sense as Free Software geeks: they share a recursive public devoted to achieving a moral and technical order in which openness and modifiability are core values ("If we are successful, we will disappear"). The implication is that the existing model and infrastructure for the publication of textbooks is of a different moral and technical order, and thus that Connexions needs to innovate not only the technology (the source code or the openness of the system) or the legal arrangements (licenses) but also the very norms and forms of textbook writing itself (coordination and, eventually, a movement). If publication once implied the appearance of reliable, final texts—even if the knowledge therein could be routinely contested by writing more texts and reviews and critiques—Connexions implies the denaturalization of not knowledge per se, but of the process whereby that knowledge is stabilized and rendered reliable, trustworthy.

A keyword for the transformation of textbook writing is *community*, as in the tagline of the Connexions project: "Sharing Knowledge and Building Communities." *Building* implies that such communities do not yet exist and that the technology will enable them; however, Connexions began with the assumption that there exist standard academic practices and norms of creating teaching materials. As a result, Connexions both enables these practices and norms, by facilitating a digital version of the textbook, and intervenes in them, by creating a different process for creating a textbook. Communities are both assumed and desired. Sometimes they are real (a group of DSP engineers, networked around Rich and others who work in his subspecialty), and sometimes they are imagined (as when in the process of grant writing we claim that the most important component of the success of the project is the "seeding" of scholarly communities). Communities, furthermore, are not audiences or consumers, and sometimes not even students or learners. They are imagined to be active, creative producers and users of teaching materials, whether for teaching or for the further creation of such materials. The structure of the community has little to do with issues of governance, solidarity, or pedagogy, and much more to do with a set of relationships that might obtain with respect to the creation of teaching materials—a community of collaborative production or collaborative debugging, as in the modulation of forms of coordination, modulated to include the activity of creating teaching materials.

Agency and Structure in Connexions

One of the most animated whiteboard conversations I remember having with Brent and Ross concerned difference between the possible "roles" that a Connexions user might occupy and the implications this could have for both the technical features of the system and the social norms that Connexions attempts to maintain and replicate. Most software systems are content to designate only "users," a generic name-and-password account that can be given a set of permissions (and which has behind it a long and robust tradition in computer-operating-system and security research). Users are users, even if they may have access to different programs and files. What Connexions needed was a way to designate that the same person might have two different *exogenous* roles: a user might be the author, but not the owner of the content, and vice versa. For instance, perhaps Rice University maintains the copyright for a work, but the author is credited for its creation. Such a situation—known, in legal terms, as "work for hire"—is routine in some universities and most corporations. So while the author is generally given the freedom and authority to create and modify the text as he or she sees fit, the university asserts copyright ownership in order to retain the right to commercially exploit the work. Such a situation is far from settled and is, of course, politically fraught, but the Connexions system, in order to be useful at all to anyone, needed to accommodate this fact. Taking an oppositional political stand would render the system useless in too many cases or cause it to become precisely the kind of authorless, creditless system as Wikipedia—a route not desired by many academics. In a perfectly open world all Connexions modules might each have identical authors and owners, but pragmatism demands that the two roles be kept separate.

Furthermore, there are many people involved every day in the creation of academic work who are neither the author nor the owner: graduate students and undergraduates, research scientists, technicians, and others in the grand, contested, complex academic ecology. In some disciplines, all contributors may get authorship credit and some of them may even share ownership, but often many of those who do the work get mentioned only in acknowledgments, or not at all. Again, although the impulse of the creators of Connexions might be to level the playing field and allow only one kind of user, the fact of the matter is that academics simply would not use

REUSE, MODIFICATION, NORMS

such a system.[13] The need for a role such as "maintainer" (which might also include "editor"), which was different from author or owner, thus also presented itself.

As Brent, Ross, and I stared at the whiteboard, the discovery of the need for multiple exogenous roles hit all of us in a kind of slow-motion shockwave. It was not simply that the content needed to have different labels attached to it to keep track of these people in a database—something deeper was at work: the law and the practice of authorship actually dictated, to a certain extent, what the software itself should look like. All of sudden, the questions were preformatted, so to speak, by the law and by certain kinds of practices that had been normalized and thus were nearly invisible: who should have permission to change what? Who will have permission to add or drop authors? Who will be allowed to make what changes, and who will have the *legal right* to do so and who the moral or customary right? What implications follow from the choices the designers make and the choices we present to authors or maintainers?

The Creative Commons licenses were key to revealing many of these questions. The licenses were in themselves modulations of Free Software licenses, but created with people like artists, musicians, scholars, and filmmakers in mind. Without them, the content in Connexions would be unlicensed, perhaps intended to be in the public domain, but ultimately governed by copyright statutes that provided no clear answers to any of these questions, as those statutes were designed to deal with older media and a different publication process. Using the Creative Commons licenses, on the other hand, meant that the situation of the content in Connexions became well-defined enough, in a legal sense, to be used as a constraint in defining the structure of the software system. The license itself provided the map of the territory by setting parameters for things such as distribution, modification, attribution, and even display, reading, or copying.

For instance, when the author and owner are different, it is not at all obvious who should be given credit. Authors, especially academic authors, expect to be given credit (which is often all they get) for an article or a textbook they have written, yet universities often retain ownership of those textbooks, and ownership would seem to imply a legal right to be identified as both owner and author (e.g., Forrester Research reports or UNESCO reports, which hide the identity of authors). In the absence of any licenses, such a scenario has no obvious solution or depends entirely on the specific context. However, the Creative Commons licenses specified the meaning of attribution and the requirement to maintain the copyright notice, thus outlining a procedure that gave the Connexions designers fixed constraints against which to measure how they would implement their system.

A positive result of such constraints is that they allow for a kind of institutional flexibility that would not otherwise be possible. Whether a university insists on expropriating copyright or allows scholars to keep their copyrights, both can use Connexions. Connexions is more "open" than traditional textbook publishing because it allows a greater number of heterogeneous contributors to participate, but it is also more "open" than something like Wikipedia, which is ideologically committed to a single definition of authorship and ownership (anonymous, reciprocally licensed collaborative creation by authors who are also the owners of their work). While Wikipedia makes such an ideological commitment, it cannot be used by institutions that have made the decision to operate as expropriators of content, or even in cases wherein authors willingly allow someone else to take credit. If authors and owners must be identical, then either the author is identified as the owner, which is illegal in some cases, or the owner is identified as the author, a situation no academic is willing to submit to.

The need for multiple roles also revealed other peculiar and troubling problems, such as the issue of giving an "identity" to long-dead authors whose works are out of copyright. So, for instance, a piece by A. E. Housman was included as a module for a class, and while it is clear that Housman is the author, the work is no longer under copyright, so Housman is no longer the copyright holder (nor is the society which published it in 1921). Yet Connexions requires that a copyright be attached to each module to allow it to be licensed openly. This particular case, of a dead author, necessitated two interesting interventions. Someone has to actually create an account for Housman and also issue the work as an "edition" or derivative under a new copyright. In this case, the two other authors are Scott McGill and Christopher Kelty. A curious question arose in this context: should we be listed both as authors and owners (and maintainers), or only as owners and maintainers? And if someone uses the module in a new context (as they have the right to do, under the license), will they be required to give attribution only to Housman, or also to McGill and Kelty as well? What rights to ownership do McGill and Kelty have over the digital version of the public-domain text by Housman?[14]

The discussion of roles circulated fluidly across concepts like law (and legal licenses), norms, community, and identity. Brent and Ross and others involved had developed sophisticated imaginations of how Connexions would fit into the existing ecology of academia, constrained all the while by both standard goals, like usability and efficiency, and by novel legal licenses and concerns about the changing practices of authors and scholars. The question, for instance, of how a module *can* be used (technically, legally) is often confused with, or difficult to disentangle from, how a module *should* be used (technically, legally, or, more generally, "socially"—with usage shaped by the community who uses it). In order to make sense of this, Connexions programmers and participants like myself are prone to using the language of custom and norm, and the figure of community, as in "the customary norms of a scholarly community."

From Law and Technology to Norm

The meaning of publication in Connexions and the questions about roles and their proper legal status emerged from the core concern with reuse, which is the primary modulation of Free Software that Connexions carries out: the modulation of the meaning of source code to include textbook writing. What makes source code such a central component of Free Software is the manner in which it is shared and transformed, not the technical features of any particular language or program. So the modulation of source code to include textbooks is not just an attempt to make textbooks exact, algorithmic, or digital, but an experiment in sharing textbook writing in a similar fashion.

This modulation also affects the other components: it creates a demand for openness in textbook creation and circulation; it demands new kinds of copyright licenses (the Creative Commons licenses); and it affects the meaning of coordination among scholars, ranging from explicit forms of collaboration and co-creation to the entire spectrum of uses and reuses that scholars normally make of their peers' works. It is this modulation of coordination that leads to the second core concern of Connexions: that of the existence of "norms" of scholarly creation, use, reuse, publication, and circulation.

Since software programmers and engineers are prone to thinking about things in concrete, practical, and detailed ways, discussions of creation, use, and circulation are rarely conducted at the level of philosophical abstraction. They are carried out on whiteboards, using diagrams.

The whiteboard diagram transcribed in figure 8 was precipitated by a fairly precise question: "When is the reuse of something in a module (or of an entire module) governed by 'academic norms' and when is it subject to the legal constraints of the licenses?" For someone to quote a piece of text from one module in another is considered *normal* practice and thus shouldn't involve concerns about legal rights and duties to fork the module (create a new modified version, perhaps containing only the section cited, which is something legal licenses explicitly allow). But what if someone borrows, say, all of the equations in a module about information theory and uses them to illustrate a very different point in a different module. Does he or she have either a normal or a legal right to do so? Should the equations be cited? What should that citation look like? What if the equations are particularly hard to mark-up in the MathML language and therefore represent a significant investment in time on the part of the original author? Should the law govern this activity, or should norms?

There is a natural tendency among geeks to answer these questions solely with respect to the law; it is, after all, highly codified and seemingly authoritative on such issues. However, there is often no need to engage the law, because of the presumed consensus ("academic norms") about how to proceed, even if those norms conflict with the law. But these norms are nowhere codified, and this makes geeks (and, increasingly, academics themselves) uneasy. As in the case of a requirement of attribution, the constraints of a written license are perceived to be much more stable and reliable than those of culture, precisely because culture is what remains contested and contestable. So the idea of creating a new "version" of a text is easier to understand when it is clearly circumscribed as a legally defined "derivative work." The Connexions software was therefore implemented in such a way that the legal right to create a derived work (to fork a module) could be done with the press of

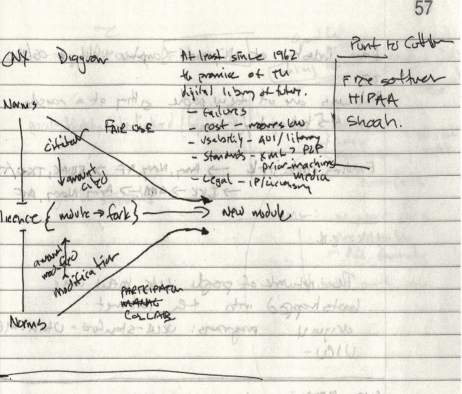

8. Whiteboard diagram: the cascade of reuse in Connexions. Conception by Ross Reedstrom, Brent Hendricks, and Christopher Kelty. Transcribed in the author's fieldnotes, 2003.

a button: a distinct module is automatically created, and it retains the name of the original author and the original owner, but now also includes the new author's name as author and maintainer. That new author can proceed to make any number of changes.

But is forking always necessary? What if the derivative work contains only a few spelling corrections and slightly updated information? Why not change the existing module (where such changes would be more akin to issuing a new edition), rather than create a legally defined derivative work? Why not simply suggest the changes to the original author? Why not *collaborate*? While a legal license gives people the right to do all of these things without ever consulting the person who licensed it, there may well be occasions when it makes much more sense to ignore those rights in favor of other norms. The answers to these questions depend a great deal on the kind and the intent of the reuse. A refined version of the whiteboard diagram, depicted in figure 9, attempts to capture the various kinds of reuse and their intersection with laws, norms, and technologies.

The center of the diagram contains a list of different kinds of imaginable reuses, arrayed from least interventionist at the top to most interventionist at the bottom, and it implies that as the intended transformations become more drastic, the likelihood of collaboration with the original author decreases. The arrow on the left indicates the legal path from cultural norms to protected fair uses; the arrow on the right indicates the technical path from built-in legal constraints based on the licenses to software tools that make collaboration (according to presumed scholarly norms) easier than the alternative (exercising the legal right to make a derivative work). With the benefit of hindsight, it seems that the arrows on either side should actually be a circle that connect laws, technologies, and norms in a chain of influence and constraint, since it is clear in retrospect that the norms of authorial practice have actually changed (or at least have been made explicit) based on the existence of licenses and the types of tools available (such as blogs and Wikipedia).

The diagram can best be understood as a way of representing, to Connexions itself (and its funders), the experiment under way with the components of Free Software. By modulating source code to include the writing of scholarly textbooks, Connexions made visible the need for new copyright licenses appropriate to this content; by making the system Internet-based and relying on open standards such as XML and Open Source components, Connexions also modulated the concept of openness to include textbook publication; and by making the system possible as an open repository of freely licensed textbook modules, Connexions made visible the changed conditions of coordination, not just between two collaborating authors, but within the entire system of publication, citation, use, reuse, borrowing, building on, plagiarizing, copying, emulating, and so on. Such changes to coordination may or may not take hold. For many scholars, they pose an immodest challenge to a working system that has developed over centuries, but for others they represent the removal of arbitrary constraints that prevent

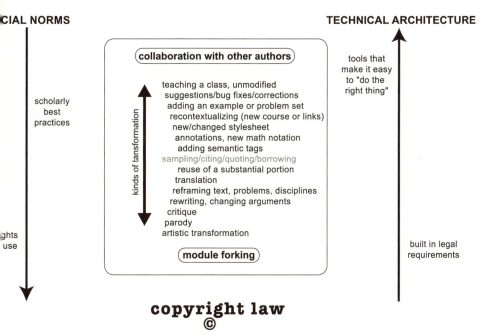

collaboration with other authors

tools that
make it easy
to "do the
right thing"

teaching a class, unmodified
suggestions/bug fixes/corrections
adding an example or problem set
recontextualizing (new course or links)
new/changed stylesheet
annotations, new math notation
adding semantic tags
sampling/citing/quoting/borrowing
reuse of a substantial portion
translation
reframing text, problems, disciplines
rewriting, changing arguments
critique
parody
artistic transformation

kinds of transformation

scholarly
best
practices

ghts
use

built in legal
requirements

module forking

copyright law
©

9. Whiteboard diagram transformed: forms of reuse in Connexions.
Conception by Christopher Kelty, 2004.

novel and innovative forms of knowledge creation and association
rendered possible in the last thirty to forty years (and especially in
the last ten). For some, these modulations might form the basis for
a final modulation—a Free Textbooks movement—but as yet no
such movement exists.

In the case of shared software source code, one of the principal
reasons for sharing it was to reuse it: to build on it, to link to it, to
employ it in ways that made building more complex objects into an
easier task. The very design philosophy of UNIX well articulates the
necessity of modularity and reuse, and the idea is no less powerful
in other areas, such as textbooks. But just as the reuse of software is
not simply a feature of software's technical characteristics, the idea
of "reusing" scholarly materials implies all kinds of questions that
are not simply questions of recombining texts. The ability to share
source code—and the ability to create complex software based on
it—requires modulations of both the legal meaning of software,
as in the case of EMACS, and the organizational form, as in the

emergence of Free Software projects other than the Free Software Foundation (the Linux kernel, Perl, Apache, etc.).

In the case of textbook reuse (but only *after* Free Software), the technical and the legal problems that Connexions addresses are relatively well specified: what software to use, whether to use XML, the need for an excellent user interface, and so on. However, the organizational, cultural, or practical meaning of reuse is not yet entirely clear (a point made by figures 8 and 9). In many ways, the recognition that there are cultural norms among academics mirrors the (re)discovery of norms and ethics among Free Software hackers.[15] But the label "cultural norms" is a mere catch-all for a problem that is probably better understood as a mixture of con-crete technical, organizational, and legal questions and as more or less abstract social imaginaries through which a particular kind of material order is understood and pursued—the creation of a re-cursive public. How do programmers, lawyers, engineers, and Free Software advocates (and anthropologists) "figure out" how norms work? How do they figure out ways to operationalize or make use of them? How do they figure out how to change them? How do they figure out how to create new norms? They do so through the modulations of existing practices, guided by imaginaries of moral and technical order. Connexions does not tend toward becoming Free Software, but it does tend toward becoming a recursive public with respect to textbooks, education, and the publication of peda-gogical techniques and knowledge. The problematic of creating an independent, autonomous public is thus the subterranean ground of both Free Software and Connexions.

To some extent, then, the matter of reuse raises a host of questions about the borders and boundaries in and of academia. Brent, Ross, and I assumed at the outset that communities have both borders and norms, and that the two are related. But, as it turns out, this is not a safe assumption. At neither the technical nor the legal level is the use of the software restricted to academics—indeed, there is no feasible way to do that and still offer it on the Internet—nor does anyone involved wish it to be so restricted. However, there is an implicit sense that the people who will contribute content will primarily be academics and educators (just as Free Software participants are expected, but not required to be programmers). As figure 9 makes clear, there may well be tremendous variation in the kinds of reuse that people wish to make, even within academia.

REUSE, MODIFICATION, NORMS

Scholars in the humanities, for instance, are loath to even imagine others creating derivative works with articles they have written and can envision their work being used only in the conventional manner of being read, cited, and critiqued. Scholars in engineering, biology, or computer science, on the other hand, may well take pleasure in the idea or act of reuse, if it is adequately understood to be a "scientific result" or a suitably stable concept on which to build.[16] Reuse can have a range of different meanings depending not only on whether it is used by scholars or academics, but within that heterogeneous group itself.

The Connexions software does not, however, enforce disciplinary differences. If anything it makes very strong and troubling claims that knowledge is knowledge and that disciplinary constraints are arbitrary. Thus, for instance, if a biologist wishes to transform a literary scholar's article on Darwin's tropes to make it reflect current evolutionary theory, he or she could do so; it is entirely possible, both legally and technically. The literary scholar could react in a number of ways, including outrage that the biologist has misread or misunderstood the work or pleasure in seeing the work refined. Connexions adheres rigorously to its ideas of openness in this regard; it neither encourages nor censures such behavior.

By contrast, as figure 9 suggests, the relationship between these two scholars can be governed either by the legal specification of rights contained in the licenses (a privately ordered legal regime dependent on a national-cum-global statutory regime) or by the customary means of collaboration enabled, perhaps enhanced, by software tools. The former is the domain of the state, the legal profession, and a moral and technical order that, for lack of a better word, might be called modernity. The latter, however, is the domain of the cultural, the informal, the practical, the interpersonal; it is the domain of ethics (prior to its modernization, perhaps) and of *tradition*.

If figure 9 is a recapitulation of modernity and tradition (what better role for an anthropologist to play!), then the presumptive boundaries around "communities" define which groups possess which norms. But the very design of Connexions—its technical and legal exactitude—immediately brings a potentially huge variety of traditions into conflict with one another. Can the biologist and the literary scholar be expected to occupy the same universe of norms? Does the fact of being academics, employees of a university, or readers of Darwin ensure this sharing of norms? How are the boundaries policed and the norms communicated and reinforced?

The problem of reuse therefore raises a much broader and more complex question: do norms actually exist? In particular, do they exist independent of the particular technical, legal, or organizational practice in which groups of people exist—outside the coordinated infrastructure of scholarship and science? And if Connexions raises this question, can the same question not also be asked of the elaborate system of professions, disciplines, and organizations that coordinate the scholarship of different communities? Are these norms, or are they "technical" and "legal" practices? What difference does formalization make? What difference does bureaucratization make?[17]

The question can also be posed this way: should norms be understood as historically changing constructs or as natural features of human behavior (regular patterns, or conventions, which emerge *inevitably* wherever human beings interact). Are they a feature of changing institutions, laws, and technologies, or do they form and persist in the same way wherever people congregate? Are norms features of a "calculative agency," as Michael Callon puts it, or are they features of the evolved human mind, as Marc Hauser argues?[18]

The answer that my informants give, in practice, concerning the mode of existence of cultural norms is neither. On the one hand, in the Connexions project the question of the mode of existence of academic norms is unanswered; the basic assumption is that certain actions are captured and constrained neither by legal constraints nor technical barriers, and that it takes people who know or study "communities" (i.e., nonlegal and nontechnical constraints) to figure out what those actions may be. On some days, the project is modestly understood to enable academics to do what they do faster and better, but without fundamentally changing anything about the practice, institutions, or legal relations; on other days, however, it is a radically transformative project, changing how people think about creating scholarly work, a project that requires educating people and potentially "changing the culture" of scholarly work, including its technology, its legal relations, and its practices.

In stark contrast (despite the very large degree of simpatico), the principal members of Creative Commons answer the question of the existence of norms quite differently than do those in Connexions:

they assert that norms not only change but are manipulated and/or channeled by the modulation of technical and legal practices (this is the novel version of law and economics that Creative Commons is founded on). Such an assertion leaves very little for norms or for culture; there may be a deep evolutionary role for rule following or for choosing socially sanctioned behavior over socially unacceptable behavior, but the real action happens in the legal and technical domains. In Creative Commons the question of the existence of norms is answered firmly in the phrase coined by Glenn Brown: "punt to culture." For Creative Commons, norms are a prelegal and pretechnical substrate upon which the licenses they create operate. Norms *must* exist for the strategy employed in the licenses to make sense—as the following story illustrates.

On the Nonexistence of Norms in the
Culture of No Culture

More than once, I have found myself on the telephone with Glenn Brown, staring at notes, a diagram, or some inscrutable collection of legalese. Usually, the conversations wander from fine legal points to music and Texas politics to Glenn's travels around the globe. They are often precipitated by some previous conversation and by Glenn's need to remind himself (and me) what we are in the middle of creating. Or destroying. His are never simple questions. While the Connexions project started with a repository of scholarly content in need of a license, Creative Commons started with licenses in need of particular kinds of content. But both projects required participants to delve into the details of both licenses and the structure of digital content, which qualified me, for both projects, as the intermediary who could help explore these intersections. My phone conversations with Glenn, then, were much like the whiteboard conversations at Connexions: filled with a mix of technical and legal terminology, and conducted largely in order to give Glenn the sense that he had cross-checked his plans with someone presumed to know better. I can't count the number of times I have hung up the phone or left the conference room wondering, "Have I just sanctioned something mad?" Yet rarely have I felt that my interventions served to do more than confirm suspicions or derail already unstable arguments.

In one particular conversation—the "punt to culture" conversation—I found myself bewildered by a sudden understanding of the process of writing legal licenses and of the particular assumptions about human behavior that need to be present in order to imagine creating these licenses or ensuring that they will be beneficial to the people who will use them.

These discussions (which often included other lawyers) happened in a kind of hypothetical space of legal imagination, a space highly structured by legal concepts, statutes, and precedents, and one extraordinarily carefully attuned to the fine details of semantics. A core aspect of operating within this imagination is the distinction between law as an abstract semantic entity and law as a practical fact that people may or may not deal with. To be sure, not all lawyers operate this way, but the warrant for thinking this way comes from no less eminent an authority than Oliver Wendell Holmes, for whom the "Path of Law" was always from practice to abstract rule, and not the reverse.[19] The opposition is unstable, but I highlight it here because it was frequently used as a *strategy* for constructing precise legal language. The ability to imagine the difference between an abstract rule designating legality and a rule encountered in practice was a first step toward seeing how the language of the rule should be constructed.

I helped write, read, and think about the first of the Creative Commons licenses, and it was through this experience that I came to understand how the crafting of legal language works, and in particular how the mode of existence of cultural or social norms relates to the crafting of legal language. Creative Commons licenses are not a familiar legal entity, however. They are modulations of the Free Software license, but they differ in important ways.

The Creative Commons licenses allow authors to grant the use of their work in about a dozen different ways—that is, the license itself comes in versions. One can, for instance, require attribution, prohibit commercial exploitation, allow derivative or modified works to be made and circulated, or some combination of all these. These different combinations actually create different licenses, each of which grants intellectual-property rights under slightly different conditions. For example, say Marshall Sahlins decides to write a paper about how the Internet is cultural; he copyrights the paper ("© 2004 Marshall Sahlins"), he requires that any use of it or any copies of it maintain the copyright notice and the attribution of authorship (these can be different), and he furthermore allows for commercial use of the paper. It would then be legal for a publishing house to take the paper off Sahlins's Linux-based Web server and publish it in a collection without having to ask permission, as long as the paper remains unchanged and he is clearly and unambiguously listed as author of the paper. The publishing house would not get any rights to the work, and Sahlins would not get any royalties. If he had specified noncommercial use, the publisher would instead have needed to contact him and arrange for a separate license (Creative Commons licenses are nonexclusive), under which he could demand some share of revenue and his name on the cover of the book.[20] But say he was, instead, a young scholar seeking only peer recognition and approbation—then royalties would be secondary to maximum circulation. Creative Commons allows authors to assert, as its members put it, "some rights reserved" or even "no rights reserved."

But what if Sahlins had chosen a license that allowed modification of his work. This would mean that I, Christopher Kelty, whether in agreement with or in objection to his work, could download the paper, rewrite large sections of it, add in my own baroque and idiosyncratic scholarship, and write a section that purports to debunk (or, what could amount to the same, augment) Sahlins's arguments. I would then be legally entitled to re-release the paper as "© 2004 Marshall Sahlins, with modifications © 2007 Christopher Kelty," so long as Sahlins is identified as the author of the paper. The nature or extent of the modifications is not legally restricted, but both the original and the modified version would be legally attributed to Sahlins (even though he would own only the first paper).

In the course of a number of e-mails, chat sessions, and phone conversations with Glenn, I raised this example and proposed that the licenses needed a way to account for it, since it seemed to me entirely possible that were I to produce a modified work that so distorted Sahlins's original argument that he did not want to be associated with the modified paper, then he should have the right also to repudiate his identification as author. Sahlins should, legally speaking, be able to ask me to remove his name from all subsequent versions of my misrepresentation, thus clearing his good name and providing me the freedom to continue sullying mine into obscurity. After hashing it out with the expensive Palo Alto legal firm that was officially drafting the licenses, we came up with text that said:

"If You create a Derivative Work, upon notice from any Licensor You must, to the extent practicable, remove from the Derivative Work any reference to such Licensor or the Original Author, as requested."

The bulk of our discussion centered around the need for the phrase, "to the extent practicable." Glenn asked me, "How is the original author supposed to monitor *all* the possible uses of her name? How will she enforce this clause? Isn't it going to be difficult to remove the name from every copy?" Glenn was imagining a situation of strict adherence, one in which the presence of the name on the paper was the same as the reputation of the individual, regardless of who actually read it. On this theory, until all traces of the author's name were expunged from each of these teratomata circulating in the world, there could be no peace, and no rest for the wronged.

I paused, then gave the kind of sigh meant to imply that I had come to my hard-won understandings of culture through arduous dissertation research: "It probably won't need to be strictly enforced in all cases—only in the significant ones. Scholars tend to respond to each other only in very circumscribed cases, by writing letters to the editor or by sending responses or rebuttals to the journal that published the work. It takes a lot of work to really police a reputation, and it differs from discipline to discipline. Sometimes, drastic action might be needed, usually not. There is so much misuse and abuse of people's arguments and work going on all the time that people only react when they are directly confronted with serious abuses. And even so, it is only in cases of negative criticism or misuse that people need respond. When a scholar uses someone's work approvingly, but incorrectly, it is usually considered petulant (at best) to correct them publicly."

"In short," I said, leaning back in my chair and acting the part of expert, "it's like, you know, c'mon—it isn't *all* law, there are a bunch of, you know, informal rules of civility and stuff that govern that sort of thing."

Then Glenn said., "Oh, okay, well that's when we punt to culture."

When I heard this phrase, I leaned too far back and fell over, joyfully stunned. Glenn had managed to capture what no amount of fieldwork, with however many subjects, could have. Some combination of American football, a twist of Hobbes or Holmes, and a lived understanding of what exactly these copyright licenses are meant to achieve gave this phrase a luminosity I usually associate only with Balinese cock-fights. It encapsulated, almost as a slogan, a very precise explanation of what Creative Commons had undertaken. It was not a theory Glenn proposed with this phrase, but a *strategy* in which a particular, if vague, theory of culture played a role.

For those unfamiliar, a bit of background on U.S. football may help. When two teams square off on the football field, the offensive team gets four attempts, called "downs," to move the ball either ten yards forward or into the end zone for a score. The first three downs usually involve one of two strategies: run or pass, run or pass. On the fourth down, however, the offensive team must either "go for it" (run or pass), kick a field goal (if close enough to the end zone), or "punt" the ball to the other team. Punting is a somewhat disappointing option, because it means giving up possession of the ball to the other team, but it has the advantage of putting the other team as far back on the playing field as possible, thus decreasing its likelihood of scoring.

To "punt to culture," then, suggests that copyright licenses try three times to *legally* restrict what a user or consumer of a work can make of it. By using the existing federal intellectual-property laws and the rules of license and contract writing, copyright licenses articulate to people what they can and cannot do with that work according to law. While the licenses do not (they cannot) *force* people, in any tangible sense, to do one thing or another, they can use the language of law and contract to warn people, and perhaps obliquely, to threaten them. If the licenses end up silent on a point—if there is no "score," to continue the analogy—then it's time to punt to culture. Rather than make more law, or call in the police, the license *strategy* relies on culture to fill in the gaps with people's own understandings of what is right and wrong, beyond the law. It operationalizes a theory of culture, a theory that emphasizes the sovereignty of nonstate customs and the diversity of systems of cultural norms. Creative Commons would prefer that its licenses remain legally minimalist. It would much prefer to assume—indeed, the licenses implicitly require—the robust, powerful existence of this multifarious, hetero-physiognomic, and formidable opponent to the law with neither uniform nor mascot, hunched at the far end of the field, preparing to, so to speak, clean law's clock.

Creative Commons's "culture" thus seems to be a somewhat vague mixture of many familiar theories. Culture is an unspecified but finely articulated set of given, evolved, designed, informal, practiced, habitual, local, social, civil, or historical norms that are expected to govern the behavior of individuals in the absence of a state, a court, a king, or a police force, at one of any number of scales. It is not monolithic (indeed, my self-assured explanation concerned only the norms of "academia"), but assumes a diversity beyond enumeration. It employs elements of relativism—*any* culture should be able to trump the legal rules. It is not a hereditary biological theory, but one that assumes historical contingency and arbitrary structures.

Certainly, whatever culture is, it is *separate from law*. Law is, to borrow Sharon Traweek's famous phrase, "a culture of no culture" in this sense. It is not the cultural and normative practices of legal scholars, judges, lawyers, legislators, and lobbyists that determine what laws will look like, but their careful, expert, *noncultural* ratiocination. In this sense, punting to culture implies that laws are the result of human design, whereas culture is the result of human action, but not of human design. Law is systematic and tractable; culture may have a deep structure, but it is intractable to human design. It can, however, be channeled and tracked, nudged or guided, *by law*.

Thus, Lawrence Lessig, one of the founders of Creative Commons has written extensively about the "regulation of social meaning," using cases such as those involving the use or nonuse of seatbelts or whether or not to allow smoking in public places. The decision not to wear a seatbelt, for instance, may have much more to do with the contextual meaning of putting on a seatbelt (don't you trust the cab driver?) than with either the existence of the seatbelt (or automatic seatbelts, for that matter) or with laws demanding their use. According to Lessig, the best law can do in the face of custom is to *change the meaning* of wearing the seatbelt: to give the refusal a dishonorable rather than an honorable meaning. Creative Commons licenses are based on a similar assumption: the law is relatively powerless in the face of entrenched academic or artistic customs, and so the best the licenses can do is channel the *meaning* of sharing and reuse, of copyright control or infringement. As Glenn explained in the context of a discussion about a license that would allow music sampling.

We anticipate that the phrase "as appropriate to the medium, genre, and market niche" might prompt some anxiety, as it leaves things relatively undefined. But there's more method here than you might expect: The definition of "sampling" or "collage" varies across different media. Rather than try to define all possible scenarios (including ones that haven't happened yet)—which would have the effect of restricting the types of re-uses to a limited set—we took the more laissez faire approach.

This sort of deference to community values—think of it as "punting to culture"—is very common in everyday business and contract law. The idea is that when lawyers have trouble defining the specialized terms of certain subcultures, they should get out of the way and let those subcultures work them out. It's probably not a surprise Creative Commons likes this sort of notion a lot.[21]

As in the case of reuse in Connexions, sampling in the music world can imply a number of different, perhaps overlapping, customary meanings of what is acceptable and what is not. For Connexions, the trick was to differentiate the cases wherein collaboration should be encouraged from the cases wherein the legal right to "sample"—to fork or to create a derived work—was the appropriate course of action. For Creative Commons, the very structure of the licenses attempts to capture this distinction as such and to allow for individuals to make determinations about the meaning of sampling themselves.[22]

At stake, then, is the construction of both technologies and legal licenses that, as Brent and Rich would assert, "make it easy for users to do the right thing." The "right thing," however, is precisely what goes unstated: the moral and technical order that guides the design of both licenses and tools. Connexions users are given tools that facilitate citation, acknowledgment, attribution, and certain kinds of reuse instead of tools that privilege anonymity or facilitate proliferation or encourage nonreciprocal collaborations. By the same token, Creative Commons licenses, while legally binding, are created with the aim of changing norms: they promote attribution and citation; they promote fair use and clearly designated uses; they are written to give users flexibility to decide what kinds of things should be allowed and what kinds shouldn't. Without a doubt, the "right thing" is right for some people and not for others—and it is thus political. But the criteria for what is right are not merely political; the criteria are what constitute the affinity of these geeks in the first place, what makes them a recursive public. They see in these instruments the possibility for the creation of authentic publics whose role is to stand outside power, outside markets, and to participate in sovereignty, and through this participation to produce liberty without sacrificing stability.

Conclusion

What happens when geeks modulate the practices that make up Free Software? What is the intuition or the cultural significance of Free Software that makes people want to emulate and modulate it? Creative Commons and Connexions modulate the practices of Free Software and extend them in new ways. They change the meaning of shared source code to include shared nonsoftware, and they try to apply the practices of license writing, coordination, and openness to new domains. At one level, such an activity is fascinating simply because of what it reveals: in the case of Connexions, it reveals the problem of determining the finality of a work. How should the authority, stability, and reliability of knowledge be assessed when work can be rendered permanently modifiable? It is an activity that reveals the complexity of the system of authorization and evaluation that has been built in the past.

The intuition that Connexions and Creative Commons draw from Free Software is an intuition about the authority of knowledge, about a reorientation of knowledge and power that demands a *response*. That response needs to be technical and legal, to be sure, but it also needs to be *public*—a response that defines the meaning of finality publicly and openly and makes modifiability an irreversible aspect of the process of stabilizing knowledge. Such a commitment is incompatible with the provision of stable knowledge by unaccountable private parties, whether individuals or corporations or governments, or by technical fiat. There must always remain the possibility that someone can question, change, reuse, and modify according to their needs.

Conclusion

The Cultural Consequences of Free Software

Free Software is changing. In all aspects it looks very different from when I started, and in many ways the Free Software described herein is not the Free Software readers will encounter if they turn to the Internet to find it. But how could it be otherwise? If the argument I make in *Two Bits* is at all correct, then modulation must constantly be occurring, for experimentation never seeks its own conclusion. A question remains, though: in changing, does Free Software and its kin preserve the imagination of moral and technical order that created it? Is the recursive public something that survives, orders, or makes sense of these changes? Does Free Software exist for more than its own sake?

In *Two Bits* I have explored not only the history of Free Software but also the question of where such future changes will have come from. I argue for seeing continuity in certain practices of everyday life precisely because the Internet and Free Software pervade everyday life to a remarkable, and growing, degree. Every day, from here to there, new projects and ideas and tools and goals emerge everywhere out of the practices that I trace through Free Software: Connexions and Creative Commons, open access, Open Source synthetic biology, free culture, access to knowledge (a2k), open cola, open movies, science commons, open business, Open Source yoga, Open Source democracy, open educational resources, the One Laptop Per Child project, to say nothing of the proliferation of wiki-everything or the "peer production" of scientific data or consumer services—all new responses to a widely felt reorientation of knowledge and power.[1] How is one to know the difference between all these things? How is one to understand the cultural significance and consequence of them? Can one distinguish between projects that promote a form of public sphere that can direct the actions of our society versus those that favor corporate, individual, or hierarchical control over decision making?

Often the first response to such emerging projects is to focus on the promises and ideology of the people involved. On the one hand, claiming to be open or free or public or democratic is something nearly everyone does (including unlikely candidates such as the defense intelligence agencies of the United States), and one should therefore be suspicious and critical of all such claims.[2] While such arguments and ideological claims are important, it would be a grave mistake to focus only on these statements. The "movement"—the ideological, critical, or promissory aspect—is just one component of Free Software and, indeed, the one that has come last, after the other practices were figured out and made legible, replicable, and modifiable. On the other hand, it is easy for geeks and Free Software advocates to denounce emerging projects, to say, "But that isn't *really* Open Source or Free Software." And while it may be tempting to fix the definition of Free Software once and for all in order to ensure a clear dividing line between the true sons and the carpetbaggers, to do so would reduce Free Software to mere repetition without difference, would sacrifice its most powerful and distinctive attribute: its responsive, emergent, *public* character.

But what questions should one ask? Where should scholars or curious onlookers focus their attention in order to see whether or not a recursive public is at work? Many of these questions are simple, practical ones: are software and networks involved at any level? Do the participants claim to understand Free Software or Open Source, either in their details or as an inspiration? Is intellectual-property law a key problem? Are participants trying to coordinate each other through the Internet, and are they trying to take advantage of voluntary, self-directed contributions of some kind? More specifically, are participants *modulating* one of these practices? Are they thinking about something in terms of source code, or source and binary? Are they changing or creating new forms of licenses, contracts, or privately ordered legal arrangements? Are they experimenting with forms of coordinating the voluntary actions of large numbers of unevenly distributed people? Are the people who are contributing aware of or actively pursuing questions of ideology, distinction, movement, or opposition? Are these practices recognized as something that creates the possibility for affinity, rather than simply arcane "technical" practices that are too complex to understand or appreciate?

In the last few years, talk of "social software" or "Web 2.0" has dominated the circuit of geek and entrepreneur conferences and discussions: Wikipedia, MySpace, Flickr, and YouTube, for example. For instance, there are scores and scores of "social" music sites, with collaborative rating, music sharing, music discovery, and so forth. Many of these directly use or take inspiration from Free Software. For all of them, intellectual property is a central and dominating concern. Key to their novelty is the leveraging and coordinating of massive numbers of people along restricted lines (i.e., music preferences that guide music discovery). Some even advocate or lobby for free(er) access to digital music. But they are not (yet) what I would identify as recursive publics: most of them are commercial entities whose structure and technical specifications are closely guarded and not open to modification. While some such entities may deal in freely licensed content (for instance, Creative Commons–licensed music), few are interested in allowing strangers to participate in, modulate, or modify the system as such; they are interested in allowing users to become consumers in more and more sophisticated ways, and not necessarily in facilitating a public culture of music. They want information and knowledge to be free, to be sure, but not necessarily the infrastructure that makes that information available and knowledge possible. Such entities lack the "recursive" commitment.

By contrast, some corners of the open-access movement are more likely to meet this criteria. As the appellation suggests, participants see it as a movement, not a corporate or state entity, a movement founded on practices of copyleft and the modulation of Free Software licensing ideas. The use of scientific data and the tools for making sense of open access are very often at the heart of controversy in science (a point often reiterated by science and technology studies), and so there is often an argument about not only the availability of data but its reuse, modification, and modulation as well. Projects like the BioBricks Foundation (biobricks.org) and new organizations like the Public Library of Science (plos.org) are committed to both availability and certain forms of collective modification. The commitment to becoming a recursive public, however, raises unprecedented issues about the nature of quality, reliability, and finality of scientific data and results—questions that will reverberate throughout the sciences as a result.

Farther afield, questions of "traditional heritage" claims, the compulsory licensing of pharmaceuticals, or new forms of "crowdsourcing" in labor markets are also open to analysis in the terms I offer in *Two Bits*.[3] Virtual worlds like Second Life, "a 3D digital world imagined, created, and owned by its residents," are increasingly laboratories for precisely the kinds of questions raised here: such worlds are far less virtual than most people realize, and the experiments conducted there far more likely to migrate into the so-called real world before we know it—including both economic and democratic experiments.[4] How far will Second Life go in facilitating a recursive public sphere? Can it survive both as a corporation and as a "world"? And of course, there is the question of the "blogosphere" as a public sphere, as a space of opinion and discussion that is radically open to the voices of massive numbers of people. Blogging gives the lie to conventional journalism's self-image as the public sphere, but it is by no means immune to the same kinds of problematic dynamics and polarizations, no more "rational-critical" than FOX News, and yet . . .

Such examples should indicate the degree to which *Two Bits* is focused on a much longer time span than simply the last couple of years and on much broader issues of political legitimacy and cultural change. Rather than offer immediate policy prescriptions or seek to change the way people think about an issue, I have approached *Two Bits* as a work of history and anthropology, making it less immediately applicable in the hopes that it is more lastingly usable. The stories I have told reach back at least forty years, if not longer. While it is clear that the Internet as most people know it is only ten to fifteen years old, it has been "in preparation" since at least the late 1950s. Students in my classes—especially hip geeks deep in Free Software apprenticeship—are bewildered to learn that the arguments and usable pasts they are rehearsing are refinements and riffs on stories that are as old or older than their parents. This deeper stability is where the cultural significance of Free Software lies: what difference does Free Software *today* introduce with respect to knowledge and power *yesterday*?

Free Software is a response to a problem, in much the same way that the Royal Society in the sixteenth century, the emergence of a publishing industry in the eighteenth century, and the institutions of the public sphere in the eighteenth and nineteenth centuries were responses. They responded to the collective challenge of creating regimes of governance that required—and encouraged—reliable empirical knowledge as a basis for their political legitimacy. Such political legitimacy is not an eternal or theoretical problem; it is a problem of constant real-world practice in creating the infrastructures by which individuals come to inhabit and understand their own governance, whether by states, corporations, or machines. If power seeks consent of the governed—and especially the consent of the democratic, self-governing kind that has become the global dominant ideal since the seventeenth century—it must also seek to ensure the stability and reliability of the knowledge on which that consent is propped.

Debates about the nature and history of publics and public spheres have served as one of the main arenas for this kind of questioning, but, as I hope I have shown here, it is a question not only of public spheres but of practices, technologies, laws, and movements, of going concerns which undergo modulation and experimentation in accord with a social imagination of order both moral and technical. "Recursive public" as a concept is not meant to replace that of public sphere. I intend neither for actors nor really for many scholars to find it generally applicable. I would not want to see it suddenly discovered everywhere, but principally in tracking the transformation, proliferation, and differentiation of Free Software and its derivatives.

Several threads from the three parts of *Two Bits* can now be tied together. The detailed descriptions of Free Software and its modulations should make clear that (1) the reason the Internet looks the way it does is due to the work of figuring out Free Software, both before and after it was recognized as such; (2) neither the Internet nor the computer is the cause of a reorientation of knowledge and power, but both are tools that render possible modulations of settled practices, modulations that reveal a much older problem regarding the legitimacy of the means of circulation and production of knowledge; (3) Free Software is not an ethical stance, but a practical response to the revelation of these older problems; and (4) the best way to understand this response is to see it as a kind of public sphere, a recursive public that is specific to the technical and moral imaginations of order in the contemporary world of geeks.

It is possible now to return to the practical and political meaning of the "singularity" of the Internet, that is, to the fact that there is only one Internet. This does not mean that there are no other networks, but only that the Internet is a singular entity and not an instance of a general type. How is it that the Internet is open in the same way to everyone, whether an individual or a corporate or a national entity? How has it become extensible (and, by extension, defensible) by and to everyone, regardless of their identity, locale, context, or degree of power?

The singularity of the Internet is both an ontological and an epistemological fact; it is a feature of the Internet's technical configurations and modes of ordering the actions of humans and machines by protocols and software. But it is also a feature of the technical and moral imaginations of the people who build, manage, inhabit, and expand the Internet. Ontologically, the creation and dissemination of standardized protocols, and novel standard-setting processes are at the heart of the story. In the case of the Internet, differences in *standards-setting processes* are revealed clearly in the form of the famous Request for Comments system of creating, distributing, and modifying Internet protocols. The RFC system, just as much as the Geneva-based International Organization for Standards, reveal the fault lines of international legitimacy in complex societies dependent on networks, software, and other high-tech forms of knowledge production, organization, and governance. The legitimacy of standards has massive significance for the abilities of individual actors to participate in their own recursive publics, whether they be publics that address software and networks or those that address education and development. But like the relationship between "law on the books" and "law in action," standards depend on the coordinated action and order of human practices.

What's more, the seemingly obvious line between a legitimate standard and a marketable product based on these standards causes nothing but trouble. The case of open systems in the 1980s high-end computer industry demonstrates how the logic of standardization is not at all clearly distinguished from the logic of the market. The open-systems battles resulted in novel forms of cooperation-within-competition that sought both standardization and competitive advantage at the same time. Open systems was an attempt to achieve a kind of "singularity," not only for a network but for a market infrastructure as well. Open systems sought ways to reform technologies and markets in tandem. What it ignored was the legal structure of intellectual property. The failure of open systems reveals the centrality of the moral and technical order of intellectual property—to both technology and markets—and shows how a reliance on this imagination of order literally renders impossible the standardization of singular market infrastructure. By contrast, the success of the Internet as a market infrastructure and as a singular entity comes in part because of the recognition of the limitations of the intellectual-property system—and Free Software in the 1990s was the main experimental arena for trying out alternatives.

The singularity of the Internet rests in turn on a counterintuitive multiplicity: the multiplicity of the UNIX operating system and its thousands of versions and imitations and reimplementations. UNIX is a great example of how novel, unexpected kinds of order can emerge from high-tech practices. UNIX is neither an academic (gift) nor a market phenomenon; it is a hybrid model of sharing that emerged from a very unusual technical and legal context. UNIX demonstrates how structured practices of sharing produce their own kind of order. Contrary to the current scholarly consensus that Free Software and its derivatives are a kind of "shadow economy" (a "sharing" economy, a "peer production" economy, a "noncommercial" economy), UNIX was never entirely outside of the mainstream market. The meanings of sharing, distribution, and profitability are related to the specific technical, legal, and organizational context. Because AT&T was prevented from commercializing UNIX, because UNIX users were keen to expand and adapt it for their own uses, and because its developers were keen to encourage and assist in such adaptations, UNIX proliferated and differentiated in ways that few commercial products could have. But it was never "free" in any sense. Rather, in combination with open systems, it set the stage for what "free" could come to mean in the 1980s and 1990s. It was a nascent recursive public, confronting the technical and legal challenges that would come to define the practices of Free Software. To suggest that it represents some kind of "outside" to a functioning economic market based in money is to misperceive how transformative of markets UNIX and the Internet (and Free Software) have been. They have initiated an imagination of moral and technical order that is not at all opposed to ideologies of market-based governance. Indeed, if anything, what UNIX and Free Software represent is an imagination of how to change an *entire market-based governance structure*—not just specific markets in things—to include a form of public sphere, a check on the power of existing authority.

UNIX and Open Systems should thus be seen as early stages of a collective technical experiment in transforming our imaginations of order, especially of the moral order of publics, markets, and self-governing peoples. The continuities and the gradualness of the change are more apparent in these events than any sudden rupture or discontinuity that the "invention of the Internet" or the passing of new intellectual-property laws might suggest. The "reorientation of knowledge and power" is more dance than earthquake; it is stratified in time, complex in its movements, and takes an experimental form whose concrete traces are the networks, infrastructures, machines, laws, and standards left in the wake of the experiments.

Availability, reusability, and modifiability are at the heart of this reorientation. The experiments of UNIX and open systems would have come to nothing if they had not also prompted a concurrent experimentation with intellectual-property law, of which the copyleft license is the central and key variable. Richard Stallman's creation of GNU EMACS and the controversy over propriety that it engendered was in many ways an attempt to deal with exactly the same problem that UNIX vendors and open-systems advocates faced: how to build extensibility into the software market—except that Stallman never saw it as a market. For him, software was and is part of the human itself, constitutive of our very freedom and, hence, inalienable. Extending software, through collective mutual aid, is thus tantamount to vitality, progress, and self-actualization. But even for those who insist on seeing software as mere product, the problem of extensibility remains. Standardization, standards processes, and market entry all appear as political problems as soon as extensibility is denied—and thus the legal solution represented by copyleft appears as an option, even though it raises new and troubling questions about the nature of competition and profitability.

New questions about competition and profitability have emerged from the massive proliferation of hybrid commercial and academic forms, forms that bring with them different traditions of sharing, credit, reputation, control, creation, and dissemination of knowledge and products that require it. The new economic demands on the university—all too easily labeled neoliberalization or corporatization—mirror changing demands on industry that it come to look more like universities, that is, that it give away more, circulate more, and cooperate more. The development of UNIX, in its details, is a symptom of these changes, and the success of Free Software is an unambiguous witness to them.

The proliferation of hybrid commercial-academic forms in an era of modifiability and reusability, among the debris of standards, standards processes, and new experiments in intellectual property, results in a playing field with a thousand different games, all of which revolve around renewed experimentation with coordination, collaboration, adaptability, design, evolution, gaming, playing, worlds, and worlding. These games are indicative of the triumph of the American love of entrepreneurialism and experimentalism; they relinquish the ideals of planning and hierarchy almost absolutely in favor of a kind of embedded, technically and legally complex anarchism. It is here that the idea of a public reemerges: the ambivalence between relinquishing control absolutely and absolute distrust of government by the few. A powerful public is a response, and a solution, so long as it remains fundamentally independent of control by the few. Hence, a commitment, widespread and growing, to a recursive public, an attempt to maintain and extend the kinds of independent, authentic, autotelic public spheres that people encounter when they come to an understanding of how Free Software and the Internet have evolved.

The open-access movement, and examples like Connexions, are attempts at maintaining such publics. Some are conceived as bulwarks against encroaching corporatization, while others see themselves as novel and innovative, but most share some of the practices hashed out in the evolution of Free Software and the Internet. In terms of scholarly publishing and open access, the movement has reignited discussions of ethics, norms, and *method*. The Mertonian ideals are in place once more, this time less as facts of scientific method than as goals. The problem of stabilizing collective knowledge has moved from being an inherent feature of science to being a problem that needs our attention. The reorientation of knowledge and power and the proliferation of hybrid commercial-academic entities in an era of massive dependence on scientific knowledge and information leads to a question about the stabilization of that knowledge.

Understanding how Free Software works and how it has developed along with the Internet and certain practices of legal and cultural critique may be essential to understanding the reliable foundation of knowledge production and circulation on which we still seek to ground legitimate forms of governance. Without Free Software, the only response to the continuing forms of excess we associate with illegitimate, unaccountable, unjust forms of governance might just be mute cynicism. With it, we are in possession of a range of practical tools, structured responses and clever ways of working through our complexity toward the promises of a shared imagination of legitimate and just governance. There is no doubt room for critique—and many scholars will demand it—but scholarly critique will have to learn how to sit, easily or uneasily, with Free Software *as critique*. Free Software can also exclude, just as any public or public sphere can, but this is not, I think, cause for resistance, but cause for joining. The alternative would be to create no new rules, no new practices, no new procedures—that is, to have what we already have. Free Software does not belong to geeks, and it is not the only form of becoming public, but it is one that will have a profound structuring effect on any forms that follow.

....
: Notes
:..................................

Introduction

Throughout this volume, some messages referenced are cited by their "Message-ID," which should allow anyone interested to access the original messages through Google Groups (http://groups.google.com).

1 A Note on Terminology: There is still debate about how to refer to Free Software, which is also known as Open Source Software. The scholarly community has adopted either FOSS or FLOSS (or F/LOSS): the former stands for the Anglo-American Free and Open Source Software; the latter stands for the continental Free, Libre and Open Source Software. *Two Bits* sticks to the simple term *Free Software* to refer to all of these things, except where it is specifically necessary to differentiate two or more names, or to specify people or events so named. The reason is primarily aesthetic and political, but *Free Software* is also the older term, as well as the one that includes issues of moral and social order. I explain in chapter 3 why there are two terms.

2 Michael M. J. Fischer, "Culture and Cultural Analysis as Experimental Systems."

3 So, for instance, when a professional society founded on charters and ideals for membership and qualification speaks as a public, it represents its members, as when the American Medical Association argues for or against changes to Medicare. However, if a new group—say, of nurses—seeks not only to participate in this discussion—which may be possible, even welcomed—but to *change the structure of representation* in order to give themselves status equal to doctors, this change is impossible, for it goes against the very aims and principles of the society. Indeed, the nurses will be urged to form their own society, not to join that of the doctors, a proposition which gives the lie to the existing structures of power. By contrast, a public is an entity that is less controlled and hence more agonistic, such that nurses might join, speak, and insist on changing the terms of debate, just as patients, scientists, or homeless people might. Their success, however, depends entirely on the force with which their actions transform the focus and terms of the public. Concepts of the public sphere have been roundly critiqued in the last twenty years for presuming that such "equality of access" is sufficient to achieve representation, when in fact other contextual factors (race, class, sex) inherently weight the representative power of different participants. But these are two different and overlapping problems: one cannot solve the problem of pernicious, invisible forms of inequality unless one first solves the problem of ensuring a certain kind of structural publicity. It is precisely the focus on maintaining publicity for a recursive public, over against massive and powerful corporate and governmental attempts to restrict it, that I locate as the central struggle of Free Software. Gender certainly influences who gets heard within Free Software, for example, but it is a mistake to focus on this inequality at the expense of the larger, more threatening form of political failure that Free Software addresses. And I think there are plenty of geeks—man, woman and animal—who share this sentiment.

4 Wikipedia is perhaps the most widely known and generally familiar example of what this book is about. Even though it is not identified as such, it is in fact a Free Software project and a "modulation" of Free Software as I describe it here. The non–technically inclined reader might keep Wikipedia in mind as an example with which to follow the argument of this book. I will return to it explicitly in part 3. However, for better or for worse, there will be no discussion of pornography.

5 Although the term *public* clearly suggests *private* as its opposite, Free Software is not anticommercial. A very large amount of money, both real and notional, is involved in the creation of Free Software. The term *re-*

cursive market could also be used, in order to emphasize the importance (especially during the 1990s) of the economic features of the practice. The point is not to test whether Free Software is a "public" or a "market," but to construct a concept adequate to the practices that constitute it.

6 See, for example, Warner, *Publics and Counterpublics*, 67–74.

7 Habermas, *The Structural Transformation of the Public Sphere*, esp. 27–43.

8 Critiques of the demand for availability and the putatively inherent superiority of transparency include Coombe and Herman, "Rhetorical Virtues" and "Your Second Life?"; Christen, "Gone Digital"; and Anderson and Bowery, "The Imaginary Politics of Access to Knowledge."

9 This description of Free Software could also be called an "assemblage." The most recent source for this is Rabinow, *Anthropos Today*. The language of thresholds and intensities is most clearly developed by Manuel DeLanda in *A Thousand Years of Non-linear History* and in *Intensive Science and Virtual Philosophy*. The term *problematization*, from Rabinow (which he channels from Foucault), is a synonym for the phrase "reorientation of knowledge and power" as I use it here.

10 See Kelty, "Culture's Open Sources."

11 The genealogy of the term *commons* has a number of sources. An obvious source is Garrett Hardin's famous 1968 article "The Tragedy of the Commons." James Boyle has done more than anyone to specify the term, especially during a 2001 conference on the public domain, which included the inspired guest-list juxtaposition of the appropriation-happy musical collective Negativland and the dame of "commons" studies, Elinor Ostrom, whose book *Governing the Commons* has served as a certain inspiration for thinking about commons versus public domains. Boyle, for his part, has ceaselessly pushed the "environmental" metaphor of speaking for the public domain as environmentalists of the 1960s and 1970s spoke for the environment (see Boyle, "The Second Enclosure Movement and the Construction of the Public Domain" and "A Politics of Intellectual Property"). The term *commons* is useful in this context precisely because it distinguishes the "public domain" as an imagined object of pure public transaction and coordination, as opposed to a "commons," which can consist of privately owned things/spaces that are managed in such a fashion that they effectively function like a "public domain" is imagined to (see Boyle, "The Public Domain"; Hess and Ostrom, *Understanding Knowledge as a Commons*).

12 Marcus and Fischer, *Anthropology as Cultural Critique*; Marcus and Clifford, *Writing Culture*; Fischer, *Emergent Forms of Life and the Anthropological Voice*; Marcus, *Ethnography through Thick and Thin*; Rabinow, *Essays on the Anthropology of Reason* and *Anthropos Today*.

13 The language of "figuring out" has its immediate source in the work of Kim Fortun, "Figuring Out Ethnography." Fortun's work refines two other sources, the work of Bruno Latour in *Science in Action* and that of Hans-Jorg Rheinberger in *Towards History of Epistemic Things*. Latour describes the difference between "science made" and "science in the making" and how the careful analysis of new objects can reveal how they come to be. Rheinberger extends this approach through analysis of the detailed practices involved in figuring out a new object or a new process—practices which participants cannot quite name or explain in precise terms until after the fact.

14 Raymond, *The Cathedral and the Bazaar*.

15 The literature on "virtual communities," "online communities," the culture of hackers and geeks, or the social study of information technology offers important background information, although it is not the subject of this book. A comprehensive review of work in anthropology and related disciplines is Wilson and Peterson, "The Anthropology of Online Communities." Other touchstones are Miller and Slater, *The Internet*; Carla Freeman, *High Tech and High Heels in the Global Economy*; Hine, *Virtual Ethnography*; Kling, *Computerization and Controversy*; Star, *The Cultures of Computing*; Castells, *The Rise of the Network Society*; Boczkowski, *Digitizing the News*. Most social-science work in information technology has dealt with questions of inequality and the so-called digital divide, an excellent overview being DiMaggio et al., "From Unequal Access to Differentiated Use." Beyond works in anthropology and science studies, a number of works from various other disciplines have recently taken up similar themes, especially Adrian MacKenzie, *Cutting Code*; Galloway, *Protocol*; Hui Kyong Chun, *Control and Freedom*; and Liu, *Laws of Cool*. By contrast, if social-science studies of information technology are set against a background of historical and ethnographic studies of "figuring out" problems of specific information technologies, software, or networks, then the literature is sparse. Examples of anthropology and science studies of figuring out include Barry, *Political Machines*; Hayden, *When Nature Goes Public*; and Fortun, *Advocating Bhopal*. Matt Ratto has also portrayed this activity in Free Software in his dissertation, "The Pressure of Openness."

16 In addition to Abbate and Salus, see Norberg and O'Neill, *Transforming Computer Technology*; Naughton, *A Brief History of the Future*; Hafner, *Where Wizards Stay Up Late*; Waldrop, *The Dream Machine*; Segaller, *Nerds 2.0.1*. For a classic autodocumentation of one aspect of the Internet, see Hauben and Hauben, *Netizens*.

17 Kelty, "Culture's Open Sources"; Coleman, "The Social Construction of Freedom"; Ratto, "The Pressure of Openness"; Joseph Feller et al., *Per-*

spectives on Free and Open Source Software; see also http://freesoftware.mit
.edu/, organized by Karim Lakhani, which is a large collection of work on
Free Software projects. Early work in this area derived both from the writings of practitioners such as Raymond and from business and management
scholars who noticed in Free Software a remarkable, surprising set of seem-
ing contradictions. The best of these works to date is Steven Weber, *The
Success of Open Source*. Weber's conclusions are similar to those presented
here, and he has a kind of cryptoethnographic familiarity (that he does not
explicitly avow) with the actors and practices. Yochai Benkler's *Wealth of
Networks* extends and generalizes some of Weber's argument.

18 Max Weber, "Objectivity in the Social Sciences and Social Policy," 68.

19 Despite what might sound like a "shoot first, ask questions later"
approach, the design of this project was in fact conducted according to
specific methodologies. The most salient is actor-network theory: Latour,
Science in Action; Law, "Technology and Heterogeneous Engineering";
Callon, "Some Elements of a Sociology of Translation"; Latour, *Pandora's
Hope*; Latour, *Re-assembling the Social*; Callon, *Laws of the Markets*; Law and
Hassard, *Actor Network Theory and After*. Ironically, there have been no
actor-network studies of networks, which is to say, of particular information and communication technologies such as the Internet. The confusion
of the word *network* (as an analytical and methodological term) with that
of *network* (as a particular configuration of wires, waves, software, and
chips, or of people, roads, and buses, or of databases, names, and diseases)
means that it is necessary to always distinguish *this-network-here* from
any-network-whatsoever. My approach shares much with the ontological
questions raised in works such as Law, *Aircraft Stories*; Mol, *The Body Multiple*; Cussins, "Ontological Choreography"; Charis Thompson, *Making Parents*; and Dumit, *Picturing Personhood*.

20 I understand a concern with scientific infrastructure to begin with
Steve Shapin and Simon Schaffer in *Leviathan and the Air Pump*, but the
genealogy is no doubt more complex. It includes Shapin, *The Social History
of Truth*; Biagioli, *Galileo, Courtier*; Galison, *How Experiments End* and *Image and Logic*; Daston, *Biographies of Scientific Objects*; Johns, *The Nature
of the Book*. A whole range of works explore the issue of scientific tools
and infrastructure: Kohler, *Lords of the Fly*; Rheinberger, *Towards a History of Epistemic Things*; Landecker, *Culturing Life*; Keating and Cambrosio, *Biomedical Platforms*. Bruno Latour's "What Rules of Method for the
New Socio-scientific Experiments" provides one example of where science
studies might go with these questions. Important texts on the subject of
technical infrastructures include Walsh and Bayma, "Computer Networks
and Scientific Work"; Bowker and Star, *Sorting Things Out*; Edwards, *The*

Closed World; Misa, Brey, and Feenberg, *Modernity and Technology*; Star and Ruhleder, "Steps Towards an Ecology of Infrastructure."

21 Dreyfus, *On the Internet*; Dean, "Why the Net Is Not a Public Sphere."

22 In addition, see Lippmann, *The Phantom Public*; Calhoun, *Habermas and the Public Sphere*; Latour and Weibel, *Making Things Public*. The debate about social imaginaries begins alternately with Benedict Anderson's *Imagined Communities* or with Cornelius Castoriadis's *The Imaginary Institution of Society*; see also Chatterjee, "A Response to Taylor's 'Modes of Civil Society'"; Gaonkar, "Toward New Imaginaries"; Charles Taylor, "Modes of Civil Society" and *Sources of the Self*.

1. Geeks and Recursive Publics

1 For the canonical story, see Levy, *Hackers*. *Hack* referred to (and still does) a clever use of technology, usually unintended by the maker, to achieve some task in an elegant manner. The term has been successfully redefined by the mass media to refer to computer users who break into and commit criminal acts on corporate or government or personal computers connected to a network. Many self-identified hackers insist that the criminal element be referred to as *crackers* (see, in particular, the entries on "Hackers," "Geeks" and "Crackers" in The Jargon File, http://www.catb .org/~esr/jargon/, also published as Raymond, *The New Hackers' Dictionary*). On the subject of definitions and the cultural and ethical characteristics of hackers, see Coleman, "The Social Construction of Freedom," chap. 2.

2 One example of the usage of *geek* is in Star, *The Cultures of Computing*. Various denunciations (e.g., Barbrook and Cameron, "The California Ideology"; Borsook, *Technolibertarianism*) tend to focus on journalistic accounts of an ideology that has little to do with what hackers, geeks, and entrepreneurs actually make. A more relevant categorical distinction than that between hackers and geeks is that between geeks and technocrats; in the case of technocrats, the "anthropology of technocracy" is proposed as the study of the limits of technical rationality, in particular the forms through which "planning" creates "gaps in the form that serve as 'targets of intervention'" (Riles, "Real Time," 393). Riles's "technocrats" are certainly not the "geeks" I portray here (or at least, if they are, it is only in their frustrating day jobs). Geeks do have libertarian, specifically Hayekian or Feyerabendian leanings, but are more likely to see technical failures not as failures of planning, but as bugs, inefficiencies, or occasionally as the products of human hubris or stupidity that is born of a faith in planning.

3 See The Geek Code, http://www.geekcode.com/.

4 Geeks are also identified often by the playfulness and agility with which they manipulate these labels and characterizations. See Michael M. J. Fischer, "Worlding Cyberspace" for an example.

5 Taylor, *Modern Social Imaginaries*, 86.

6 On the subject of imagined communities and the role of information technologies in imagined networks, see Green, Harvey, and Knox, "Scales of Place and Networks"; and Flichy, *The Internet Imaginaire*.

7 Taylor, *Modern Social Imaginaries*, 32.

8 Ibid., 33–48. Taylor's history of the transition from feudal nobility to civil society to the rise of republican democracies (however incomplete) is comparable to Foucault's history of the birth of biopolitics, in *La naissance de la biopolitique*, as an attempt to historicize governance with respect to its theories and systems, as well as within the material forms it takes.

9 Ricoeur, *Lectures on Ideology and Utopia*, 2.

10 Geertz, "Ideology as a Cultural System"; Mannheim, *Ideology and Utopia*. Both, of course, also signal the origin of the scientific use of the term proximately with Karl Marx's "German Ideology" and more distantly in the Enlightenment writings of Destutt de Tracy.

11 Geertz, "Ideology as a Cultural System," 195.

12 Ibid., 208–13.

13 The depth and the extent of this issue is obviously huge. Ricoeur's *Lectures on Ideology and Utopia* is an excellent analysis to the problem of ideology prior to 1975. Terry Eagleton's books *The Ideology of the Aesthetic* and *Ideology: An Introduction* are Marxist explorations that include discussions of hegemony and resistance in the context of artistic and literary theory in the 1980s. Slavoj Žižek creates a Lacanian-inspired algebraic system of analysis that combines Marxism and psychoanalysis in novel ways (see Žižek, *Mapping Ideology*). There is even an attempt to replace the concept of ideology with a metaphor of "software" and "memes" (see Balkin, *Cultural Software*). The core of the issue of ideology as a practice (and the vicissitudes of materialism that trouble it) are also at the heart of works by Pierre Bourdieu and his followers (on the relationship of ideology and hegemony, see Laclau and Mouffe, *Hegemony and Socialist Strategy*). In anthropology, see Comaroff and Comaroff, *Ethnography and the Historical Imagination*.

14 Ricoeur, *Lectures on Ideology and Utopia*, 10.

15 Taylor, *Modern Social Imaginaries*, 23.

16 Ibid., 25.

17 Ibid., 26–27.

18 Ibid., 28.

19 The question of gender plagues the topic of computer culture. The gendering of hackers and geeks and the more general exclusion of women in computing have been widely observed by academics. I can do no more here than direct readers to the increasingly large and sophisticated literature on the topic. See especially Light, "When Computers Were Women"; Turkle, *The Second Self* and *Life on the Screen*. With respect to Free Software, see Nafus, Krieger, Leach, "Patches Don't Have Gender." More generally, see Kirkup et al., *The Gendered Cyborg*; Downey, *The Machine in Me*; Faulkner, "Dualisms, Hierarchies and Gender in Engineering"; Grint and Gill, *The Gender-Technology Relation*; Helmreich, *Silicon Second Nature*; Herring, "Gender and Democracy in Computer-Mediated Communication"; Kendall, "'Oh No! I'm a NERD!'"; Margolis and Fisher, *Unlocking the Clubhouse*; Green and Adam, *Virtual Gender*; P. Hopkins, *Sex/Machine*; Wajcman, *Feminism Confronts Technology* and "Reflections on Gender and Technology Studies"; and Fiona Wilson, "Can't Compute, Won't Compute." Also see the novels and stories of Ellen Ullman, including *Close to the Machine* and *The Bug: A Novel*.

20 Originally coined by Steward Brand, the phrase was widely cited after it appeared in Barlow's 1994 article "The Economy of Ideas."

21 On the genesis of "virtual communities" and the role of Steward Brand, see Turner, "Where the Counterculture Met the New Economy."

22 Warner, "Publics and Counterpublics," 51.

23 Ibid., 51–52. See also Warner, *Publics and Counterpublics*, 69.

24 The rest of this message can be found in the Silk-list archives at http://groups.yahoo.com/group/silk-list/message/2869 (accessed 18 August 2006). The reference to "Fling" is to a project now available at http://fling.sourceforge.net/ (accessed 18 August 2006). The full archives of Silk-list can be found at http://groups.yahoo.com/group/silk-list/ and the full archives of the FoRK list can be found at http://www.xent.com/mailman/listinfo/fork/.

25 Vinge, "The Coming Technological Singularity."

26 Moore's Law—named for Gordon Moore, former head of Intel—states that the speed and capacity of computer central processing units (CPUs) doubles every eighteen months, which it has done since roughly 1970. Metcalfe's Law—named for Robert Metcalfe, inventor of Ethernet—states that the utility of a network equals the square of the number of users, suggesting that the number of things one can do with a network increases exponentially as members are added linearly.

27 This quotation from the 1990s is attributed to Electronic Frontier Foundation's founder and "cyber-libertarian" John Gilmore. Whether there is any truth to this widespread belief expressed in the statement is not clear. On the one hand, the protocol to which this folklore refers—the general system of "message switching" and, later, "packet switching" invented by Paul Baran at RAND Corporation—does seem to lend itself to robustness (on this history, see Abbate, *Inventing the Internet*). However, it is not clear that nuclear threats were the only reason such robustness was a design goal; simply to ensure communication in a distributed network was necessary in itself. Nonetheless, the story has great currency as a myth of the nature and structure of the Internet. Paul Edwards suggests that both stories are true ("Infrastructure and Modernity," 216–20, 225n13).

28 Lessig, *Code and Other Laws of Cyberspace*. See also Gillespie, "Engineering a Principle" on the related history of the "end to end" design principle.

29 This is constantly repeated on the Internet and attributed to David Clark, but no one really knows where or when he stated it. It appears in a 1997 interview of David Clark by Jonathan Zittrain, the transcript of which is available at http://cyber.law.harvard.edu/jzfallsem//trans/clark/ (accessed 18 August 2006).

30 Ashish "Hash" Gulhati, e-mail to Silk-list mailing list, 9 September 2000, http://groups.yahoo.com/group/silk-list/message/3125.

31 Eugen Leitl, e-mail to Silk-list mailing list, 9 September 2000, http://groups.yahoo.com/group/silk-list/message/3127. Python is a programming language. Mojonation was a very promising peer-to-peer application in 2000 that has since ceased to exist.

32 In particular, this project focuses on the Transmission Control Protocol (TCP), the User Datagram Protocol (UDP), and the Domain Name System (DNS). The first two have remained largely stable over the last thirty years, but the DNS system has been highly politicized (see Mueller, *Ruling the Root*).

33 On Internet standards, see Schmidt and Werle, *Coordinating Technology*; Abbate and Kahin, *Standards Policy for Information Infrastructure*.

2. Reformers, Polymaths, Transhumanists

1 Foucault, "What Is Enlightenment," 319.

2 Stephenson, *In the Beginning Was the Command Line*.

3 Message-ID: tht55.221960$701.2930569@news4.giganews.com.

4 The Apple-Microsoft conflict was given memorable expression by Umberto Eco in a widely read piece that compared the Apple user interface to Catholicism and the PC user interface to Protestantism ("La bustina di Minerva," *Espresso*, 30 September 1994, back page).

5 One entry on Wikipedia differentiates religious wars from run-of-the-mill "flame wars" as follows: "Whereas a flame war is usually a particular spate of flaming against a non-flamy background, a holy war is a drawn-out disagreement that may last years or even span careers" ("Flaming [Internet]," http://en.wikipedia.org/wiki/Flame_war [accessed 16 January 2006]).

6 Message-ID: 369tva$8l0@csnews.cs.colorado.edu.

7 Message-ID: c1dz4.145472$mb.2669517@news6.giganews.com. It should be noted, in case the reader is unsure how serious this is, that EGCS stood for Extended GNU Compiler System, not Ecumenical GNU Compiler Society.

8 "Martin Luther, Meet Linus Torvalds," *Salon*, 12 November 1998, http://archive.salon.com/21st/feature/1998/11/12feature.html (accessed 5 February 2005).

9 See http://www.stallman.org/saint.html (accessed 5 February 2005) and http://www.dina.kvl.dk/~abraham/religion/ (accessed 5 February 2005). On EMACS, see chapter 6.

10 Message-ID: 6ms27l$6e1@bgtnsc01.worldnet.att.net. In one very humorous case the comparison is literalized "Microsoft acquires Catholic Church" (Message-ID: gaijin-870804300-dragonwing@sec.lia.net).

11 Paul Fusco, "The Gospel According to Joy," *New York Times*, 27 March 1988, Sunday Magazine, 28.

12 See, for example, Matheson, *The Imaginative World of the Reformation*. There is rigorous debate about the relation of print, religion, and capitalism: one locus classicus is Eisenstein's *The Printing Press as an Agent of Change*, which was inspired by McLuhan, *The Gutenberg Galaxy*. See also Ian Green, *Print and Protestantism in Early Modern England* and *The Christian's ABCs*; Chadwick, *The Early Reformation on the Continent*, chaps. 1–3.

13 Crain, *The Story of A*, 16–17.

14 Ibid., 20–21.

15 At a populist level, this was captured by John Perry Barlow's "Declaration of Independence of the Internet," http://homes.eff.org/~barlow/Declaration-Final.html.

16 Foucault, "What Is Enlightenment," 309–10.

17 Ibid., 310.

18 Ibid., 310.

19 Adrian Gropper, interview by author, 28 November 1998.

20 Adrian Gropper, interview by author, 28 November 1998.

21 Sean Doyle, interview by author, 30 March 1999.

22 Feyerabend, *Against Method*, 215–25.

23 One of the ways Adrian discusses innovation is via the argument of the Harvard Business School professor Clayton Christensen's *The Innovator's Dilemma*. It describes "sustaining vs. disruptive" technologies as less an issue of how technologies work or what they are made of, and more an issue of how their success and performance are measured. See Adrian Gropper, "The Internet as a Disruptive Technology," *Imaging Economics*, December 2001, http://www.imagingeconomics.com/library/200112-10.asp (accessed 19 September 2006).

24 On kinds of civic duty, see Fortun and Fortun, "Scientific Imaginaries and Ethical Plateaus in Contemporary U.S. Toxicology."

25 There is, in fact, a very specific group of people called transhumanists, about whom I will say very little. I invoke the label here because I think certain aspects of transhumanism are present across the spectrum of engineers, scientists, and geeks.

26 See the World Transhumanist Association, http://transhumanism .org/ (accessed 1 December 2003) or the Extropy Institute, http://www .extropy.org/ (accessed 1 December 2003). See also Doyle, *Wetwares*, and Battaglia, "For Those Who Are Not Afraid of the Future," for a sidelong glance.

27 Huxley, *New Bottles for New Wine*, 13–18.

28 The computer scientist Bill Joy wrote a long piece in *Wired* warning of the outcomes of research conducted without ethical safeguards and the dangers of eugenics in the past, "Why the Future Doesn't Need Us," *Wired* 8.4 [April 2000], http://www.wired.com/wired/archive/8.04/joy .html (accessed 27 June 2005).

29 Vinge, "The Coming Technological Singularity."

30 Eugen Leitl, e-mail to Silk-list mailing list, 16 May 2000, http://groups .yahoo.com/group/silk-list/message/2410.

31 Eugen Leitl, e-mail to Silk-list mailing list, 7 August 2000, http://groups .yahoo.com/group/silk-list/message/2932.

32 Friedrich A. Hayek, *Law, Legislation and Liberty*, 1:20.

3. The Movement

1 For instance, Richard Stallman writes, "The Free Software movement and the Open Source movement are like two political camps within the free software community. Radical groups in the 1960s developed a reputation for factionalism: organizations split because of disagreements on details of strategy, and then treated each other as enemies. Or at least, such is the

image people have of them, whether or not it was true. The relationship between the Free Software movement and the Open Source movement is just the opposite of that picture. We disagree on the basic principles, but agree more or less on the practical recommendations. So we can and do work together on many specific projects. We don't think of the Open Source movement as an enemy. The enemy is proprietary software" ("Why 'Free Software' Is Better than 'Open Source,'" GNU's Not Unix! http://www.gnu .org/philosophy/free-software-for-freedom.html [accessed 9 July 2006]). By contrast, the Open Source Initiative characterizes the relationship as follows: "How is 'open source' related to 'free software'? The Open Source Initiative is a marketing program for free software. It's a pitch for 'free software' because it works, not because it's the only right thing to do. We're selling freedom on its merits" (http://www.opensource.org/advocacy/faq .php [accessed 9 July 2006]). There are a large number of definitions of Free Software: canonical definitions include Richard Stallman's writings on the Free Software Foundation's Web site, www.fsf.org, including the "Free Software Definition" and "Confusing Words and Phrases that Are Worth Avoiding." From the Open Source side there is the "Open Source Definition" (http://www.opensource.org/licenses/). Unaffiliated definitions can be found at www.freedomdefined.org.

2 Moody, *Rebel Code*, 193.

3 Frank Hecker, quoted in Hamerly and Paquin, "Freeing the Source," 198.

4 See Moody, *Rebel Code*, chap. 11, for a more detailed version of the story.

5 Bruce Perens, "The Open Source Definition," 184.

6 Steven Weber, *The Success of Open Source*.

7 "Netscape Announces Plans to Make Next-Generation Communicator Source Code Available Free on the Net," Netscape press release, 22 January 1998, http://wp.netscape.com/newsref/pr/newsrelease558.html (accessed 25 Sept 2007).

8 On the history of software development methodologies, see Mahoney, "The Histories of Computing(s)" and "The Roots of Software Engineering."

9 Especially good descriptions of what this cycle is like can be found in Ullman, *Close to the Machine* and *The Bug*.

10 Jamie Zawinski, "resignation and postmortem," 31 March 1999, http://www.jwz.org/gruntle/nomo.html.

11 Ibid.

12 Ibid.

13 Ibid.

14 Ibid.

15 "Open Source Pioneers Meet in Historic Summit," press release, 14 April 1998, O'Reilly Press, http://press.oreilly.com/pub/pr/796.

16 See Hamerly and Paquin, "Freeing the Source." The story is elegantly related in Moody, *Rebel Code*, 182–204. Raymond gives Christine Petersen of the Foresight Institute credit for the term *open source*.

17 From Raymond, *The Cathedral and the Bazaar*. The changelog is available online only: http://www.catb.org/~esr/writings/cathedral-bazaar/cathedral-bazaar/.

18 Josh McHugh, "For the Love of Hacking," *Forbes*, 10 August 1998, 94–100.

19 On social movements—the closest analog, developed long ago—see Gerlach and Hine, *People, Power, Change*, and Freeman and Johnson, *Waves of Protest*. However, the Free Software and Open Source Movements do not have "causes" of the kind that conventional movements do, other than the perpetuation of Free and Open Source Software (see Coleman, "Political Agnosticism"; Chan, "Coding Free Software"). Similarly, there is no single development methodology that would cover only Open Source. Advocates of Open Source are all too willing to exclude those individuals or organizations who follow the same "development methodology" but *do not use a Free Software license*—such as Microsoft's oft-mocked "shared-source" program. The list of licenses approved by both the Free Software Foundation and the Open Source Initiative is substantially the same. Further, the Debian Free Software Guidelines and the "Open Source Definition" are almost identical (compare http://www.gnu.org/philosophy/license-list.html with http://www.opensource.org/licenses/ [both accessed 30 June 2006]).

20 It is, in the terms of Actor Network Theory, a process of "enrollment" in which participants find ways to rhetorically align—and to disalign—their interests. It does not constitute the substance of their interest, however. See Latour, *Science in Action*; Callon, "Some Elements of a Sociology of Translation."

21 Coleman, "Political Agnosticism."

22 See, respectively, Raymond, *The Cathedral and the Bazaar*, and Williams, *Free as in Freedom*.

23 For example, Castells, *The Internet Galaxy*, and Weber, *The Success of Open Source* both tell versions of the same story of origins and development.

4. Sharing Source Code

1 "Sharing" source code is not the only kind of sharing among geeks (e.g., informal sharing to communicate ideas), and UNIX is not the only shared software. Other examples that exhibit this kind of proliferation (e.g., the LISP programming language, the TeX text-formatting system) are as ubiquitous as UNIX today. The inverse of my argument here is that selling produces a different kind of order: many products that existed in much larger numbers than UNIX have since disappeared because they were never ported or forked; they are now part of dead-computer museums and collections, if they have survived at all.

2 The story of UNIX has not been told, and yet it has been told hundreds of thousands of times. Every hacker, programmer, computer scientist, and geek tells a version of UNIX history—a usable past. Thus, the sources for this chapter include these stories, heard and recorded throughout my fieldwork, but also easily accessible in academic work on Free Software, which enthusiastically participates in this potted-history retailing. See, for example, Steven Weber, *The Success of Open Source*; Castells, *The Internet Galaxy*; Himanen, *The Hacker Ethic*; Benkler, *The Wealth of Networks*. To date there is but one detailed history of UNIX—*A Quarter Century of UNIX*, by Peter Salus—which I rely on extensively. Matt Ratto's dissertation, "The Pressure of Openness," also contains an excellent analytic history of the events told in this chapter.

3 The intersection of UNIX and TCP/IP occurred around 1980 and led to the famous switch from the Network Control Protocol (NCP) to the Transmission Control Protocol/Internet Protocol that occurred on 1 January 1983 (see Salus, *Casting the Net*).

4 Light, "When Computers Were Women"; Grier, *When Computers Were Human*.

5 There is a large and growing scholarly history of software: Wexelblat, *History of Programming Languages* and Bergin and Gibson, *History of Programming Languages 2* are collected papers by historians and participants. Key works in history include Campbell-Kelly, *From Airline Reservations to Sonic the Hedgehog*; Akera and Nebeker, *From 0 to 1*; Hashagen, Keil-Slawik, and Norberg, *History of Computing—Software Issues*; Donald A. MacKenzie, *Mechanizing Proof*. Michael Mahoney has written by far the most about the early history of software; his relevant works include "The Roots of Software Engineering," "The Structures of Computation," "In Our Own Image," and "Finding a History for Software Engineering." On UNIX in particular, there is shockingly little historical work. Martin Campbell-Kelly and William Aspray devote a mere two pages in their general history *Computer*. As early as 1978, Ken Thompson and Dennis Ritchie were reflecting on the "history" of UNIX in "The UNIX Time-Sharing System: A Retrospective." Ritchie maintains a Web site that contains a valuable collection of early documents and his own reminiscences (http://www

.cs.bell-labs.com/who/dmr/). Mahoney has also conducted interviews with the main participants in the development of UNIX at Bell Labs. These interviews have not been published anywhere, but are drawn on as background in this chapter (interviews are in Mahoney's personal files).

6 Turing, "On Computable Numbers." See also Davis, *Engines of Logic*, for a basic explanation.

7 Sharing programs makes sense in this period only in terms of user groups such as SHARE (IBM) and USE (DEC). These groups were indeed sharing source code and sharing programs they had written (see Akera, "Volunteerism and the Fruits of Collaboration"), but they were constituted around specific machines and manufacturers; brand loyalty and customization were familiar pursuits, but sharing source code across dissimilar computers was not.

8 See Waldrop, *The Dream Machine*, 142–47.

9 A large number of editors were created in the 1970s; Richard Stallman's EMACS and Bill Joy's vi remain the most well known. Douglas Engelbart is somewhat too handsomely credited with the creation of the interactive computer, but the work of Butler Lampson and Peter Deutsch in Berkeley, as well as that of the Multics team, Ken Thompson, and others on early on-screen editors is surely more substantial in terms of the fundamental ideas and problems of manipulating text files on a screen. This story is largely undocumented, save for in the computer-science literature itself. On Engelbart, see Bardini, *Bootstrapping*.

10 See Campbell-Kelly, *From Airline Reservations to Sonic the Hedgehog*.

11 Ibid., 107.

12 Campbell-Kelly and Aspray, *Computer*, 203–5.

13 Ultimately, the Department of Justice case against IBM used bundling as evidence of monopolistic behavior, in addition to claims about the creation of so-called Plug Compatible Machines, devices that were reverse-engineered by meticulously constructing both the mechanical interface and the software that would communicate with IBM mainframes. See Franklin M. Fischer, *Folded, Spindled, and Mutilated*; Brock, *The Second Information Revolution*.

14 The story of this project and the lessons Brooks learned are the subject of one of the most famous software-development handbooks, *The Mythical Man-Month*, by Frederick Brooks.

15 The computer industry has always relied heavily on trade secret, much less so on patent and copyright. Trade secret also produces its own form of order, access, and circulation, which was carried over into the early software industry as well. See Kidder, *The Soul of a New Machine* for a classic account of secrecy and competition in the computer industry.

16 On time sharing, see Lee et al., "Project MAC." Multics makes an appearance in nearly all histories of computing, the best resource by far being Tom van Vleck's Web site http://www.multicians.org/.

17 Some widely admired technical innovations (many of which were borrowed from Multics) include: the hierarchical file system, the command shell for interacting with the system; the decision to treat everything, including external devices, as the same kind of entity (a file), the "pipe" operator which allowed the output of one tool to be "piped" as input to another tool, facilitating the easy creation of complex tasks from simple tools.

18 Salus, *A Quarter Century of UNIX*, 33–37.

19 Campbell-Kelly, *From Airline Reservations to Sonic the Hedgehog*, 143.

20 Ritchie's Web site contains a copy of a 1974 license (http://cm.bell-labs .com/cm/cs/who/dmr/licenses.html) and a series of ads that exemplify the uneasy positioning of UNIX as a commercial product (http://cm.bell-labs .com/cm/cs/who/dmr/unixad.html). According to Don Libes and Sandy Ressler, "The original licenses were source licenses. . . . [C]ommercial institutions paid fees on the order of $20,000. If you owned more than one machine, you had to buy binary licenses for every additional machine [i.e., you were not allowed to copy the source and install it] you wanted to install UNIX on. They were fairly pricey at $8000, considering you couldn't resell them. On the other hand, educational institutions could buy source licenses for several hundred dollars—just enough to cover Bell Labs' administrative overhead and the cost of the tapes" (*Life with UNIX*, 20–21).

21 According to Salus, this licensing practice was also a direct result of Judge Thomas Meaney's 1956 antitrust consent decree which required AT&T to reveal and to license its patents for nominal fees (*A Quarter Century of UNIX*, 56); see also Brock, *The Second Information Revolution*, 116–20.

22 Even in computer science, source code was rarely formally shared, and more likely presented in the form of theorems and proofs, or in various idealized higher-level languages such as Donald Knuth's MIX language for presenting algorithms (Knuth, *The Art of Computer Programming*). Snippets of actual source code are much more likely to be found in printed form in handbooks, manuals, how-to guides, and other professional publications aimed at training programmers.

23 The simultaneous development of the operating system and the norms for creating, sharing, documenting, and extending it are often referred to as the "UNIX philosophy." It includes the central idea that one should build on the ideas (software) of others (see Gancarz, *The Unix Philosophy* and *Linux and the UNIX Philosophy*). See also Raymond, *The Art of UNIX Programming*.

24 Bell Labs threatened the nascent *UNIX NEWS* newsletter with trade-mark infringement, so "USENIX" was a concession that harkened back to the original USE users' group for DEC machines, but avoided explicitly using the name UNIX. Libes and Ressler, *Life with UNIX*, 9.

25 Salus, *A Quarter Century of Unix*, 138.

26 Ibid., emphasis added.

27 Ken Thompson and Dennis Ritchie, "The Unix Operating System," *Bell Systems Technical Journal* (1974).

28 Greg Rose, quoted in Lions, *Commentary*, n.p.

29 Lions, *Commentary*, n.p.

30 Ibid.

31 Tanenbaum's two most famous textbooks are *Operating Systems* and *Computer Networks*, which have seen three and four editions respectively.

32 Tanenbaum was not the only person to follow this route. The other acknowledged giant in the computer-science textbook world, Douglas Comer, created Xinu and Xinu-PC (UNIX spelled backwards) in *Operating Systems Design* in 1984.

33 McKusick, "Twenty Years of Berkeley Unix," 32.

34 Libes and Ressler, *Life with UNIX*, 16–17.

35 A recent court case between the Utah-based SCO—the current owner of the legal rights to the original UNIX source code—and IBM raised yet again the question of how much of the original UNIX source code exists in the BSD distribution. SCO alleges that IBM (and Linus Torvalds) inserted SCO-owned UNIX source code into the Linux kernel. However, the incredibly circuitous route of the "original" source code makes these claims hard to ferret out: it was developed at Bell Labs, licensed to multiple universities, used as a basis for BSD, sold to an earlier version of the company SCO (then known as the Santa Cruz Operation), which created a version called Xenix in cooperation with Microsoft. See the diagram by Eric Lévénez at http://www.levenez.com/unix/. For more detail on this case, see www .groklaw.com.

36 See Vinton G. Cerf and Robert Kahn, "A Protocol for Packet Network Interconnection." For the history, see Abbate, *Inventing the Internet*; Norberg and O'Neill, *A History of the Information Techniques Processing Office*. Also see chapters 1 and 5 herein for more detail on the role of these protocols and the RFC process.

37 Waldrop, *The Dream Machine*, chaps. 5 and 6.

38 The exception being a not unimportant tool called Unix to Unix Copy Protocol, or uucp, which was widely used to transmit data by phone and formed the bases for the creation of the Usenet. See Hauben and Hauben, *Netizens*.

39 Salus, *A Quarter Century of UNIX*, 161.

40 *TCP/IP Digest* 1.6 (11 November 1981) contains Joy's explanation of Berkeley's intentions (Message-ID: anews.aucbvax.5236).

41 See Andrew Leonard, "BSD Unix: Power to the People, from the Code," *Salon*, 16 May 2000, http://archive.salon.com/tech/fsp/2000/05/16/chapter_2_part_one/.

42 Norberg and O'Neill, *A History of the Information Techniques Processing Office*, 184–85. They cite Comer, *Internetworking with TCP/IP*, 6 for the figure.

5. Conceiving Open Systems

1 Quoted in Libes and Ressler, *Life with UNIX*, 67, and also in Critchley and Batty, *Open Systems*, 17. I first heard it in an interview with Sean Doyle in 1998.

2 *Moral* in this usage signals the "moral and social order" I explored through the concept of social imaginaries in chapter 1. Or, in the Scottish Enlightenment sense of Adam Smith, it points to the right organization and relations of exchange among humans.

3 There is, of course, a relatively robust discourse of open systems in biology, sociology, systems theory, and cybernetics; however, that meaning of *open systems* is more or less completely distinct from what *openness* and *open systems* came to mean in the computer industry in the period book-ended by the arrivals of the personal computer and the explosion of the Internet (ca. 1980–93). One relevant overlap between these two meanings can be found in the work of Carl Hewitt at the MIT Media Lab and in the interest in "agorics" taken by K. Eric Drexler, Bernardo Huberman, and Mark S. Miller. See Huberman, *The Ecology of Computation*.

4 Keves, "Open Systems Formal Evaluation Process," 87.

5 General Motors stirred strong interest in open systems by creating, in 1985, its Manufacturing Automation Protocol (MAP), which was built on UNIX. At the time, General Motors was the second-largest purchaser of computer equipment after the government. The Department of Defense and the U.S. Air Force also adopted and required POSIX-compliant UNIX systems early on.

6 Paul Fusco, "The Gospel According to Joy," *New York Times*, 27 March 1988, Sunday Magazine, 28.

7 "Dinosaur" entry, The Jargon File, http://catb.org/jargon/html/D/dinosaur.html.

8 Crichtley and Batty, *Open Systems*, 10.

9 An excellent counterpoint here is Paul Edwards's *The Closed World*, which clearly demonstrates the appeal of a thoroughly and hierarchically controlled system such as the Semi-Automated Ground Environment (SAGE) of the Department of Defense against the emergence of more "green world" models of openness.

10 Crichtley and Batty, *Open Systems*, 13.

11 McKenna, *Who's Afraid of Big Blue?* 178, emphasis added. McKenna goes on to suggest that computer companies can differentiate themselves by adding services, better interfaces, or higher reliability—ironically similar to arguments that the Open Source Initiative would make ten years later.

12 Richard Stallman, echoing the image of medieval manacled wretches, characterized the blind spot thus: "Unix does not give the user any more legal freedom than Windows does. What they mean by 'open systems' is that you can mix and match components, so you can decide to have, say, a Sun chain on your right leg and some other company's chain on your left leg, and maybe some third company's chain on your right arm, and this is supposed to be better than having to choose to have Sun chains on all your limbs, or Microsoft chains on all your limbs. You know, I don't care whose chains are on each limb. What I want is not to be chained by anyone" ("Richard Stallman: High School Misfit, Symbol of Free Software, MacArthur-certified Genius," interview by Michael Gross, Cambridge, Mass., 1999, 5, http://www.mgross.com/MoreThgsChng/interviews/stallman1.html).

13 A similar story can be told about the emergence, in the late 1960s and early 1970s, of manufacturers of "plug-compatible" devices, peripherals that plugged into IBM machines (see Takahashi, "The Rise and Fall of the Plug Compatible Manufacturers"). Similarly, in the 1990s the story of browser compatibility and the World Wide Web Consortium (W3C) standards is another recapitulation.

14 McKenna, *Who's Afraid of Big Blue?* 178.

15 Pamela Gray, *Open Systems*.

16 Eric Raymond, "Origins and History of Unix, 1969–1995," The Art of UNIX Programming, http://www.faqs.org/docs/artu/ch02s01.html #id2880014.

17 Libes and Ressler, *Life with UNIX*, 22. Also noted in Tanenbaum, "The UNIX Marketplace in 1987," 419.

18 Libes and Ressler, *Life with UNIX*, 67.

19 A case might be made that a third definition, the ANSI standard for the C programming language, also covered similar ground, which of course it would have had to in order to allow applications written on one

operating system to be compiled and run on another (see Gray, *Open Systems*, 55–58; Libes and Ressler, *Life with UNIX*, 70–75).

20 "AT&T Deal with Sun Seen," *New York Times*, 19 October 1987, D8.

21 Thomas C. Hayesdallas, "AT&T's Unix Is a Hit at Last, and Other Companies Are Wary," *New York Times*, 24 February 1988, D8.

22 "Unisys Obtains Pacts for Unix Capabilities," *New York Times*, 10 March 1988, D4.

23 Andrew Pollack, "Computer Gangs Stake Out Turf," *New York Times*, 13 December 1988, D1. See also Evelyn Richards, "Computer Firms Get a Taste of 'Gang Warfare,'" *Washington Post*, 11 December 1988, K1; Brit Hume, "IBM, Once the Bully on the Block, Faces a Tough New PC Gang," *Washington Post*, 3 October 1988, E24.

24 "What Is Unix?" The Unix System, http://www.unix.org/what_is_unix/history_timeline.html.

25 "About the Open Group," The Open Group, http://www.opengroup.org/overview/vision-mission.htm.

26 "What Is Unix?" The Unix System, http://www.unix.org/what_is_unix/history_timeline.html.

27 Larry McVoy was an early voice, within Sun, arguing for solving the open-systems problem by turning to Free Software. Larry McVoy, "The Sourceware Operating System Proposal," 9 November 1993, http://www.bitmover.com/lm/papers/srcos.html.

28 The distinction between a protocol, an implementation and a standard is important: *Protocols* are descriptions of the precise terms by which two computers can communicate (i.e., a dictionary and a handbook for communicating). An *implementation* is the creation of software that uses a protocol (i.e., actually does the communicating; thus two implementations using the same protocol should be able to share data. A *standard* defines which protocol should be used by which computers, for what purposes. It may or may not define the protocol, but will set limits on changes to that protocol.

29 The advantages of such an unplanned and unpredictable network have come to be identified in hindsight as a design principle. See Gillespie, "Engineering a Principle" for an excellent analysis of the history of "end to end" or "stupid" networks.

30 William Broad, "Global Network Split as Safeguard," *New York Times*, 5 October 1983, A13.

31 See the incomparable *BBS: The Documentary*, DVD, directed by Jason Scott (Boston: Bovine Ignition Systems, 2005), http://www.bbsdocumentary.com/.

32 Grier and Campbell, "A Social History of Bitnet and Listserv 1985–1991."

33 On Usenet, see Hauben and Hauben, *Netizens.* See also Pfaffenberger, "'A Standing Wave in the Web of Our Communications.'"

34 Schmidt and Werle, *Coordinating Technology,* chap. 7.

35 See, for example, Martin, *Viewdata and the Information Society.*

36 There is little information on the development of open systems; there is, however, a brief note from William Stallings, author of perhaps the most widely used textbook on networking, at "The Origins of OSI," http://williamstallings.com/Extras/OSI.html.

37 Brock, *The Second Information Revolution* is a good introductory source for this conflict, at least in its policy outlines. The Federal Communications Commission issued two decisions (known as "Computer 1" and "Computer 2") that attempted to deal with this conflict by trying to define what counted as voice communication and what as data.

38 Brock, *The Second Information Revolution,* chap. 10.

39 Drake, "The Internet Religious War."

40 Malamud, *Exploring the Internet;* see also Michael M. J. Fischer, "Worlding Cyberspace."

41 The usable past of Giordano Bruno is invoked by Malamud to signal the heretical nature of his own commitment to openly publishing standards that ISO was opposed to releasing. Bruno's fate at the hands of the Roman Inquisition hinged in some part on his acceptance of the Copernican cosmology, so he has been, like Galileo, a natural figure for revolutionary claims during the 1990s.

42 Abbate, *Inventing the Internet;* Salus, *Casting the Net;* Galloway, *Protocol;* and Brock, *The Second Information Revolution.* For practitioner histories, see Kahn et al., "The Evolution of the Internet as a Global Information System"; Clark, "The Design Philosophy of the DARPA Internet Protocols."

43 Kahn et al., "The Evolution of the Internet as a Global Information System," 134–140; Abbate, *Inventing the Internet,* 114–36.

44 Kahn and Cerf, "A Protocol for Packet Network Intercommunication," 637.

45 Clark, "The Design Philosophy of the DARPA Internet Protocols," 54–55.

46 RFCs are archived in many places, but the official site is RFC Editor, http://www.rfc-editor.org/.

47 RFC Editor, RFC 2555, 6.

48 Ibid., 11.

49 This can be clearly seen, for instance, by comparing the various editions of the main computer-networking textbooks: cf. Tanenbaum, *Computer Networks,* 1st ed. (1981), 2d ed. (1988), 3d ed. (1996), and 4th ed. (2003); Stallings, *Data and Computer Communications,* 1st ed. (1985), 2d ed. (1991),

3d ed. (1994), 4th ed. (1997), and 5th ed. (2004); and Comer, *Internetworking with TCP/IP* (four editions between 1991 and 1999).

50 Sunshine, *Computer Network Architectures and Protocols*, 5.

51 The structure of the IETF, the Internet Architecture Board, and the ISOC is detailed in Comer, *Internetworking with TCP/IP*, 8–13; also in Schmidt and Werle, *Coordinating Technology*, 53–58.

52 Message-ID: 673c43e160cia758@sluvca.slu.edu. See also Berners-Lee, *Weaving the Web*.

6. Writing Copyright Licenses

1 The legal literature on Free Software expands constantly and quickly, and it addresses a variety of different legal issues. Two excellent starting points are Vetter, "The Collaborative Integrity of Open-Source Software" and "'Infectious' Open Source Software."

2 Coleman, "The Social Construction of Freedom."

3 "The GNU General Public Licence, Version 2.0," http://www.gnu.org /licenses/old-licenses/gpl-2.0.html.

4 All existing accounts of the hacker ethic come from two sources: from Stallman himself and from the colorful and compelling chapter about Stallman in Steven Levy's *Hackers*. Both acknowledge a prehistory to the ethic. Levy draws it back in time to the MIT Tech Model Railroad Club of the 1950s, while Stallman is more likely to describe it as reaching back to the scientific revolution or earlier. The stories of early hackerdom at MIT are avowedly Edenic, and in them hackers live in a world of uncontested freedom and collegial competition—something like a writer's commune without the alcohol or the brawling. There are stories about a printer whose software needed fixing but was only available under a nondisclosure agreement; about a requirement to use passwords (Stallman refused, chose <return> as his password, and hacked the system to encourage others to do the same); about a programming war between different LISP machines; and about the replacement of the Incompatible Time-Sharing System with DEC's TOPS-20 ("Twenex") operating system. These stories are oft-told usable pasts, but they are not representative. Commercial constraints have always been part of academic life in computer science and engineering: hardware and software were of necessity purchased from commercial manufacturers and often controlled by them, even if they offered "academic" or "educational" licenses.

5 Delanda, "Open Source."

6 Dewey, *Liberalism and Social Action*.

7 *Copyright Act of 1976*, Pub. L. No. 94–553, 90 Stat. 2541, enacted 19 October 1976; and *Copyright Amendments*, Pub. L. No. 96–517, 94 Stat. 3015, 3028 (amending §101 and §117, title 17, *United States Code*, regarding computer programs), enacted 12 December 1980. All amendments since 1976 are listed at http://www.copyright.gov/title17/92preface.html.

8 The history of the copyright and software is discussed in Litman, *Digital Copyright*; Cohen et al., *Copyright in a Global Information Economy*; and Merges, Menell, and Lemley, *Intellectual Property in the New Technological Age*.

9 See Wayner, *Free for All*; Moody, *Rebel Code*; and Williams, *Free as in Freedom*. Although this story could be told simply by interviewing Stallman and James Gosling, both of whom are still alive and active in the software world, I have chosen to tell it through a detailed analysis of the Usenet and Arpanet archives of the controversy. The trade-off is between a kind of incomplete, fly-on-the-wall access to a moment in history and the likely revisionist retellings of those who lived through it. All of the messages referenced here are cited by their "Message-ID," which should allow anyone interested to access the original messages through Google Groups (http://groups.google.com).

10 Eugene Ciccarelli, "An Introduction to the EMACS Editor," MIT Artificial Intelligence Laboratory, AI Lab Memo no. 447, 1978, 2.

11 Richard Stallman, "EMACS: The Extensible, Customizable Self-documenting Display Editor," MIT Artificial Intelligence Laboratory, AI Lab Memo no. 519a, 26 March 1981, 19. Also published as Richard M. Stallman, "EMACS: The Extensible, Customizable Self-documenting Display Editor," *Proceedings of the ACM SIGPLAN SIGOA Symposium on Text Manipulation*, 8–10 June (ACM, 1981), 147–56.

12 Richard Stallman, "EMACS: The Extensible, Customizable Self-documenting Display Editor," MIT Artificial Intelligence Laboratory, AI Lab Memo no. 519a, 26 March 1981, 24.

13 Richard M. Stallman, "EMACS Manual for ITS Users," MIT Artificial Intelligence Laboratory, AI Lab Memo no. 554, 22 October 1981, 163.

14 Back in January of 1983, Steve Zimmerman had announced that the company he worked for, CCA, had created a commercial version of EMACS called CCA EMACS (Message-ID: 385@yetti.uucp). Zimmerman had not written this version entirely, but had taken a version written by Warren Montgomery at Bell Labs (written for UNIX on PDP-11s) and created the version that was being used by programmers at CCA. Zimmerman had apparently distributed it by ftp at first, but when CCA determined that it might be worth something, they decided to exploit it commercially, rather than letting Zimmerman distribute it "freely." By Zimmerman's own account, this whole procedure required ensuring that there was nothing left of the original code by Warren Montgomery that Bell Labs owned (Message-ID: 730@masscomp.uucp).

15 Message-ID for Gosling: bnews.sri-arpa.865.

16 The thread starting at Message-ID: 969@sdcsvax.uucp contains one example of a discussion over the difference between public-domain and commercial software.

17 In particular, a thread discussing this in detail starts at Message-ID: 172@encore.uucp and includes Message-ID: 137@osu-eddie.UUCP, Message-ID: 1127@godot.uucp, Message-ID: 148@osu-eddie.uucp.

18 Message-ID: bnews.sri-arpa.988.

19 Message-ID: 771@mit-eddie.uucp, announced on net.unix-wizards and net.usoft.

20 Message-ID: 771@mit-eddie.uucp.

21 Various other people seem to have conceived of a similar scheme around the same time (if the Usenet archives are any guide), including Guido Van Rossum (who would later become famous for the creation of the Python scripting language). The following is from Message-ID: 5568@mcvax.uucp:

/* This software is copyright (c) Mathematical Centre, Amsterdam,
 * 1983.
 * Permission is granted to use and copy this software, but not for
 * profit,
 * and provided that these same conditions are imposed on any person
 * receiving or using the software.
 */

22 For example, Message-ID: 6818@brl-tgr.arpa.

23 Stallman, "The GNU Manifesto." Available at GNU's Not Unix!, http://www.gnu.org/gnu/manifesto.html.

24 The main file of the controversy was called display.c. A version that was modified by Chris Torek appears in net.sources, Message-ID: 424@umcp-cs.uucp. A separate example of a piece of code written by Gosling bears a note that claims he had declared it public domain, but did not "include the infamous Stallman anti-copyright clause" (Message-ID: 78@tove.uucp).

25 Message-ID: 7773@ucbvax.arpa.

26 Message-ID: 11400007@inmet.uucp.

27 Message-ID: 717@masscomp.uucp.

28 Message-ID: 4421@mit-eddie.uucp.

29 Message-ID: 4486@mit-eddie.uucp. Stallman also recounts this version of events in "RMS Lecture at KTH (Sweden)," 30 October 1986, http://www.gnu.org/philosophy/stallman-kth.html.

30 Message-ID: 2334@sun.uucp.

31 Message-ID: 732@masscomp.uucp.

32 Message-ID: 103@unipress.uucp.

33 With the benefit of hindsight, the position that software could be in the public domain also seems legally uncertain, given that the 1976 changes to USC§17 abolished the requirement to register and, by the same token, to render uncertain the status of code contributed to Gosling and incorporated into GOSMACS.

34 Message-ID: 18@megatest. Note here the use of "once proud hacker ethic," which seems to confirm the perpetual feeling that the ethic has been compromised.

35 Message-ID: 287@mit-athena.uucp.

36 Message-ID: 4559@mit-eddie.uucp.

37 Message-ID: 4605@mit-eddie.uucp.

38 Message-ID: 104@unipress.uucp.

39 Joaquim Martillo, Message-ID: 287@mit-athena.uucpp: "Trying to forbid RMS from using discarded code so that he must spend time to reinvent the wheel supports his contention that 'software hoarders' are slowing down progress in computer science."

40 *Diamond V. Diehr,* 450 U.S. 175 (1981), the Supreme Court decision, forced the patent office to grant patents on software. Interestingly, software patents had been granted much earlier, but went either uncontested or unenforced. An excellent example is patent 3,568,156, held by Ken Thompson, on regular expression pattern matching, granted in 1971.

41 Calvin Mooers, in his 1975 article "Computer Software and Copyright," suggests that the IBM unbundling decision opened the doors to thinking about copyright protection.

42 Message-ID: 933@sdcrdcf.uucp.

43 Gosling's EMACS 264 (Stallman copied EMACS 84) is registered with the Library of Congress, as is GNU EMACS 15.34. Gosling's EMACS Library of Congress registration number is TX-3–407–458, registered in 1992. Stallman's registration number is TX-1–575–302, registered in May 1985. The listed dates are uncertain, however, since there are periodic re-registrations and updates.

44 This is particularly confusing in the case of "dbx." Message-ID: 4437@mit-eddie.uucp, Message-ID: 6238@shasta.arpa, and Message-ID: 730@masscomp.uucp.

45 Message-ID: 4489@mit-eddie.uucp.

46 A standard practice well into the 1980s, and even later, was the creation of so-called clean-room versions of software, in which new programmers and designers who had not seen the offending code were hired to re-implement it in order to avoid the appearance of trade-secret violation. Copyright law is a strict liability law, meaning that ignorance does not absolve the infringer, so the practice of "clean-room engineering" seems not to have been as successful in the case of copyright, as the meaning of infringement remains murky.

47 Message-ID: 730@masscomp.uucp. AT&T was less concerned about copyright infringement than they were about the status of their trade secrets. Zimmerman quotes a statement (from Message-ID: 108@emacs.uucp) that he claims indicates this: "Beginning with CCA EMACS version 162.36z, CCA EMACS no longer contained any of the code from Mr. Montgomery's EMACS, or any methods or concepts which would be known only by programmers familiar with BTL [Bell Labs] EMACS of any version." The statement did not mention copyright, but implied that CCA EMACS did not contain any AT&T trade secrets, thus preserving their software's trade-secret status. The fact that EMACS was a conceptual design—a particular kind of interface, a LISP interpreter, and extensibility—that was very widely imitated had no apparent bearing on the legal status of these secrets.

48 CONTU Final Report, http://digital-law-online.info/CONTU/contu1 .html (accessed 8 December 2006).

49 The cases that determine the meaning of the 1976 and 1980 amendments begin around 1986: *Whelan Associates, Inc. v. Jaslow Dental Laboratory, Inc., et al.*, U.S. Third Circuit Court of Appeals, 4 August 1986, 797 F.2d 1222, 230 USPQ 481, affirming that "structure (or sequence or organization)" of software is copyrightable, not only the literal software code; *Computer Associates International, Inc. v. Altai, Inc.*, U.S. Second Circuit Court of Appeals, 22 June 1992, 982 F.2d 693, 23 USPQ 2d 1241, arguing that the structure test in Whelan was not sufficient to determine infringement and thus proposing a three-part "abstraction-filiation-comparison" test; *Apple Computer, Inc. v. Microsoft Corp*, U.S. Ninth Circuit Court of Appeals, 1994, 35 F.3d 1435, finding that the "desktop metaphor" used in Macintosh and Windows was not identical and thus did not constitute infringement; *Lotus Development Corporation v. Borland International, Inc.* (94–2003), 1996, 513 U.S. 233, finding that the "look and feel" of a menu interface was not copyrightable.

50 The relationship between the definition of *source* and *target* befuddles software law to this day, one of the most colorful examples being the DeCSS case. See Coleman, "The Social Construction of Freedom," chap. 1: Gallery of CSS Descramblers, http://www.cs.cmu.edu/~dst/DeCSS/gallery/.

51 An interesting addendum here is that the manual for EMACS was also released at around the same time as EMACS 16 and was available as a TeX file. Stallman also attempted to deal with the paper document in the same fashion (see Message-ID: 4734@mit-eddie. uucp, 19 July 1985), and this would much later become a different and trickier issue that would result in the GNU Free Documentation License.

52 Message-ID: 659@umcp-cs.uucp.

53 Message-ID: 8605202356.aa12789@ucbvax.berkeley.edu.

54 See Coleman, "The Social Construction of Freedom," chap. 6, on the Debian New Maintainer Process, for an example of how induction into a Free Software project stresses the legal as much as the technical, if not more.

55 For example, Message-ID: 5745@ucbvax.arpa.

56 See Message-ID: 8803031948.aa01085@venus.berkeley.edu.

7. Coordinating Collaborations

1 Research on coordination in Free Software forms the central core of recent academic work. Two of the most widely read pieces, Yochai Benkler's "Coase's Penguin" and Steven Weber's *The Success of Open Source*, are directed at classic research questions about collective action. Rishab Ghosh's "Cooking Pot Markets" and Eric Raymond's *The Cathedral and the Bazaar* set many of the terms of debate. Josh Lerner's and Jean Tirole's "Some Simple Economics of Open Source" was an early contribution. Other important works on the subject are Feller et al., *Perspectives on Free and Open Source Software*; Tuomi, *Networks of Innovation*; Von Hippel, *Democratizing Innovation*.

2 On the distinction between adaptability and adaptation, see Federico Iannacci, "The Linux Managing Model," http://opensource.mit.edu/papers/iannacci2.pdf. Matt Ratto characterizes the activity of Linux-kernel developers as a "culture of re-working" and a "design for re-design," and captures the exquisite details of such a practice both in coding and in the discussion between developers, an activity he dubs the "pressure of openness" that "results as a contradiction between the need to maintain productive collaborative activity and the simultaneous need to remain open to new development directions" ("The Pressure of Openness," 112–38).

3 Linux is often called an operating system, which Stallman objects to on the theory that a kernel is only one part of an operating system. Stallman suggests that it be called GNU/Linux to reflect the use of GNU operating-system tools in combination with the Linux kernel. This not-so-subtle ploy to take credit for Linux reveals the complexity of the distinctions. The kernel is at the heart of hundreds of different "distributions"—such as Debian, Red Hat, SuSe, and Ubuntu Linux—all of which also use GNU tools, but

which are often collections of software larger than just an operating system. Everyone involved seems to have an intuitive sense of what an operating system is (thanks to the pedagogical success of UNIX), but few can draw any firm lines around the object itself.

4 Eric Raymond directed attention primarily to Linux in *The Cathedral and the Bazaar*. Many other projects preceded Torvalds's kernel, however, including the tools that form the core of both UNIX and the Internet: Paul Vixie's implementation of the Domain Name System (DNS) known as BIND; Eric Allman's sendmail for routing e-mail; the scripting languages perl (created by Larry Wall), python (Guido von Rossum), and tcl/tk (John Ousterhout); the X Windows research project at MIT; and the derivatives of the original BSD UNIX, FreeBSD and OpenBSD. On the development model of FreeBSD, see Jorgensen, "Putting It All in the Trunk" and "Incremental and Decentralized Integration in FreeBSD." The story of the genesis of Linux is very nicely told in Moody, *Rebel Code*, and Williams, *Free as in Freedom*; there are also a number of papers—available through Free/Opensource Research Community, http://freesoftware.mit.edu/—that analyze the development dynamics of the Linux kernel. See especially Ratto, "Embedded Technical Expression" and "The Pressure of Openness." I have conducted much of my analysis of Linux by reading the Linux Kernel Mailing List archives, http://lkml.org. There are also annotated summaries of the Linux Kernel Mailing List discussions at http://kerneltraffic.org.

5 Howard Rheingold, *The Virtual Community*. On the prehistory of this period and the cultural location of some key aspects, see Turner, *From Counterculture to Cyberculture*.

6 Julian Dibbell's "A Rape in Cyberspace" and Sherry Turkle's *Life on the Screen* are two classic examples of the detailed forms of life and collaborative ethical creation that preoccupied denizens of these worlds.

7 The yearly influx of students to the Usenet and Arpanet in September earned that month the title "the longest month," due to the need to train new users in use and etiquette on the newsgroups. Later in the 1990s, when AOL allowed subscribers access to the Usenet hierarchy, it became known as "eternal September." See "September that Never Ended," Jargon File, http://catb.org/~esr/jargon/html/S/September-that-never-ended.html.

8 Message-ID: 1991aug25.205708.9541@klaava.helsinki.fi.

9 Message-ID: 12595@star.cs.vu.nl.

10 Indeed, initially, Torvalds's terms of distribution for Linux were more restrictive than the GPL, including limitations on distributing it for a fee or for handling costs. Torvalds eventually loosened the restrictions and switched to the GPL in February 1992. Torvalds's release notes for Linux 0.12 say, "The Linux copyright will change: I've had a couple of requests to make it compatible with the GNU copyleft, removing the 'you may not distribute it for money' condition. I agree. I propose that the copyright be changed so that it conforms to GNU—pending approval of the persons who have helped write code. I assume this is going to be no problem for anybody: If you have grievances ('I wrote that code assuming the copyright would stay the same') mail me. Otherwise The GNU copyleft takes effect as of the first of February. If you do not know the gist of the GNU copyright—read it" (http://www.kernel.org/pub/linux/kernel/Historic/old-versions/RELNOTES-0.12).

11 Message-ID: 12667@star.cs.vu.nl.

12 Message-ID: 12595@star.cs.vu.nl. Key parts of the controversy were reprinted in Dibona et al. *Open Sources.*

13 Steven Weber, *The Success of Open Source,* 164.

14 Quoted in Zack Brown, "Kernel Traffic #146 for 17Dec2001," Kernel Traffic, http://www.kerneltraffic.org/kernel-traffic/kt20011217_146 .html; also quoted in Federico Iannacci, "The Linux Managing Model," http://opensource.mit.edu/papers/iannacci2.pdf.

15 Message-ID: 673c43e160C1a758@sluvca.slu.edu. See also, Berners-Lee, *Weaving the Web.*

16 The original Apache Group included Brian Behlendorf, Roy T. Fielding, Rob Harthill, David Robinson, Cliff Skolnick, Randy Terbush, Robert S. Thau, Andrew Wilson, Eric Hagberg, Frank Peters, and Nicolas Pioch. The mailing list new-httpd eventually became the Apache developers list. The archives are available at http://mail-archives.apache.org/mod_mbox/httpd-dev/ and cited hereafter as "Apache developer mailing list," followed by sender, subject, date, and time.

17 For another version of the story, see Moody, *Rebel Code,* 127–28. The official story honors the Apache Indian tribes for "superior skills in warfare strategy and inexhaustible endurance." Evidence of the concern of the original members over the use of the name is clearly visible in the archives of the Apache project. See esp. Apache developer mailing list, Robert S. Thau, Subject: The political correctness question . . . , 22 April 1995, 21:06 EDT.

18 Mockus, Fielding, and Herbsleb, "Two Case Studies of Open Source Software Development," 3.

19 Apache developer mailing list, Andrew Wilson, Subject: Re: httpd patch B5 updated, 14 March 1995, 21:49 GMT.

20 Apache developer mailing list, Rob Harthill, Subject: Re: httpd patch B5 updated, 14 March 1995, 15:10 MST.

21 Apache developer mailing list, Cliff Skolnick, Subject: Process (please read), 15 March 1995, 3:11 PST; and Subject: Patch file format, 15 March 1995, 3:40 PST.

22 Apache developer mailing list, Rob Harthill, Subject: patch list vote, 15 March 1995, 13:21:24 MST.

23 Apache developer mailing list, Rob Harthill, Subject: apache-0.2 on hyperreal, 18 March 1995, 18:46 MST.

24 Apache developer mailing list, Cliff Skolnick, Subject: Re: patch list vote, 21 March 1995, 2:47 PST.

25 Apache developer mailing list, Paul Richards, Subject: Re: vote counting, 21 March 1995, 22:24 GMT.

26 Roy T. Fielding, "Shared Leadership in the Apache Project."

27 Apache developer mailing list, Robert S. Thau, Subject: Re: 0.7.2b, 7 June 1995, 17:27 EDT.

28 Apache developer mailing list, Robert S. Thau, Subject: My Garage Project, 12 June 1995, 21:14 GMT.

29 Apache developer mailing list, Rob Harthill, Subject: Re: Shambhala, 30 June 1995, 9:44 MDT.

30 Apache developer mailing list, Rob Harthill, Subject: Re: Shambhala, 30 June 1995, 14:50 MDT.

31 Apache developer mailing list, Rob Harthill, Subject: Re: Shambhala, 30 June 1995, 16:48 GMT.

32 Gabriella Coleman captures this nicely in her discussion of the tension between the individual virtuosity of the hacker and the corporate populism of groups like Apache or, in her example, the Debian distribution of Linux. See Coleman, The Social Construction of Freedom.

33 Apache developer mailing list, Robert S. Thau, Subject: Re: Shambhala, 1 July 1995, 14:42 EDT.

34 A slightly different explanation of the role of modularity is discussed in Steven Weber, *The Success of Open Source*, 173–75.

35 Tichy, "RCS."

36 See Steven Weber, *The Success of Open Source*, 117–19; Moody, *Rebel Code*, 172–78. See also Shaikh and Cornford, "Version Management Tools."

37 Linus Torvalds, "Re: [PATCH] Remove Bitkeeper Documentation from Linux Tree," 20 April 2002, http://www.uwsg.indiana.edu/hypermail/linux/kernel/0204.2/1018.html. Quoted in Shaikh and Cornford, "Version Management Tools."

38 Andrew Orlowski, "'Cool it, Linus'—Bruce Perens," *Register*, 15 April 2005, http://www.theregister.co.uk/2005/04/15/perens_on_torvalds/page2.html.

39 Similar debates have regularly appeared around the use of non-free compilers, non-free debuggers, non-free development environments, and so forth. There are, however, a large number of people who write and promote Free Software that runs on proprietary operating systems like Macintosh and Windows, as well as a distinction between tools and formats. So, for instance, using Adobe Photoshop to create icons is fine so long as they are in standard open formats like PNG or JPG, and not proprietary forms such as GIF or photoshop.

40 Quoted in Jeremy Andrews, "Interview: Larry McVoy," Kernel Trap, 28 May 2002, http://Kerneltrap.org/node/222.

41 Steven Weber, *The Success of Open Source*, 132.

42 Raymond, *The Cathedral and the Bazaar*.

43 Gabriella Coleman, in "The Social Construction of Freedom," provides an excellent example of a programmer's frustration with font-lock in EMACS, something that falls in between a bug and a feature. The programmer's frustration is directed at the stupidity of the design (and implied designers), but his solution is not a fix, but a work-around—and it illustrates how debugging does not always imply collaboration.

44 Dan Wallach, interview, 3 October 2003.

45 Mitchell Waldrop's *The Dream Machine* details the family history well.

8. "If We Succeed, We Will Disappear"

1 In January 2005, when I first wrote this analysis, this was true. By April 2006, the Hewlett Foundation had convened the Open Educational Resources "movement" as something that would transform the production and circulation of textbooks like those created by Connexions. Indeed, in Rich Baraniuk's report for Hewlett, the first paragraph reads: "A grassroots movement is on the verge of sweeping through the academic world. The *open education movement* is based on a set of intuitions that are shared by a remarkably wide range of academics: that knowledge should be free and open to use and re-use; that collaboration should be easier, not harder; that people should receive credit and kudos for contributing to education and research; and that concepts and ideas are linked in unusual and surprising ways and not the simple linear forms that textbooks present. Open education promises to fundamentally change the way authors, instructors, and students interact worldwide" (Baraniuk and King, "Connexions"). (In a nice confirmation of just how embedded participation can become in anthropology, Baraniuk cribbed the second sentence from something I had written two years earlier as part of a description of what I thought Connexions hoped to achieve.) The "movement" as such still does not quite exist, but the momentum for it is clearly part of the actions that Hewlett hopes to achieve.

2 See Chris Beam, "Fathom.com Shuts Down as Columbia Withdraws," *Columbia Spectator*, 27 January 2003, http://www.columbiaspectator.com/. Also see David Noble's widely read critique, "Digital Diploma Mills."

3 "Provost Announces Formation of Council on Educational Technology," *MIT Tech Talk*, 29 September 1999, http://web.mit.edu/newsoffice/1999/council-0929.html.

4 The software consists of a collection of different Open Source Software cobbled together to provide the basic platform (the Zope and Plone content-management frameworks, the PostGresQL database, the python programming language, and the cvs version-control software).

5 The most significant exception has been the issue of tools for authoring content in XML. For most of the life of the Connexions project, the XML mark-up language has been well-defined and clear, but there has been no way to write a module in XML, short of directly writing the text and the tags in a text editor. For all but a very small number of possible users, this feels too much like programming, and they experience it as too frustrating to be worth it. The solution (albeit temporary) was to encourage users to make use of a proprietary XML editor (like a word processor, but capable of creating XML content). Indeed, the Connexions project's devotion to openness was tested by one of the most important decisions its participants made: to pursue the creation of an Open Source XML text editor in order to provide access to completely open tools for creating completely open content.

6 Boyle, "Mertonianism Unbound," 14.

7 The movement is the component that remains unmodulated: there is no "free textbook" movement associated with Connexions, even though many of the same arguments that lead to a split between Free Software and Open Source occur here: the question of whether the term *free* is confusing, for example, or the role of for-profit publishers or textbook companies. In the end, most (though not all) of the Connexions staff and many of its users are content to treat it as a useful tool for composing novel kinds of digital educational material—not as a movement for the liberation of educational content.

8 Boyle, "Conservatives and Intellectual Property."

9 Lessig's output has been prodigious. His books include *Code and Other Laws of Cyber Space, The Future of Ideas, Free Culture*, and *Code: Version 2.0*. He has also written a large number of articles and is an active blogger (http://www.lessig.org/blog/).

10 There were few such projects under way, though there were many in the planning stages. Within a year, the Public Library of Science had launched itself, spearheaded by Harold Varmus, the former director of the National Institutes of Health. At the time, however, the only other large scholarly project was the MIT Open Course Ware project, which, although it had already agreed to use Creative Commons licenses, had demanded a peculiar one-off license.

11 The fact that I organized a workshop to which I invited "informants" and to which I subsequently refer as research might strike some, both in anthropology and outside it, as wrong. But it is precisely the kind of occasion I would argue has become central to the problematics of method in cultural anthropology today. On this subject, see Holmes and Marcus, "Cultures of Expertise and the Management of Globalization." Such strategic and seemingly ad hoc participation does not exclude one from attempting to later disentangle oneself from such participation, in order to comment on the value and significance, and especially to offer critique. Such is the attempt to achieve objectivity in social science, an objectivity that goes beyond the basic notions of bias and observer-effect so common in the social sciences. "Objectivity" in a broader social sense includes the observation of the conceptual linkages that both precede such a workshop (constituted the need for it to happen) and follow on it, independent of any particular meeting. The complexity of mobilizing objectivity in discussions of the value and significance of social or economic phenomena was well articulated a century ago by Max Weber, and problems of method in the sense raised by him seem to me to be no less fraught today. See Max Weber, "Objectivity in the Social Sciences."

12 *Suntrust v. Houghton Mifflin Co.*, U.S. Eleventh Circuit Court of Appeals, 2001, 252 F. 3d 1165.

13 Neil Strauss, "An Uninvited Bassist Takes to the Internet," *New York Times*, 25 August 2002, sec. 2, 23.

14 Indeed, in a more self-reflective moment, Glenn once excitedly wrote to me to explain that what he was doing was "code-switching" and that he thought that geeks who constantly involved themselves in technology, law, music, gaming, and so on would be prime case studies for a code-switching study by anthropologists.

15 See Kelty, "Punt to Culture."

16 Lessig, "The New Chicago School."

17 Hence, Boyle's "Second Enclosure Movement" and "copyright conservancy" concepts (see Boyle, "The Second Enclosure Movement"; Bollier, *Silent Theft*). Perhaps the most sophisticated and compelling expression of the institutional-economics approach to understanding Free Software is the work of Yochai Benkler, especially "Sharing Nicely" and "Coase's Penguin." See also Benkler, *Wealth of Networks*.

18 Steven Weber's *The Success of Open Source* is exemplary.

19 Carrington and King, "Law and the Wisconsin Idea."

20 In particular, Glenn Brown suggested Oliver Wendell Holmes as a kind of origin point both for critical legal realism and for law and economics, a kind of filter through which lawyers get both their Nietzsche and their liberalism (see Oliver Wendell Holmes, "The Path of the Law"). Glenn's opinion was that what he called "punting to culture" (by which he meant writing minimalist laws which allow social custom to fill in the details) descended more or less directly from the kind of legal reasoning embodied in Holmes: "Note that [Holmes] is probably best known in legal circles for arguing that questions of morality be removed from legal analysis and left to the field of ethics. this is what makes him the godfather of both the posners of the world, and the crits, and the strange hybrids like lessig" (Glenn Brown, personal communication, 11 August 2003).

9. Reuse, Modification, Norms

1 Actor-network theory comes closest to dealing with such "ontological" issues as, for example, airplanes in John Law's *Aircraft Stories*, the disease atheroscleroris in Annemarie Mol's *The Body Multiple*, or in vitro fertilization in Charis Thompson's *Making Parents*. The focus here on finality is closely related, but aims at revealing the temporal characteristics of highly modifiable kinds of knowledge-objects, like textbooks or databases, as in Geoffrey Bowker's *Memory Practices in the Sciences*.

2 Merton, "The Normative Structure of Science."

3 See Johns, *The Nature of the Book*; Eisenstein, *The Printing Press as an Agent of Change*; McLuhan, *The Gutenberg Galaxy* and *Understanding Media*; Febvre and Martin, *The Coming of the Book*; Ong, *Ramus, Method, and the Decay of Dialogue*; Chartier, *The Cultural Uses of Print in Early Modern France* and *The Order of Books*; Kittler, *Discourse Networks 1800/1900* and *Gramophone, Film, Typewriter*.

4 There is less communication between the theorists and historians of copyright and authorship and those of the book; the former are also rich in analyses, such as Jaszi and Woodmansee, *The Construction of Authorship*; Mark Rose, *Authors and Owners*; St. Amour, *The Copyrights*; Vaidhyanathan, *Copyrights and Copywrongs*.

5 Eisenstein, *The Printing Press as an Agent of Change*. Eisenstein's work makes direct reference to McLuhan's thesis in *The Gutenberg Galaxy*, and Latour relies on these works and others in "Drawing Things Together."

6 Johns, *The Nature of the Book*, 19–20.

7 On this subject, cf. Pablo Boczkowski's study of the digitization of newspapers, *Digitizing the News*.

8 *Conventional* here is actually quite historically proximate: the system creates a pdf document by translating the XML document into a LaTeX document, then into a pdf document. LaTeX has been, for some twenty years, a standard text-formatting and typesetting language used by some

sectors of the publishing industry (notably mathematics, engineering, and computer science). Were it not for the existence of this standard from which to bootstrap, the Connexions project would have faced a considerably more difficult challenge, but much of the infrastructure of publishing has already been partially transformed into a computer-mediated and -controlled system whose final output is a printed book. Later in Connexions's lifetime, the group coordinated with an Internet-publishing startup called Qoop.com to take the final step and make Connexions courses available as print-on-demand, cloth-bound textbooks, complete with ISBNs and back-cover blurbs.

9 See Johns, *The Nature of the Book*; Warner, *The Letters of the Republic*.

10 On fixity, see Eisenstein's *The Printing Press as an Agent of Change* which cites McLuhan's *The Gutenberg Galaxy*. The stability of texts is also questioned routinely by textual scholars, especially those who work with manuscripts and complicated *varoria* (for an excellent introduction, see Bornstein and Williams, *Palimpsest*). Michel Foucault's "What Is an Author?" addresses a related but orthogonal problematic and is unconcerned with the relatively sober facts of a changing medium.

11 A salient and recent point of comparison can be found in the form of Lawrence Lessig's "second edition" of his book *Code*, which is titled *Code: Version 2.0* (*version* is used in the title, but *edition* is used in the text). The first book was published in 1999 ("ancient history in Internet time"), and Lessig convinced the publisher to make it available as a wiki, a collaborative Web site which can be directly edited by anyone with access. The wiki was edited and updated by hordes of geeks, then "closed" and reedited into a second edition with a new preface. It is a particularly tightly controlled example of collaboration; although the wiki and the book were freely available, the modification and transformation of them did not amount to a simple free-for-all. Instead, Lessig leveraged his own authority, his authorial voice, and the power of Basic Books to create something that looks very much like a traditional second edition, although it was created by processes unimaginable ten years ago.

12 The most familiar comparison is Wikipedia, which was started after Connexions, but grew far more quickly and dynamically, largely due to the ease of use of the system (a bone of some contention among the Connexions team). Wikipedia has come under assault primarily for being unreliable. The suspicion and fear that surround Wikipedia are similar to those that face Connexions, but in the case of Wikipedia entries, the commitment to openness is stubbornly meritocratic: any article can be edited by anyone at anytime, and it matters not how firmly one is identified as an expert by rank, title, degree, or experience—a twelve year old's knowledge of the Peloponnesian War is given the same access and status as an eighty-year-old classicist's. Articles are not owned by individuals, and

all work is pseudonymous and difficult to track. The range of quality is therefore great, and the mainstream press has focused largely on whether Wikipedia is more or less reliable than conventional encyclopedias, not on the process of knowledge production. See, for instance, George Johnson, "The Nitpicking of the Masses vs. the Authority of the Experts," *New York Times*, 3 January 2006, Late Edition—Final, F2; Robert McHenry, "The Faith-based Encyclopedia," *TCS Daily*, 15 November 2004, http://www .techcentralstation.com/111504A.html.

13 Again, a comparison with Wikipedia is apposite. Wikipedia is, morally speaking, and especially in the persona of its chief editor, Jimbo Wales, totally devoted to merit-based equality, with users getting no special designation beyond the amount and perceived quality of the material they contribute. Degrees or special positions of employment are anathema. It is a quintessentially American, anti-intellectual-fueled, Horatio Alger–style approach in which the slate is wiped clean and contributors are given a chance to prove themselves independent of background. Connexions, by contrast, draws specifically from the ranks of intellectuals or academics and seeks to replace the infrastructure of publishing. Wikipedia is interested only in creating a better encyclopedia. In this respect, it is transhumanist in character, attributing its distinctiveness and success to the advances in technology (the Internet, wiki, broadband connections, Google). Connexions on the other hand is more polymathic, devoted to intervening into the already complexly constituted organizational practice of scholarship and academia.

14 An even more technical feature concerned the issue of the order of authorship. The designers at first decided to allow Connexions to simply display the authors in alphabetical order, a practice adopted by some disciplines, like computer science. However, in the case of the Housman example this resulted in what looked like a module authored principally by me, and only secondarily by A. E. Housman. And without the ability to explicitly designate order of authorship, many disciplines had no way to express their conventions along these lines. As a result, the system was redesigned to allow users to designate the order of authorship as well.

15 I refer here to Eric Raymond's "discovery" that hackers possess unstated norms that govern what they do, in addition to the legal licenses and technical practices they engage in (see Raymond, "Homesteading the Noosphere"). For a critique and background on hacker ethics and norms, see Coleman, "The Social Construction of Freedom."

16 Bruno Latour's *Science in Action* makes a strong case for the centrality of "black boxes" in science and engineering for precisely this reason.

17 I should note, in my defense, that my efforts to get my informants to read Max Weber, Ferdinand Tönnies, Henry Maine, or Emile Durkheim

proved far less successful than my creation of nice Adobe Illustrator diagrams that made explicit the reemergence of issues addressed a century ago. It was not for lack of trying, however.

18 Callon, *The Laws of the Markets*; Hauser, *Moral Minds*.

19 Oliver Wendell Holmes, "The Path of Law."

20 In December 2006 Creative Commons announced a set of licenses that facilitate the "follow up" licensing of a work, especially one initially issued under a noncommercial license.

21 Message from the cc-sampling mailing list, Glenn Brown, Subject: BACKGROUND: "AS APPROPRIATE TO THE MEDIUM, GENRE, AND MARKET NICHE," 23 May 2003, http://lists.ibiblio.org/pipermail/cc-sampling/2003-May/000004.html.

22 Sampling offers a particularly clear example of how Creative Commons differs from the existing practice and infrastructure of music creation and intellectual-property law. The music industry has actually long recognized the fact of sampling as something musicians do and has attempted to deal with it by making it an explicit economic practice; the music industry thus encourages sampling by facilitating the sale between labels and artists of rights to make a sample. Record companies will negotiate prices, lengths, quality, and quantity of sampling and settle on a price.

This practice is set opposite the assumption, also codified in law, that the public has a right to a fair use of copyrighted material without payment or permission. Sampling a piece of music might seem to fall into this category of use, except that one of the tests of fair use is that the use not impact any existing market for such uses, and the fact that the music industry has effectively created a market for the buying and selling of samples means that sampling now routinely falls outside the fair uses codified in the statute, thus removing sampling from the domain of fair use. Creative Commons licenses, on the other hand, say that owners should be able to designate their material as "sample-able," to give permission ahead of time, and by this practice to encourage others to do the same. They give an "honorable" meaning to the practice of sampling for free, rather than the dishonorable one created by the industry. It thus becomes a war over the meaning of norms, in the law-and-economics language of Creative Commons and its founders.

Conclusion

1 See http://cnx.org, http://www.creativecommons.org, http://www.earlham.edu/~peters/fos/overview.htm, http://www.biobricks.org, http://www.freebeer.org, http://freeculture.org, http://www.cptech.org/

a2k, http://www.colawp.com/colas/400/cola467_recipe.html, http://www.elephantsdream.org, http://www.sciencecommons.org, http://www.plos.org, http://www.openbusiness.cc, http://www.yogaunity.org, http://osdproject.com, http://www.hewlett.org/Programs/Education/oer/, and http://olpc.com.

2 See Clive Thompson, "Open Source Spying," *New York Times Magazine*, 3 December 2006, 54.

3 See especially Christen, "Tracking Properness" and "Gone Digital"; Brown, *Who Owns Native Culture?* and "Heritage as Property." Crowdsourcing fits into other novel forms of labor arrangements, ranging from conventional outsourcing and off-shoring to newer forms of bodyshopping and "virtual migration" (see Aneesh, *Virtual Migration*; Xiang, *"Global Bodyshopping"*).

4 Golub, "Copyright and Taboo"; Dibbell, *Play Money.*

Bibliography

Abbate, Janet. *Inventing the Internet*. Cambridge, Mass: MIT Press, 1999.

Abbate, Janet, and Brian Kahin, eds. *Standards Policy for Information Infrastructure*. Cambridge, Mass.: MIT Press, 1995.

Abelson, Harold, and Gerald J. Sussman. *The Structure and Interpretation of Computer Programs*. Cambridge, Mass.: MIT Press, 1985.

Akera, Atsushi. "Volunteerism and the Fruits of Collaboration: The IBM User Group SHARE." *Technology and Culture* 42.4 (October 2001): 710–736.

Akera, Atsushi, and Frederik Nebeker, eds. *From 0 to 1: An Authoritative History of Modern Computing*. New York: Oxford University Press, 2002.

Anderson, Benedict. *Imagined Communities: Reflections on the Origins and Spread of Nationalism*. London: Verso, 1983.

Anderson, Jane, and Kathy Bowery. "The Imaginary Politics of Access to Knowledge." Paper presented at the Contexts of Invention Conference, Cleveland, Ohio, 20–23 April 2006.

Aneesh, A. *Virtual Migration: The Programming of Globalization*. Durham, N.C.: Duke University Press, 2006.

Arendt, Hannah. *The Human Condition*. 2d ed. University of Chicago Press, 1958.

Balkin, Jack. *Cultural Software: A Theory of Ideology*. New Haven, Conn.: Yale University Press, 1998.

Baraniuk, Richard, and W. Joseph King. "Connexions: Sharing Knowledge and Building Communities." *Sloan-C Review: Perspectives in Quality Online Education* 4.9 (September 2005): 8. http://www.aln.org/publications/view/v4n9/coverv4n9.htm.

Barbrook, Richard, and Andy Cameron. "The California Ideology." *Science as Culture* 26 (1996): 44–72.

Bardini, Thierry. *Bootstrapping: Douglas Engelbart, Co-evolution and the Origins of Personal Computing*. Stanford, Calif.: Stanford University Press, 2001.

Barlow, John Perry. "The Economy of Ideas." *Wired* 2.3 (March 1994).

Barry, Andrew. *Political Machines: Governing a Technological Society*. London: Athlone Press, 2001.

Battaglia, Deborah. "'For Those Who Are Not Afraid of the Future': Raëlian Clonehood in the Public Sphere." In *E.T. Culture: Anthropology in Outerspaces*, ed. Deborah Battaglia, 149–79. Durham, N.C.: Duke University Press, 2005.

Benkler, Yochai. "Coase's Penguin, or Linux and the Nature of the Firm." *Yale Law Journal* 112.3 (2002): 369–446.

———. "Sharing Nicely: On Shareable Goods and the Emergence of Sharing as a Modality of Economic Production." *Yale Law Journal* 114.2 (2004): 273–358.

———. *The Wealth of Networks: How Social Production Transforms Markets and Freedom*. New Haven, Conn.: Yale University Press, 2006.

Bergin, Thomas J., Jr., and Richard G. Gibson Jr., eds. *History of Programming Languages 2*. New York: Association for Computing Machinery Press, 1996.

Berners-Lee, Tim, with Mark Fischetti. *Weaving the Web: The Original Design and Ultimate Destiny of the World Wide Web by Its Inventor*. San Francisco: Harper San Francisco, 1999.

Biagioli, Mario. *Galileo, Courtier: The Practice of Science in the Culture of Absolutism*. Chicago: University of Chicago Press, 1993.

Boczkowski, Pablo. *Digitizing the News: Innovation in Online Newspapers*. Cambridge, Mass.: MIT Press, 2004.

Bollier, David. *Silent Theft: The Private Plunder of Our Common Wealth*. New York: Routledge, 2002.

Bornstein, George, and Ralph G. Williams, eds. *Palimpsest: Editorial Theory in the Humanities*. Ann Arbor: University of Michigan Press, 1993.

Borsook, Paulina. *Cyberselfish: A Critical Romp through the Terribly Libertarian Culture of High Tech*. New York: Public Affairs, 2000.

Bowker, Geoffrey. *Memory Practices in the Sciences*. Cambridge, Mass.: MIT Press, 2006.

Bowker, Geoffrey C., and Susan Leigh Star. *Sorting Things Out: Classification and Its Consequences*. Cambridge, Mass.: MIT Press, 1999.

Boyle, James. "Conservatives and Intellectual Property." *Engage* 1 (April 2000): 83. http://www.law.duke.edu/boylesite/Federalist.htm.

———. "Mertonianism Unbound? Imagining Free, Decentralized Access to Most Cultural and Scientific Material." In *Understanding Knowledge as a Common: From Theory to Practice*, ed. Charlotte Hess and Elinor Ostrom, 123–44. Cambridge, Mass.: MIT Press, 2006. http://www.james-boyle .com/mertonianism.pdf.

———. "A Politics of Intellectual Property: Environmentalism for the Net?" *Duke Law Journal* 47.1 (October 1997): 87–116.

———, ed. "The Public Domain." Special issue, *Law and Contemporary Problems* 66.1–2 (winter–spring 2003).

———. "The Second Enclosure Movement and the Construction of the Public Domain." In James Boyle, ed., "The Public Domain," special issue, *Law and Contemporary Problems* 66.1–2 (winter–spring 2003): 33–74.

Brock, Gerald. *The Second Information Revolution*. Cambridge, Mass.: Harvard University Press, 2003.

Brooks, Frederick. *The Mythical Man-month: Essays on Software Engineering*. Reading, Mass.: Addison-Wesley, 1975.

Brown, Michael. "Heritage as Property." In *Property in Question: Value Transformation in the Global Economy*, ed. Katherine Verdery and Caroline Humphrey, 49–68. Oxford: Berg, 2004.

———. *Who Owns Native Culture?* Cambridge, Mass.: Harvard University Press, 2003.

Calhoun, Craig, ed. *Habermas and the Public Sphere*. Cambridge, Mass.: MIT Press, 1992.

Callon, Michel. *The Laws of the Markets*. London: Blackwell, 1998.

———. "Some Elements of a Sociology of Translation: Domestication of the Scallops and the Fishermen of St Brieuc Bay." In *Power, Action and Belief: A New Sociology of Knowledge*, ed. John Law, 196–233. London: Routledge and Kegan Paul, 1986.

Callon, Michel, Cécile Méadel, and Vololona Rabeharisoa. "The Economy of Qualities." *Economy and Society* 31.2 (May 2002): 194–217.

Campbell-Kelly, Martin. *From Airline Reservations to Sonic the Hedgehog: A History of the Software Industry*. Cambridge, Mass.: MIT Press, 2003.

Campbell-Kelly, Martin, and William Aspray. *Computer: A History of the Information Machine*. New York: Basic Books, 1996.

Carrington, Paul D., and Erika King. "Law and the Wisconsin Idea." *Journal of Legal Education* 47 (1997): 297.

Castells, Manuel. *The Internet Galaxy: Reflections on the Internet, Business and Society*. New York: Oxford University Press, 2001.

———. *The Rise of the Network Society*. Cambridge, Mass.: Blackwell, 1996.

Castoriadis, Cornelius. *The Imaginary Institution of Society*. Cambridge, Mass.: MIT Press, 1987.

Cerf, Vinton G., and Robert Kahn. "A Protocol for Packet Network Inter-connection." *IEEE Transactions on Communications* 22.5 (May 1974): 637–48.

Chadwick, Owen. *The Early Reformation on the Continent*. Oxford: Oxford University Press, 2001.

Chan, Anita. "Coding Free Software, Coding Free States: Free Software Legislation and the Politics of Code in Peru." *Anthropological Quarterly* 77.3 (summer 2004): 531–45.

Chartier, Roger. *The Cultural Uses of Print in Early Modern France*. Princeton: Princeton University Press, 1988.

———. *The Order of Books: Readers, Authors, and Libraries in Europe between the Fourteenth and Eighteenth Centuries*. Trans. Lydia G. Cochrane. Stanford, Calif.: Stanford University Press, 1994.

Chatterjee, Partha. "A Response to Taylor's 'Modes of Civil Society.'" *Public Culture* 3.1 (1990): 120–21.

Christen, Kim. "Gone Digital: Aboriginal Remix and the Cultural Commons." *International Journal of Cultural Property* 12 (August 2005): 315–45.

———. "Tracking Properness: Repackaging Culture in a Remote Australian Town." *Cultural Anthropology* 21.3 (August 2006): 416–46.

Christensen, Clayton. *The Innovator's Dilemma: When New Technologies Cause Great Firms to Fail*. Boston: Harvard Business School Press, 1997.

Chun, Wendy Hui Kyong. *Control and Freedom: Power and Paranoia in the Age of Fiber Optics*. Cambridge, Mass.: MIT Press, 2006.

Clark, David. "The Design Philosophy of the DARPA Internet Protocols." 1988. In *Computer Communications: Architectures, Protocols, and Standards*, 3d ed., ed. William Stallings, 54–62. Los Alamitos, Calif.: IEEE Computer Society Press, 1992.

Cohen, Julie, Lydia Pallas Loren, Ruth Gana Okediji, and Maureen O'Rourke, eds. *Copyright in a Global Information Economy*. Aspen, Colo.: Aspen Law and Business Publishers, 2001.

Coleman, E. Gabriella. "The Political Agnosticism of Free and Open Source Software and the Inadvertent Politics of Contrast." *Anthropological Quarterly* 77.3 (summer 2004): 507–19.

———. "The Social Construction of Freedom: Hackers, Ethics and the Liberal Tradition." Ph.D. diss., University of Chicago, 2005.

Comaroff, Jean, and John Comaroff. *Ethnography and the Historical Imagination.* Boulder, Colo.: Westview, 1992.

Comer, Douglas E. *Internetworking with TCP/IP.* 4th ed. Upper Saddle River, N.J.: Prentice Hall, 2000.

———. *Operating System Design.* 1st ed. 2 vols. Englewood Cliffs, N.J.: Prentice Hall, 1984.

Coombe, Rosemary, and Andrew Herman. "Rhetorical Virtues: Property, Speech, and the Commons on the World-Wide Web." *Anthropological Quarterly* 77.3 (summer 2004): 559–574.

———. "Your Second Life? Goodwill and the Performativity of Intellectual Property in Online Digital Gaming." *Cultural Studies* 20.2–3 (March–May 2006): 184–210.

Crain, Patricia. *The Story of A: The Alphabetization of America from* The New England Primer *to* The Scarlet Letter. Stanford, Calif.: Stanford University Press, 2000.

Critchley, Terry A., and K. C. Batty. *Open Systems: The Reality.* Englewood Cliffs, N.J.: Prentice Hall, 1993.

Cussins, Charis. "Ontological Choreography: Agency through Objectification in Infertility Clinics." *Social Studies of Science* 26.3 (1996): 575–610.

Daston, Lorraine, ed. *Biographies of Scientific Objects.* Chicago: University of Chicago Press, 2000.

Davis, Martin. *Engines of Logic: Mathematicians and the Origin of the Computer.* W. W. Norton, 2001.

Dean, Jodi. "Why the Net Is Not a Public Sphere." *Constellations* 10.1 (March 2003): 95.

DeLanda, Manuel. *Intensive Science and Virtual Philosophy.* London: Continuum Press, 2002.

———. "Open Source: A Movement in Search of a Philosophy." Paper presented to the Institute for Advanced Study, Princeton, N.J., 2001. http://www.cddc.vt.edu/host/delanda/pages/opensource.htm.

———. *A Thousand Years of Non-linear History.* New York: Zone Books, 1997.

Dewey, John. *Freedom and Culture.* 1939; repr., Amherst, N.Y.: Prometheus Books, 1989.

———. *Liberalism and Social Action.* New York: G. P. Putnam's Sons, 1935.

———. *The Public and Its Problems.* Chicago: Swallow Press, 1927; repr., Sage Books / Swallow Press, 1954.

Dibbell, Julian. *Play Money: Or, How I Quit My Day Job and Made Millions Trading Virtual Loot.* New York: Basic Books, 2006.

———. "A Rape in Cyberspace." *Village Voice* 38.51 (December 1993): 21.

Dibona, Chris, et al. *Open Sources: Voices from the Open Source Revolution.* Sebastopol, Calif.: O'Reilly Press, 1999.

DiMaggio, Paul, Esther Hargittai, C. Celeste, and S. Shafer. "From Unequal Access to Differentiated Use: A Literature Review and Agenda for Research on Digital Inequality." In *Social Inequality,* ed. Kathryn Neckerman, 355–400. New York: Russell Sage Foundation, 2004.

Downey, Gary L. *The Machine in Me: An Anthropologist Sits among Computer Engineers.* London: Routledge, 1998.

Doyle, Richard. *Wetwares: Experiments in Postvital Living.* Minneapolis: University of Minnesota Press, 2003.

Drake, William. "The Internet Religious War." *Telecommunications Policy* 17 (December 1993): 643–49.

Dreyfus, Hubert. *On the Internet.* London: Routledge, 2001.

Dumit, Joseph. *Picturing Personhood: Brain Scans and Biomedical Identity.* Princeton: Princeton University Press, 2004.

Eagleton, Terry. *Ideology: An Introduction.* London: Verso Books, 1991.

———. *The Ideology of the Aesthetic.* Cambridge, Mass.: Blackwell, 1990.

Edwards, Paul N. *The Closed World: Computers and the Politics of Discourse in the Cold War.* Cambridge, Mass.: MIT Press, 1996.

———. "Infrastructure and Modernity: Force, Time, and Social Organization in the History of Sociotechnical Systems." In *Modernity and Technology,* ed. Thomas Misa, Philip Brey, and Andrew Feenberg, 185–225. Cambridge, Mass.: MIT Press, 2003.

Eisenstein, Elizabeth. *The Printing Press as an Agent of Change: Communications and Cultural Transformations in Early Modern Europe.* 2 vols. Cambridge: Cambridge University Press, 1979.

Faulkner, W. "Dualisms, Hierarchies and Gender in Engineering." *Social Studies of Science* 30.5 (2000): 759–92.

Febvre, Lucien, and Henri-Jean Martin. *The Coming of the Book: The Impact of Printing 1450–1800.* Trans. David Gerard. 1958; repr., London: Verso, 1976.

Feller, Joseph, Brian Fitzgerald, Scott A. Hissam, and Karim R. Lakhani, eds. *Perspectives on Free and Open Source Software.* Cambridge, Mass.: MIT Press, 2005.

Feyerabend, Paul. *Against Method.* 3rd ed. 1975. London: Verso Books, 1993.

Fielding, Roy T. "Shared Leadership in the Apache Project." *Communications of the ACM* 42.4 (April 1999): 42–43.

Fischer, Franklin M. *Folded, Spindled, and Mutilated*. Cambridge, Mass.: MIT Press, 1983.

Fischer, Michael M. J. "Culture and Cultural Analysis as Experimental Systems." *Cultural Anthropology* 22.1 (Feburary 2007): 1–65.

———. *Emergent Forms of Life and the Anthropological Voice*. Durham, N.C.: Duke University Press, 2003.

———. "Worlding Cyberspace." In *Critical Anthropology Now*, ed. George Marcus, 245–304. Santa Fe, N.M.: School for Advanced Research Press, 1999.

Flichy, Patrice. *The Internet Imaginaire*. Trans. Liz Carey-Libbrecht. Cambridge, Mass.: MIT Press, 2007.

Fortun, Kim. *Advocating Bhopal: Environmentalism, Disaster, New Global Orders*. Chicago: University of Chicago Press, 2003.

———. "Figuring Out Ethnography." In *Fieldwork Isn't What It Used to Be*, ed. George Marcus and James Faubion. Chicago: University of Chicago Press, forthcoming.

Fortun, Kim, and Mike Fortun. "Scientific Imaginaries and Ethical Plateaus in Contemporary U.S. Toxicology." *American Anthropologist* 107.1 (2005): 43–54.

Foucault, Michel. *La naissance de la biopolitique: Cours au Collège de France (1978–1979)*. Paris: Gallimard / Le Seuil, 2004.

———. "What Is an Author?" In *The Foucault Reader*, ed. P. Rabinow, 101–20. New York: Pantheon Books, 1984.

———. 1997. "What Is Enlightenment?" In *Ethics*, ed. Paul Rabinow, 303–17. Vol. 2 of *The Essential Works of Foucault 1954–1984*. New York: New Press, 1997.

Freeman, Carla. *High Tech and High Heels in the Global Economy*. Durham, N.C.: Duke University Press, 2000.

Freeman, Jo, and Victoria Johnson, eds. *Waves of Protest: Social Movements since the Sixties*. Lanham, Md.: Rowman and Littlefield, 1999.

Galison, Peter. *How Experiments End*. Chicago: University of Chicago Press, 1987.

———. *Image and Logic: The Material Culture of Microphysics*. Chicago: University of Chicago Press, 1997.

Galloway, Alexander. *Protocol, or How Control Exists after Decentralization*. Cambridge, Mass.: MIT Press, 2004.

Gancarz, Mike. *Linux and the UNIX Philosophy*. Boston: Digital Press, 2003.

———. *The Unix Philosophy*. Boston: Digital Press, 1994.

Gaonkar, Dilip. "Toward New Imaginaries: An Introduction." *Public Culture* 14.1 (2002): 1–19.

Geertz, Clifford. "Ideology as a Cultural System." In *The Interpretation of Cultures*, 193–233. New York: Basic Books, 1973.

Gerlach, Luther P., and Virginia H. Hine. *People, Power, Change: Movements of Social Transformation*. Indianapolis: Bobbs-Merrill, 1970.

Ghosh, Rishab Ayer. "Cooking Pot Markets: An Economic Model for the Trade in Free Goods." *First Monday* 3.3 (1998). http://www.firstmonday.org/issues/issue3_3/ghosh/.

Gillespie, Tarleton. "Engineering a Principle: 'End to End' in the Design of the Internet." *Social Studies of Science* 36.3 (2006): 427–57.

Golub, Alex. "Copyright and Taboo." *Anthropological Quarterly* 77.3 (2004): 521–30.

Gray, Pamela. *Open Systems: A Business Strategy for the 1990s*. London: McGraw-Hill, 1991.

Green, Ellen, and Allison Adam. *Virtual Gender: Technology, Consumption and Identity*. London: Routledge, 2001.

Green, Ian. *The Christian's ABCs: Catechisms and Catechizing in England c1530–1740*. Oxford: Oxford University Press, 1996.

———. *Print and Protestantism in Early Modern England*. Oxford: Oxford University Press, 2000.

Green, Sarah, Penny Harvey, and Hannah Knox. "Scales of Place and Networks: An Ethnography of the Imperative to Connect through Information and Communication Technologies." *Current Anthropology* 46.5 (December 2005): 805–26.

Grier, David Alan. *When Computers Were Human*. Princeton: Princeton University Press, 2005.

Grier, David Alan, and Mary Campbell. "A Social History of Bitnet and Listserv 1985–1991." *IEEE Annals of the History of Computing* (April–June 2000): 32–41.

Grint, Keith, and Rosalind Gill. *The Gender-Technology Relation: Contemporary Theory and Research*. London: Taylor and Francis, 1995.

Habermas, Jürgen. *The Structural Transformation of the Public Sphere: An Inquiry into a Category of Bourgeois Society*. Trans. Thomas Burger, with the assistance of Frederick Lawrence. Cambridge, Mass.: MIT Press, 1991.

Hafner, Katie. *Where Wizards Stay Up Late: The Origins of the Internet*. New York: Simon and Schuster, 1998.

Hamerly, Jim, and Tom Paquin, with Susan Walton. "Freeing the Source." In *Open Sources: Voices from the Open Source Revolution*, by Chris Dibona et al., 197–206. Sebastopol, Calif.: O'Reilly Press, 1999.

Hardin, Garrett. "The Tragedy of the Commons." *Science* 162 (1968): 1,243–48.

Hashagen, Ulf, Reinhard Keil-Slawik, and Arthur Norberg, eds. *History of Computing—Software Issues*. Berlin: Springer Verlag, 2002.

Hauben, Michael, and Rhonda Hauben. *Netizens: On the History and Impact of Usenet and the Internet*. Los Alamitos, Calif.: IEEE Computer Society Press, 1997.

Hauser, Marc. *Moral Minds: How Nature Designed Our Universal Sense of Right and Wrong*. New York: Ecco Press, 2006.

Hayden, Cori. *When Nature Goes Public: The Making and Unmaking of Bioprospecting in Mexico*. Princeton: Princeton University Press, 2003.

Hayek, Friedrich A. *Law, Legislation and Liberty*. Vol. 1, *Rules and Order*. Chicago: University of Chicago Press, 1970.

Helmreich, Stefan. *Silicon Second Nature: Culturing Artificial Life in a Digital World*. Berkeley: University of California Press, 1998.

Herring, Susan C. "Gender and Democracy in Computer-Mediated Communication." In *Computerization and Controversy: Value Conflicts and Social Choices*, ed. Rob Kling and Charles Dunlop, 476–89. 2d ed. Orlando: Academic Press, 1995.

Hess, Charlotte, and Elinor Ostrom, eds. *Understanding Knowledge as a Common: From Theory to Practice*. Cambridge, Mass.: MIT Press, 2006.

Himanen, Pekka. *The Hacker Ethic and the Spirit of the Information Age*. New York: Random House, 2001.

Hine, Christine. *Virtual Ethnography*. London: Sage, 2000.

Holmes, Douglas, and George Marcus. "Cultures of Expertise and the Management of Globalization: Toward the Re-Functioning of Ethnography." In *Global Assemblages: Technology, Politics, and Ethics as Anthropological Problems*, ed. Aiwa Ong and Stephen J. Collier, 235–52. Boston: Blackwell, 2005.

Holmes, Oliver Wendell. "The Path of Law." *Harvard Law Review* 10 (1897): 457.

Hopkins, Patrick D., ed. *Sex/Machine: Readings in Culture, Gender and Technology*. Bloomington: Indiana University Press, 1998.

Huberman, Bernardo A., ed. *The Ecology of Computation*. Amsterdam: North-Holland, 1988.

Huxley, Julian. *New Bottles for New Wine: Essays*. New York: Harper, 1957.

Jaszi, Peter, and Martha Woodmansee, eds. *The Construction of Authorship: Textual Appropriation in Law and Literature*. Durham, N.C.: Duke University Press, 1994.

Johns, Adrian. *The Nature of the Book: Print and Knowledge in the Making*. Chicago: University of Chicago Press, 1998.

Jorgensen, Neils. "Incremental and Decentralized Integration in FreeBSD." In *Perspectives on Free and Open Source Software*, ed. Feller et al., 227–44. Cambridge, Mass.: MIT Press, 2004.

———. "Putting It All in the Trunk: Incremental Software Development in the FreeBSD Open Source Project." *Information Systems Journal* 11.4 (2001): 321–36.

Kahn, Robert, et al. "The Evolution of the Internet as a Global Information System." *International Information and Libraries Review* 29 (1997): 129–51.

Kahn, Robert, and Vint Cerf. "A Protocol for Packet Network Intercommunication." *IEEE Transactions on Communications* Com-22.5 (May 1974): 637–44.

Keating, Peter, and Alberto Cambrosio. *Biomedical Platforms: Realigning the Normal and the Pathological in Late-twentieth-century Medicine.* Cambridge, Mass.: MIT Press, 2003.

Kelty, Christopher, ed. "Culture's Open Sources." *Anthropological Quarterly* 77.3 (summer 2004): 499–506. http://aq.gwu.edu/archive/table_summer04.htm.

———. "Punt to Culture." *Anthropological Quarterly* 77.3 (summer 2004): 547–58.

Kendall, Lori. "'Oh No! I'm a NERD!' Hegemonic Masculinity on an Online Forum." *Gender and Society* 14.2 (2000): 256–74.

Keves, Brian William. "Open Systems Formal Evaluation Process." Paper presented at the USENIX Association Proceedings of the Seventh Systems Administration Conference (LISA VII), Monterey, California, 1–5 November 1993.

Kidder, Tracy. *The Soul of a New Machine.* Boston: Little, Brown, 1981.

Kirkup, Gill, Linda Janes, Kath Woodward, and Fiona Hovenden. *The Gendered Cyborg: A Reader.* London: Routledge, 2000.

Kittler, Friedrich. *Discourse Networks 1800/1900.* Trans. Michael Metteer, with Chris Cullens. 1985; repr., Stanford, Calif.: Stanford University Press, 1990.

———. *Gramophone, Film, Typewriter.* Trans. Geoffry Winthrop-Young and Michael Wutz. 1986; repr., Stanford, Calif.: Stanford University Press, 1999.

Kling, Rob. *Computerization and Controversy: Value Conflicts and Social Choices.* San Diego: Academic Press, 1996.

Knuth, Donald. *The Art of Computer Programming.* 3d ed. Reading, Mass.: Addison-Wesley, 1997.

Kohler, Robert. *Lords of the Fly: Drosophila Genetics and the Experimental Life.* Chicago: University of Chicago Press, 1994.

Laclau, Ernesto, and Chantal Mouffe. *Hegemony and Socialist Strategy*. London: Verso, 1985.

Landecker, Hannah. *Culturing Life: How Cells Became Technologies*. Cambridge, Mass.: Harvard University Press, 2007.

Latour, Bruno. "Drawing Things Together." In *Representation in Scientific Practice*, ed. Michael Lynch and Steve Woolgar, 19–68. Cambridge, Mass.: MIT Press, 1990.

———. *Pandora's Hope: Essays on the Reality of Science Studies*. Cambridge, Mass.: Harvard University Press, 1999.

———. *Re-assembling the Social: An Introduction to Actor-Network Theory*. Oxford: Oxford University Press, 2005.

———. *Science in Action: How to Follow Scientists and Engineers through Society*. Cambridge, Mass.: Harvard University Press, 1987.

———. "What Rules of Method for the New Socio-scientific Experiments." In *Experimental Cultures: Configurations between Science, Art and Technology 1830–1950, Conference Proceedings*, 123. Berlin: Max Planck Institute for the History of Science, 2001.

Latour, Bruno, and Peter Weibel, eds. *Making Things Public: Atmospheres of Democracy*. Cambridge, Mass.: MIT Press, 2005.

Law, John. *Aircraft Stories: Decentering the Object in Technoscience*. Durham, N.C.: Duke University Press, 2002.

———. "Technology and Heterogeneous Engineering: The Case of Portuguese Expansion." In *The Social Construction of Technological Systems: New Directions in the Sociology and History of Technology*, ed. W. E. Bijker, T. P. Hughes, and T. J. Pinch, 111–134. Cambridge, Mass.: MIT Press, 1987.

Law, John, and John Hassard. *Actor Network Theory and After*. Sociological *Review* Monograph. Malden, Mass.: Blackwell 1999.

Leach, James, Dawn Nafus, and Berbard Krieger. *Gender: Integrated Report of Findings*. Free/Libre and Open Source Software: Policy Support (FLOSSPOLS) D 16, 2006. http://www.jamesleach.net/downloads/FLOSSPOLS-D16-Gender_Integrated_Report_of_Findings.pdf.

Lee, J. A. N., R. M. Fano, A. L. Scherr, F. J. Corbato, and V. A. Vyssotsky. "Project MAC." *Annals of the History of Computing* 14.2 (1992): 9–42.

Lerner, Josh, and Jean Tirole. "Some Simple Economics of Open Source." *Industrial Economics* 50.2 (June 2002): 197–234.

Lessig, Lawrence. *Code: Version 2.0*. New York: Basic Books, 2006.

———. *Code and Other Laws of Cyber Space*. New York: Basic Books, 1999.

———. *Free Culture: The Nature and Future of Creativity*. New York: Penguin, 2003.

———. *The Future of Ideas: The Fate of the Commons in a Connected World*. New York: Random House, 2001.

———. "The New Chicago School." *Legal Studies* 27.2 (1998): 661–91.

Levy, Steven. *Hackers: Heroes of the Computer Revolution.* New York: Basic Books, 1984.

Libes, Don, and Sandy Ressler. *Life with UNIX: A Guide for Everyone.* Englewood Cliffs, N.J.: Prentice Hall, 1989.

Light, Jennifer. "When Computers Were Women." *Technology and Culture* 40.3 (July 1999): 455–483.

Lions, John. *Lions' Commentary on UNIX 6th Edition with Source Code.* 1977; repr., San Jose: Peer to Peer Communications, 1996.

Lippmann, Walter. *The Phantom Public.* New York: Macmillan, 1927.

Litman, Jessica. *Digital Copyright.* New York: Prometheus Books, 2001.

Liu, Alan. *The Laws of Cool: Knowledge Work and the Culture of Information.* Chicago: University of Chicago Press, 2004.

MacKenzie, Adrian. *Cutting Code: Software and Sociality.* Digital Formations Series. New York: Peter Lang, 2005.

MacKenzie, Donald A. *Mechanizing Proof: Computing, Risk, and Trust.* Cambridge, Mass.: MIT Press, 2001.

Mahoney, Michael. "Finding a History for Software Engineering." *Annals of the History of Computing* 26.1 (2004): 8–19.

———. "The Histories of Computing(s)." *Interdisciplinary Science Reviews* 30.2 (2005): 119–35.

———. "In Our Own Image: Creating the Computer." In *The Changing Image of the Sciences*, ed. Ida Stamhuis, Teun Koetsier, and Kees de Pater, 9–28. Dordrecht: Kluwer Academic Publishers, 2002.

———. "The Roots of Software Engineering." *CWI Quarterly* 3.4 (1990): 325–34.

———. "The Structures of Computation." In *The First Computers: History and Architectures*, ed. Raul Rojas and Ulf Hashagen, 17–32. Cambridge, Mass.: MIT Press, 2000.

Malamud, Carl. *Exploring the Internet: A Technical Travelogue.* Englewood Cliffs, N.J.: Prentice Hall, 1992.

Mannheim, Karl. *Ideology and Utopia: Introduction to the Sociology of Knowledge.* New York: Harcourt and Brace, 1946.

Marcus, George. *Ethnography through Thick and Thin.* Chicago: University of Chicago Press, 1998.

Marcus, George, and James Clifford. *Writing Culture: The Poetics and Politics of Ethnography.* Berkeley: University of California Press, 1986.

Marcus, George, and Michael M. J. Fischer. *Anthropology as Cultural Critique: An Experimental Moment in the Human Sciences.* Chicago: University of Chicago Press, 1986.

Margolis, Jane, and Allen Fisher. *Unlocking the Clubhouse: Women in Computing*. Cambridge, Mass.: MIT Press, 2002.

Martin, James. *Viewdata and the Information Society*. Englewood Cliffs, N.J.: Prentice Hall, 1982.

Matheson, Peter. *The Imaginative World of the Reformation*. Edinburgh, Scotland: T and T Clark, 2000.

McKenna, Regis. *Who's Afraid of Big Blue? How Companies Are Challenging IBM—and Winning*. Reading, Mass.: Addison-Wesley, 1989.

McKusick, M. Kirk. "Twenty Years of Berkeley Unix: From AT&T-owned to Freely Redistributable." In *Open Sources: Voices from the Open Source Revolution*, Chris Dibona et al., 31–46. ACM Sebastopol, Calif.: O'Reilly Press, 1999.

McLuhan, Marshall. *The Gutenberg Galaxy: The Making of Typographic Man*. Toronto: University of Toronto Press, 1966.

———. *Understanding Media: The Extensions of Man*. 1964; repr., Cambridge, Mass.: MIT Press, 1994.

Merges, Robert, Peter Menell, and Mark Lemley, eds. *Intellectual Property in the New Technological Age*. 3d ed. New York: Aspen Publishers, 2003.

Merton, Robert. "The Normative Structure of Science." In *The Sociology of Science: Theoretical and Empirical Investigations*, 267–80. Chicago: University of Chicago Press, 1973.

Miller, Daniel, and Don Slater. *The Internet: An Ethnography*. Oxford: Berg, 2000.

Misa, Thomas, Philip Brey, and Andrew Feenberg, eds. *Modernity and Technology*. Cambridge, Mass.: MIT Press, 2003.

Mockus, Audris, Roy T. Fielding, and James Herbsleb. "Two Case Studies of Open Source Software Development: Apache and Mozilla." *ACM Transactions in Software Engineering and Methodology* 11.3 (2002): 309–46.

Mol, Annemarie. *The Body Multiple: Ontology in Medical Practice*. Durham, N.C.: Duke University Press, 2002.

Moody, Glyn. *Rebel Code: Inside Linux and the Open Source Revolution*. Cambridge, Mass.: Perseus, 2001.

Mooers, Calvin. "Computer Software and Copyright." *Computer Surveys* 7.1 (March 1975): 45–72.

Mueller, Milton. *Ruling the Root: Internet Governance and the Taming of Cyberspace*. Cambridge, Mass.: MIT Press, 2004.

Naughton, John. *A Brief History of the Future: From Radio Days to Internet Years in a Lifetime*. Woodstock, N.Y.: Overlook Press, 2000.

Noble, David. "Digital Diploma Mills: The Automation of Higher Education." *First Monday* 3.1 (5 January 1998). http://www.firstmonday.org/issues/issue3_1/.

Norberg, Arthur L., and Judy O'Neill. *A History of the Information Techniques Processing Office of the Defense Advanced Research Projects Agency*. Minneapolis: Charles Babbage Institute, 1992.

———. *Transforming Computer Technology: Information Processing for the Pentagon, 1962–1986*. Baltimore: Johns Hopkins University Press, 1996.

Ong, Walter. *Ramus, Method, and the Decay of Dialogue: From the Art of Discourse to the Art of Reason*. 1983; repr., Chicago: University of Chicago Press, 2004.

Ostrom, Elinor. *Governing the Commons: The Evolution of Institutions for Collective Action*. Cambridge: Cambridge University Press, 1991.

Perens, Bruce. "The Open Source Definition." In *Open Sources: Voices from the Open Source Revolution*, Dibona et al., 171–188. Sebastopol, Calif.: O'Reilly Press, 1999. http://perens.com/OSD.html and http://www.oreilly.com/catalog/opensources/book/perens.html.

Pfaffenberger, Bryan. "'A Standing Wave in the Web of our Communications': USENet and the Socio-technical Construction of Cyberspace Values." In *From Usenet to CoWebs: Interacting with Social Information Spaces*, ed. Christopher Lueg and Danyel Fisher, 20–43. London: Springer, 2003.

Rabinow, Paul. *Anthropos Today: Reflections on Modern Equipment*. Princeton: Princeton University Press, 2003.

———. *Essays on the Anthropology of Reason*. Princeton: Princeton University Press, 1997.

Ratto, Matt. "Embedded Technical Expression: Code and the Leveraging of Functionality." *Information Society* 21.3 (July 2005): 205–13.

———. "The Pressure of Openness: The Hybrid work of Linux Free/Open Source Kernel Developers." Ph.D. diss., University of California, San Diego, 2003.

Raymond, Eric S. *The Art of UNIX Programming*. Boston: Addison-Wesley, 2004.

———. *The Cathedral and the Bazaar: Musings on Linux and Open Source by an Accidental Revolutionary*. Sebastopol, Calif.: O'Reilly Press, 2001. See esp. "Homesteading the Noosphere," 79–135.

———, ed. *The New Hackers' Dictionary*. 3d ed. Cambridge, Mass.: MIT Press, 1996.

Rheinberger, Hans-Jörg. *Towards a History of Epistemic Things: Synthesizing Proteins in the Test Tube*. Stanford, Calif.: Stanford University Press, 1997.

Rheingold, Howard. *The Virtual Community: Homesteading on the Electronic Frontier*. Rev. ed. 1993; repr., Cambridge, Mass.: MIT Press, 2000.

Ricoeur, Paul. *Lectures on Ideology and Utopia*. New York: Columbia University Press, 1986.

Riles, Annelise. "Real Time: Unwinding Technocratic and Anthropological Knowledge." *American Ethnologist* 31.3 (August 2004): 392–405.

Ritchie, Dennis. "The UNIX Time-Sharing System: A Retrospective." *Bell System Technical Journal* 57.6, pt. 2 (July–August 1978). http://cm.bell-labs .com/cm/cs/who/dmr/retroindex.html.

Rose, Mark. *Authors and Owners: The Invention of Copyright.* Cambridge, Mass.: Harvard University Press, 1995.

Salus, Peter. *Casting the Net: From ARPANET to Internet and Beyond.* Reading, Mass.: Addison-Wesley, 1995.

——. *A Quarter Century of UNIX.* Reading, Mass.: Addison-Wesley, 1994.

Schmidt, Susanne K., and Raymund Werle. *Coordinating Technology: Studies in the International Standardization of Telecommunications.* Cambridge, Mass.: MIT Press, 1998.

Segaller, Stephen. *Nerds 2.0.1: A Brief History of the Internet.* New York: TV Books, 1998.

Shaikh, Maha, and Tony Cornford. "Version Management Tools: CVS to BK in the Linux Kernel." Paper presented at the Twenty-fifth International Conference on Software Engineering—Taking Stock of the Bazaar: The Third Workshop on Open Source Software Engineering, Portland, Oregon, 3–10 May 2003.

Shapin, Steven. *The Social History of Truth: Civility and Science in Seventeenth Century England.* Chicago: University of Chicago Press, 1994.

Shapin, Steven, and Simon Schaffer. *Leviathan and the Air Pump: Hobbes, Boyle and the Experimental Life.* Princeton: Princeton University Press, 1985.

Smith, Adam. *The Theory of Moral Sentiments.* 1759; repr., Cambridge: Cambridge University Press, 2002.

Stallings, William. *Data and Computer Communications.* London: Macmillan, 1985.

Stallman, Richard. "The GNU Manifesto." *Dr. Dobb's* 10.3 (March 1985): 30–35.

St. Amour, Paul K. *The Copywrights: Intellectual Property and the Literary Imagination.* Ithaca, N.Y.: Cornell University Press, 2003.

Star, Susan Leigh, ed. *The Cultures of Computing.* Malden, Mass.: Blackwell, 1995.

Star, Susan Leigh, and Karen Ruhleder. "Steps towards an Ecology of Infrastructure: Complex Problems in Design and Access for Large-scale Collaborative Systems." *Information Systems Research* 7 (1996): 111–33.

Stephenson, Neal. *In the Beginning Was the Command Line.* New York: Avon / Perennial, 1999.

Sunshine, Carl. *Computer Network Architectures and Protocols*. 2d ed. New York: Plenum Press, 1989.

Takahashi, Shigeru. "The Rise and Fall of the Plug Compatible Manufacturers." *IEEE Annals of the History of Computing* (January–March 2005): 4–16.

Tanenbaum, Andrew. *Computer Networks*. 1st ed. Upper Saddle River, N.J.: Prentice Hall, 1981.

————. *Operating Systems: Design and Implementation*. 1st ed. Englewood Cliffs, N.J.: Prentice Hall, 1987.

————. "The UNIX Marketplace in 1987: Life, the UNIverse and Everything." In *Proceedings of the Summer 1987 USENIX Conference*, 419–24. Phoenix, Ariz.: USENIX, 1987.

Taylor, Charles. *Modern Social Imaginaries*. Durham, N.C.: Duke University Press, 2004.

————. "Modes of Civil Society." *Public Culture* 3.1 (1990): 95–132.

————. *Multiculturalism and the Politics of Recognition*. Princeton, N.J.: Princeton University Press, 1992.

————. *Sources of the Self: The Making of the Modern Identity*. Cambridge, Mass.: Harvard University Press, 1989.

Thompson, Charis. *Making Parents: The Ontological Choreography of Reproductive Technologies*. Cambridge, Mass.: MIT Press, 2005.

Thompson, Ken, and Dennis Ritchie. "The UNIX Time-Sharing System." *Communications of the ACM* 17.7 (July 1974): 365–75.

Tichy, Walter F. "RCS: A System for Version Control." *Software: Practice and Experience* 15.7 (July 1985): 637–54.

Torvalds, Linus, with David Diamond. *Just for Fun: The Story of an Accidental Revolutionary*. New York: HarperCollins, 2002.

Tuomi, Ilkka. *Networks of Innovation: Change and Meaning in the Age of the Internet*. New York: Oxford University Press, 2002.

Turing, Alan. "On Computable Numbers, with an Application to the Entscheidungsproblem." *Proceedings of the London Mathematical Society* 2.1 (1937): 230.

Turkle, Sherry. *Life on the Screen: Identity in the Age of the Internet*. New York: Simon and Schuster, 1995.

————. *The Second Self: Computers and the Human Spirit*. New York: Simon and Schuster, 1984.

Turner, Fred. *From Counterculture to Cyberculture: Stewart Brand, the Whole Earth Network, and the Rise of Digital Utopianism*. Chicago: University of Chicago Press, 2006.

————. "Where the Counterculture Met the New Economy." *Technology and Culture* 46.3 (July 2005): 485–512.

Ullman, Ellen. *The Bug: A Novel.* New York: Nan A. Talese, 2003.

———. *Close to the Machine: Technophilia and Its Discontents.* San Francisco: City Lights, 1997.

Vaidhyanathan, Siva. *Copyrights and Copywrongs; The Rise of Intellectual Property and How It Threatens Creativity.* New York: New York University Press, 2001.

Vetter, Greg R. "The Collaborative Integrity of Open-Source Software." *Utah Law Review* 2004.2 (2004): 563–700.

———. "'Infectious Open Source Software: Spreading Incentives or Promoting Resistance?" *Rutgers Law Journal* 36.1 (fall 2004): 53–162.

Vinge, Vernor. "The Coming Technological Singularity: How to Survive in the Post-Human Era" (1993). http://www.rohan.sdsu.edu/faculty/vinge/misc/singularity.html (accessed 18 August 2006).

Von Hippel, Eric. *Democratizing Innovation.* Cambridge, Mass.: MIT Press, 2005.

Wajcman, Judy. *Feminism Confronts Technology.* Cambridge: Polity, 1991.

———. "Reflections on Gender and Technology Studies: In What State Is the Art?" *Social Studies of Science* 30.3 (2000): 447–64.

Waldrop, Mitchell. *The Dream Machine: J. C. R. Licklider and the Revolution that Made Computing Personal.* New York: Viking, 2002.

Walsh, John, and Todd Bayma. "Computer Networks and Scientific Work." *Social Studies of Science* 26.3 (August 1996): 661–703.

Warner, Michael. *The Letters of the Republic: Publication and the Public Sphere in Eighteenth-century America.* Cambridge, Mass.: Harvard University Press, 1990.

———. "Publics and Counterpublics." *Public Culture* 14.1 (2002): 49–90.

———. *Publics and Counterpublics.* New York: Zone Books, 2003.

Wayner, Peter. *Free for All: How LINUX and the Free Software Movement Undercut the High-Tech Titans.* New York: Harper Business, 2000.

Weber, Max. "Objectivity in the Social Sciences and Social Policy." In *The Methodology of the Social Sciences,* trans. and ed. Edward Shils and Henry A. Finch, 50–112. New York: Free Press, 1949.

Weber, Steven. *The Success of Open Source.* Cambridge, Mass.: Harvard University Press, 2004.

Wexelblat, Richard L., ed. *History of Programming Languages.* New York: Academic Press, 1981.

Williams, Sam. *Free as in Freedom: Richard Stallman's Crusade for Free Software.* Sebastopol, Calif.: O'Reilly Press, 2002.

Wilson, Fiona. "Can't Compute, Won't Compute: Women's Participation in the Culture of Computing." *New Technology, Work and Employment* 18.2 (2003): 127–42.

Wilson, Samuel M., and Leighton C. Peterson. "The Anthropology of Online Communities." *Annual Reviews of Anthropology* 31 (2002): 449–67.

Xiang, Biao. *"Global Bodyshopping": An Indian Labor System in the Information Technology Industry*. Princeton: Princeton University Press, 2006.

Žižek, Slavoj, ed. *Mapping Ideology*. London: Verso, 1994.

Index

Artificial Intelligence Lab (AI Lab), at MIT, 184–188, 206–208
Assemblage, 313n9
AT&T, 119–120, 126–131, 160, 165, 192, 201, 216, 307; Bell Laboratories, 126–131, 230, 238; divestiture in 1984, 154, 201; version of EMACS, 194, 196; version of UNIX, 138, 153
Attribution: copyright licenses and, 265, 283–284, 286; disavowal of, 295–296
Authorship, 75, 195, 274, 282–293; copyright vs., 244n4; moral rights of, 257; order of, 346n14; ownership vs., 205, 284
Availability: open systems and, 152, 166–167, 175–176; reorientation of power and knowledge and, 10–13, 308–310, 313n8

Bangalore, 43–46
Baraniuk, Richard, 244–258, 271–272, 299, 341n1
Barlow, John Perry, 46, 55, 264
Behlendorf, Brian, 110–111; as head of Apache, 223–229
Bell Laboratories. *See* AT&T
Bentham, Jeremy, 181–182
Berkeley Systems Distribution (BSD) (version of UNIX), 111, 137–141, 153, 157, 176, 238, 327n35; FreeBSD, 219, 226, 231, 338n4
Berkman Center for Internet and Society, 259
Berlin, 19–20, 27–28, 36–38
Berners-Lee, Tim, 103, 177, 223
Bitkeeper (Source Code Management software), 232–236
Blogosphere, as recursive public, 304
Bolt, Beranek, and Newman (BBN), 139–141
Bone, Jeff, 51–61
Bowker, Geoffrey, 344n1

Boyle, James, 30, 256–268, 313n11
Brooks, Frederick, 124, 325n14
Brown, Glenn Otis, 264–268, 293–300, 343n14; on history of legal realism, 343n20
Bruno, Giordano, 170, 331n41
BSD. *See* Berkeley Systems Distribution
BSD License, 104, 140
Bugs, 106, 224, 230, 237
Burris, C. Sidney, 248

C (programming language), 119, 126, 329n19
Calvinism, 93
"Cathedral and the Bazaar," 109–110, 243, 353n1
Censorship, 51, 53, 55–56
Cerf, Vinton, 138, 172–173
Chari, Bharath, 45
Clark, David, 58, 60, 172
Coleman, Gabriella, 180, 337n54, 340n32, 341n43, 346n15
Collaboration: competition vs., 152, 158–166, 307; different meanings of, 228–229; forking vs., 287–288. *See also* coordination
Commité Consultative de Information, Technologie et Télécommunications (CCITT), 169, 175
Commons, ix, xi, 16–17, 33, 246, 313n11
Communities, 214, 252, 281; norms and 281, 285–293
comp.os.minix, 217; "Linux Is Obsolete" debate on, 218–219
Computer Corporation of America (CCA), 190–199, 201–202, 333n14, 336n47
Concurrent Versioning System (cvs), 106, 212, 229; history of, 231; Linux and 232
Connexions project, 3,11, 17, 244–246; connection to Creative

Commons, 258–259, 262–263; as "factory of knowledge, 249–251; as Free Software project, 254–255; history and genesis of, 247–258; line between content and software, 279; meaning of publication in, 275–286; model of learning in, 249, 253–254; modules in, 249–251; Open CourseWare, 252, 263; relationship to education, 251; relationship to hypertext, 266; roles in, 282–293, 345nn 12–13; stages of producing a document in, 275–277; textbooks and, 248–249, 274–282

"Content," 255–265

Control, relationship to power, 70,73

CONTU report, 202

Cooperation. *See* Collaboration

Coordination (component of Free Software), 15, 105–107, 213–215, 224–240; individual virtuosity vs. hierarchical planning, 211–212, 220–221, 223, 227–229, 237, 340n32; of Linux vs. Minix, 216–218; modulations of, 233–236, 246, 253, 257, 281, 285–288; of technology and people, 210–211

Copyleft licenses (component of Free Software), 49, 103–105, 140, 178, 179–209, 309; commercial use and, 265; Creative Commons version, 294–300; derivative uses and 265, 286, 291, 294; disavowal clause in, 295; early attempts, 334n21; forking and, 287–288; as hack of copyright law, 182–183; modulations of, 234–236, 253, 256–257, 259, 285 *See also* Copyright; General Public License; Intellectual property

Copyright, 2, 74–76, 182, 189; changes in 1976, 134, 187,

199–206; changes in 1980, 183, 187, 199–206; infringement and software, 189–209; legal definition of software and, 189–209; requirement to register, 200, 260; rights granted by, 183; software and copyrightability, 189, 199, 335n41; specificity of media and, 278–279; transfer of, 205; works for hire, 282. *See also* Copyleft licenses; Copyright infringement; Intellectual property

Copyright infringement, 199, 335nn46–47; EMACS controversy and, 193, 196–200; infringement on own invention, 198; legal threats and, 195, 205–206; permissions and, 194–206; redistribution of software as, 204

Copyright licenses. *See* Copyleft licenses; Copyright

Crain, Patricia, 74–75

Creative Commons, 3, 11, 17, 246, 258–268, 283; activism of, 261; connection to Connexions, 262–263; marketing of, 265; origin and history of 259–261, writing of licenses, 294–300

Credit, 282–284, 295–296, 309. *See also* Attribution

Critique, Free Software as, 211, 222, 235–236, 239

Cultural significance, x, 13, 245, 269, 305

Culture, 38, 44, 244; Creative Commons version of, 297–300; as experimental system, 2; law vs., 298; punting to, 293–300, 343n20

Cygnus Solutions (corporation), 99, 110

Debugging, 216–217; patching vs., 224, 228–229, 237, 341n43

Defense, Department of, 171–176, 328n5, 329n9

Defense Advanced Research Projects Agency (DARPA), 138–141, 171–172
Delanda, Manuel, 181, 329n9
Design, 80, 134, 185, 220, 337n2; and evolution, 221–222
Dewey, John, 25, 181, 241
diff (software tool), 129–130, 137, 238; history of, 230–231
Differentiation of software, 120, 137, 141–142, 155–156; in Apache, 226–229. *See also* Forking; Sharing source code
Digital Equipment Corporation (corporation), 147, 192; DECNet, 168
Digital Millennium Copyright Act (DMCA), 99
Digital signal processing (DSP), 244, 248, 272, 281
Disney, Walt, 261
Distance learning, 251–252
Distributed phenomena, ethnography of, 19–20
Domain Name System (DNS), 52, 108, 338n4
Dotcom era, 110–112
Doyle, Sean, 31–37, 43, 55, 61, 72–74, 80–81, 84–86, 150, 328n1
Dyson, Esther, 55

Economics, institutional, 343n17
Editions, print vs. electronic, 277–278. *See also* Versions
Edwards, Paul, 315n20, 329n9
Eisenstein, Elizabeth, 273, 276, 344n5
Eldred, Eric, 259–260
EMACS (text editor), 70, 180–209, 214, 256, 341n43; controversy about, 188–199, 235, 308; ersatz versions, 185–186, 189; legal status of, 193–199; modularity and extensibility of, 184–188; number of users, 218

EMACS commune, 186–188, 192, 196
End User License Agreements (EULAs), 180
Enlightenment, 64–66, 77–78, 85–86, 255, 273
Entrepreneurialism, 33–35, 81–84, 309
Ethnographic data: availability of, 21, 181, 333n9; mailing lists and, 181;
Ethnography, 19–22, 28, 114–115
Evil, 72–74, 76, 103, 150
Experiment, collective technical, 15, 98, 308
Experimentation, 206, 212, 241, 245, 268, 288; administrative reform as, 181–182. *See also* Modulations
Extensible Mark-up Language (XML), 275–276, 288, 344n8; editors, 342n5
Extropians, 86

Fair use, 200, 203, 299, 347n22
Federal Communications Commission (FCC), 331n37
Feyerabend, Paul, 85, 316n2
Figuring out, 17, 18–19, 65, 114, 144, 155, 181, 236, 263–268, 270, 286, 290, 306, 314n13, 314n15
File Transfer Protocol (ftp), 216–217
Finality, 11–12, 270–293; certainty and stability vs., 270; fixity vs., 280; in Wikipedia and Connexions, 276–277
Firefox, 98, 106. *See also* Netscape Navigator
Fixity, 273–276, 280 345n10. *See also* Finality
Folklore of Internet, 55. *See also* Usable pasts
Forking, 136–141; in Apache, 226–229; in Connexions, 286–293; in Linux, 232–233. *See also* Sharing source code

Lessig, Lawrence, 56–57, 259–268, 342n9, 344n20, 345n11; law and economics and, 267, 298; style of presentations, 262
Liberalism, classical, 53
Libertarianism, 55–56, 109, 318n27
Licensing, of UNIX, 127–128
Licklider, J. C. R., 138
Linux (Free Software project), 15, 69, 108, 111, 136, 177, 206, 211–214, 231, 248, 277, 327n35, 337nn3–4; license terms, 338n10; origins in Minix, 136, 215–217; planning vs. adaptability in, 220–222, 337n2; process of decision making, 220; Source Code Management tools and, 229–236; VGER tree and, 232
Linux Kernel Mailing List (LKML), 338n4
Lions, John, 132–135
LISP (programming language) 123, 323n1; interpreter in EMACS, 204
Luther, Martin, 66–69

Mahoney, Michael, 324n5
Mainframes, 146–147
Malamud, Carl, 170–171
Mannheim, Karl, 40–41
Massachusetts Institute of Technology (MIT), 34, 206–208, 332n4; open courseware and, 251–252
McCool, Rob, 100, 212, 223–224
McGill, Scott, 284–285
McIlroy, Douglas, 126, 230
McKenna, Regis, 151–152
McLuhan, Marshall, 276, 344n5
McVoy, Larry, 232–236, 330n27
Meaning, regulation through law, 298
Medcommons, 33–34
Mergers, 158–160
Merton, Robert, 271; Mertonian norms, 271, 280, 310
Metcalfe's Law, 52, 318n26

Mickey Mouse, 261
Microcomputers, 146–147
Microsoft, 103, 158, 237, 323n19; as Catholic Church, 70, 74–76; Internet Explorer, 73, 100–101; Windows Operating System, 132, 140, 164–165; Xenix (version of UNIX), 153–155, 190
Mikro e.V., 37
Mill, John Stuart, 53, 78
Miller, Dave, 232
Minix (operating system), 135–136, 137, 212, 215–217; goals of, 218–219. See also Tanenbaum, Andrew
Modernity, 78; tradition and, 291–292, 346n17
Modifiability, 10–13, 17, 152, 166–167, 176–177, 202, 216, 238–239, 256, 264, 285, 308–320; EMACS and, 184–188; implications for finality, 11–12, 270–286; modularity in software, 220, 228, 340n34; relation of different disciplines to, 291. See also Adaptability
Modulation: of Free Software, 16, 212, 244, 267, 301–304; practices of, 245–246, 253–258, 264
Mol, Anne-Marie, 344n1
Monopoly, 54, 120, 124, 127, 144, 151–152, 158, 165–170, 325n13
Montgomery, Warren, 194, 201, 333n14, 336n47
Moody, Glyn, 19, 232
Moore's Law, 52, 318n26
Moral and technical order, 28–29, 43, 50, 54, 57, 61, 66, 77, 90, 110, 115, 118, 131, 140–144, 150, 155, 166, 177, 185–187, 215, 236, 246, 260, 263, 291, 299, 301, 307–309; definition of, 328n2. See also Social imaginary
Mosaic (web browser), 100–101, 177, 223. See also Netscape Navigator

Packet-switching, 171–172, 318n27
Participant observation, 244, 268, 271, 343n11; writing copyright licenses as, 294–300
Pascal (programming language), 138
Patches (software), 220; debugging vs., 224; voting in software development and, 222–229
Patents on software, 199–200, 335n40
Pedagogy, 246; Minix and, 217–218; network protocols and, 175; operating systems and, 119, 141–142, 166, 215, 337n3
Peer production, 210, 307
Perens, Bruce, 105, 108, 234
perl (programming language), 108, 338n4
Permission, 283–286. *See also* Copyright infringement
Planning, 12, 60, 211, 238, 316n2. *See also* Adaptability
Polymaths, 266, 280; transhumanists vs., 77–94
Portability, of operating systems 123–125, 153–155
Porting. *See* Sharing source code
POSIX (standard) 155, 157–58, 328n5
Power, relationship to control, 70, 73. *See also* Reorientation of power and knowledge.
Practices, x, 12, 113–116, 180, 302; five components of Free Software, 3, 97–98, 181, 210–211, 245, 253, 302; norms vs., 290–292; opposed to legal changes, 199–200; stories as, 114
Pragmatism, 85
Prentice Hall, 216–217
Printing press, 273, 279
Programming, 58–59, 80, 120–123
Programming languages, 68, 119–126, 324n5
Progress, 64–65, 77, 88, 92

Proliferation of software, 120, 123, 141–142, 155, 185, 208, 308, 323n1. *See also* Forking; Sharing source code
Property. *See* Intellectual property
Proprietary systems: closed, 32; open vs., 126, 143; lock-in and, 145
Protestant Reformation, 65–76, 94, 114, 149; as usable past, 54
Protocols: distinguished from standards and implementation, 330n28; Open Systems Interconnection (OSI), 144, 166–177; TCP/IP, 55, 60, 120, 138–141, 144, 166–177, 319n32
Public, x, 3–7, 38–43, 302; autotelic and independent, 48–49, 50, 53, 61–63, 76, 290, 312n3; private vs., 312n5; self-grounding of, 47–48. *See also* Recursive public
Publication, 49–50, 54; as notional event, 275; transformation by Internet, 271–286
Public domain, 103, 119, 261–262, 335n33; commons vs., 313n11; contrasted with free, 191–193, 334n16; Creative Commons licenses and, 283; environmentalism and, 267; literary texts in, 285; meaning in EMACS controversy, 190–199. *See also* Commons
Public Library of Science (PLoS), 342n10
Public sphere: recursive public vs., 305; theories of, 4, 8, 22–23, 38–43, 211, 311n3. *See also* Public; Recursive public
python (programming language), 334n21

QED (text editor), 184

Raelians, 86
Ratto, Matt, 337n2

Parts of this book have been published elsewhere. A much earlier version of chapter 1 was published as "Geeks, Social Imaginaries and Recursive Publics," *Cultural Anthropology* 20.2 (summer 2005); chapter 6 as "The EMACS Controversy," in Mario Biagioli, Martha Woodmansee, and Peter Jaszi, eds., *Contexts of Invention* (forthcoming); and parts of chapter 9 as "Punt to Culture," *Anthropological Quarterly* 77.3.

CHRISTOPHER M. KELTY
is an assistant professor of anthropology
at Rice University.

Library of Congress Cataloging-in-
Publication Data
Kelty, Christopher M.
Two bits : the cultural significance of free
software / Christopher M. Kelty.
p. cm. — (Experimental futures)
Includes bibliographical references and index.
ISBN-13: 978-0-8223-4242-7 (cloth : alk. paper)
ISBN-13: 978-0-8223-4264-9 (pbk. : alk. paper)
1. Information society. 2. Open source
software—Social aspects. I. Title.
HM851K45 2008
303.48'33—dc22   2007049447